W9-CSF-290

Topics in
Physical Chemistry

vol 1

Edited by Deutsche Bunsen-Gesellschaft für Physikalische Chemie
Editors: H. Baumgärtel, E.U. Franck, W. Grünbein

K. Christmann

Introduction to
Surface Physical Chemistry

 Steinkopff Verlag Darmstadt
Springer-Verlag New York

Author's address:
Prof. Dr. Klaus Christmann
Institut für Physikalische
und Theoretische Chemie
der Freien Universität Berlin
Takustraße 3
1000 Berlin 33

Die Deutsche Bibliothek – CIP-Einheitsaufnahme

Christmann, Klaus:
Introduction to Surface Physical Chemistry / K. Christmann. –
Darmstadt : Steinkopff ; New York : Springer, 1991
  (Topics in Physical Chemistry ; Vol. 1)
  ISBN 3–7985–0858–5 (Steinkopff)
  ISBN 0–387–91405–6 (Springer)
NE: GT

Chemistry Editor: Dr. Maria Magdalene Nabbe – English Editor: James C. Willis
Production: Heinz J. Schäfer

Printed in Germany

Typesetting: Ernst Pendl, Heidelberg – Printing and bookbinding: Druckhaus Darmstadt

Printed on acid-free paper

Introduction to the series:
Topics in Physical Chemistry

Science continues to expand exponentially and there is no sign of a levelling off. The consequence is a fast-growing gap between the knowledge a scientist takes with him after graduation from University and today's state-of-the-art; chemists are not excluded. Today it is rather difficult to find the entrance to a new field of interest, more difficult than it was in the past. A textbook may contain the basic physical chemistry of the topic in question, and one will find numerous original papers on the topic of interest in journals. Often one may find a progress report or an "advanced" presentation on the subject, both frequently too difficult and too sophisticated for the nonspecialist to grasp an understanding of the topic. There remains a gap between the classical textbook on physical chemistry and the presentation in "advanced" articles or in original papers written for the specialist, but not for the graduate student or for the learned, but not specialized chemist.

The executive committee of the Deutsche Bunsen-Gesellschaft für Physikalische Chemie, a few years ago, decided to introduce a series "Topics in Physical Chemistry". The purpose was to help the chemist close the gap between even the finest textbook on physical chemistry and the most current research in a particular field. Fortunately, we found colleagues willing to edit this series and, fortunately, we have the cooperation of the publishers Dr. Dietrich Steinkopff Verlag; the result is this first volume of "Topics in Physical Chemistry". Other volumes will follow, two or three per year, and, in this way, we hope to build a practical library on different current fields of physical chemistry. The background provided by a modern textbook of physical chemistry will be a sufficient introduction to the specific topics. The bridge between this background and current research in the field, as published in scientific journals, will be these "Topics", not just for physical chemists, but also for the whole community of chemists, graduate students and researchers.

Anyone who wishes to suggest themes for the series, or has proposals for its improvement, or would like to contribute to the series is encouraged to contact the editors; the cooperation of the scientific community is indispensable and welcome.

Alarich Weiss   *Darmstadt, 1991*

V

Preface

The remaining years of our ending millennium are characterized by a tempestuous development of Surface Science, whose ultimate consequences are presently hard to foresee. While some of these consequences are apparent to everybody (e.g. modern information electronics would hardly be possible without the progress in device fabrication which, in turn, has required profound knowledge of surface technology) there are several other disciplines where the impact of surface physical chemistry may not be so obvious, but, nevertheless, has contributed much to the technological progress made in the past, and is expected to cause even more such benefit in the future. We only list here the classical synthetic inorganic chemistry or the technical chemistry which have both greatly benefited from a more fundamental understanding of heterogeneously catalyzed (surface) processes leading to, among others, improved industrial fabrication processes. (We selectively mention the promising attempts to model the Fischer-Tropsch reaction or the ammonia synthesis reaction, remembering also the optimization of the hydrocarbon reforming process by developing appropriate bimetallic catalyst materials). Furthermore, materials science with its considerations of corrosion, embrittlement, and fracture, as well as energy technology with its considerations of photovoltaics, hydrogen storage, or fuel cell development, must also be mentioned here. Many other important aspects remain unmentioned, because of space limitations.

The writing of this book arose from the intention of the editors to bring a new series "Topics in Physical Chemistry" into being, thus presenting a selection of current problems and research activities in the field of physical chemistry to a more generally educated scientific community. The various volumes of the series will be prepared accordingly, and it is a great pleasure to present this first such volume and to have it devoted to the physical chemistry of surfaces. This underscores once again how important this subject is, and how it is expected to stimulate other areas of fundamental and applied physical chemistry.

Toward this end a first step is certainly to awake interest in surface chemistry and physics, and in pursuing this goal, a platform of somewhat more general surface physics is presented, flavored with a variety of practical examples taken from both chemistry and physics. Correspondingly, this volume is equally interesting for students of chemistry and physics, and to chemists and physicists employed in industry, who want to gain some insight into elementary processes that play a role, for instance, in heterogeneously catalyzed reactions.

The second step then must be to provide an easy entry into the matter, and there is no doubt that this task is a very difficult one, because it requires a precise and extensive presentation of also the theoretical background, which is certainly welcome for a physicist, but may perhaps sometimes bore a chemist. On the other hand, these chemists expect a concise treatment especially of practical and experimental chemical problems which may not be too interesting for a solid state physicist. To satisfy both groups requires a balancing act between a representation based on well-defined physical (but, perhaps, "esoteric") and more ill-defined chemical (but practically, much more relevant) conditions. Of course, both representations could have been given, if it were not for practical space limitations. Hopefully, the reader will appreciate this limitation, which sometimes made it necessary to omit useful mathematical derivations of physical laws or relations and to leave out more indepth explanations, or to refer to more elementary textbooks. Therefore, we do not claim that this volume represents a textbook in surface physical chemistry, despite the fairly broad title. Nevertheless, we have tried, whenever possible, to generate a physical understanding of surface processes and, moreover, to elucidate the close relationship between classical surface physics and applied interface chemistry, in particular, as far as heterogeneous catalysis is concerned. Furthermore, we offer a relatively comprehensive list of references at the end of each chapter that includes many of the original publications, for the benefit of the interested reader. As pointed out before, one of the most dominant perspectives of surface and interface science must be assumed in semiconductor physics and technology, where, however, chemical processes become increasingly important, especially in the fabrication stage, in addition to the traditional physical methods. Despite this importance of semiconductors we do not attempt to thoroughly cover these materials and their physics in this book, rather we place the emphasis on more chemical problems. On the other hand, we intended and hope that the general scope of this book will also enable those readers who are not particularly engaged in semiconductor physics to at least understand some of the elementary problems. Here, we especially consider the adsorption process as a decisive part of each surface reaction. In today's surface science, the methodological aspect has also become a very important issue, and accordingly, we provide the reader with a composition of most frequently used and valuable surface analytical techniques. This treatment does not compete with the many specialized textbooks to the various analytical methods, but it should, nevertheless, provide an overall understanding of the physical background of these methods and the advantages or disadvantages of their exploitation in surface science.

In addition to hoping that this presentation will be widely accepted and understood I also must express my sincere thanks to all those who have helped in developing this book, in particular, to Mrs. Karin Schubert, who processed the text and drafted all the illustrations. I gratefully acknowledge the critical reading of the manuscript by Dr. Jörn Manz and Dr. Karl-Heinz Rieder, who helped to eliminate errors and to improve the text. And a special thanks goes to the publisher, Dr. Dietrich Steinkopff Verlag, Darmstadt, especially to their Chemistry Editor, Dr. Maria Magdalene Nabbe, for a very successful collaboration in making this volume possible.

Klaus Christmann   *Berlin,*
                                                                              June 1991

Contents

1 Introduction

1.1 The Importance of Surfaces and Surface Physical Chemistry

The first half of this century was governed by great discoveries in the field of particle and solid state physics that led to the foundation of quantum theory as probably the most useful concept to describe the properties of matter. With regard to chemistry, extremely important developments in reaction kinetics, complex chemistry, and synthetic organic chemistry were made at the same time. Today, this pattern has changed somewhat and it must now be obvious to any observant natural scientist that, not only biological or biochemical disciplines more and more predominate but also surface and interface phenomena steadily gain interest. Information technology, metallurgy, heterogeneous catalysis, materials science – all these disciplines make use of the physical chemistry of surfaces and interfaces.

Evidently, any interaction of solid or liquid matter with its environment (which may either be in the gaseous, liquid or solid state) can only come about via the *surface* of the respective condensed phase. Note that the term "surface" is mostly used in the context of gas-liquid or gas-solid phase boundaries; otherwise, the term "interface" is used. This interaction involves a variety of fundamental questions pertinent to physical chemistry, including:

- What is the actual topography of the surface, i.e., the geometrical location of the topmost atoms of a regular crystal? The answer should contain information about bond lengths, bond angles, long-range order of the surface atoms, as well as about possible crystallographic defects (steps, kinks, dislocations, etc.).
- Are there any structure differences between surface and bulk (relaxation, restructuring phenomena, possible surface compound formation with different crystallography)?
- What is the electron structure of the surface? Here, we expect information about the valence state of the surface atoms, the shape and direction of the electron clouds in the surface, about the formation of electron bands, surface states or, more generally, about the conducting or insulating properties of the surface region.
- Equally important is to determine the kind and number of surface atoms, that is, a chemical analysis should elucidate the surface chemical composition and possible concentration gradients perpendicular or parallel to the surface (enrichment or depletion effects, island or domain formation), a problem pertinent to alloys or any kind of mixtures.
- How do adsorption effects occur, and if they do, what is the local geometry of the adsorption site, and what is the configuration of the adsorption complex? Again, we require to know bond lengths and angles of the adsorbed species, as well as the long-range order within the adsorbate layer, including clustering and island formation.
- What are the chemical interactions between solid and liquid surfaces and an adsorbate atom or molecule (chemical-binding energies, lateral adsorbate-adsorbate interaction energies), and how do the interaction forces depend on the concentration of the adsorbate?

1

– Finally, how can chemical reactions occur between two different adsorbed species on the surface or between only one adsorbed reactant and the gas phase? Here, we search for the *reactivity* of a given system and, in particular, for the reaction mechanism.

Thus, besides these rather fundamental questions there are, of course, just as many surface-related problems in the area of "practical" physical chemistry or better technical chemistry that concern the large-scale fabrication of basic chemicals (heterogenous catalysis), automotive exhaust air pollution, corrosion, etch-pit formation of stainless steel, crack and fracture phenomena in materials science or chemical engineering, or the coating of surfaces (optical lenses and mirrors, passivation of aluminum and other metals, etc.). The list is extensive. If we simply concentrate on heterogeneous catalysis, we remember that about 70% of all basic chemicals are fabricated via catalysts. The synthesis and/or refinement of hydrocarbons is certainly one of the most important branches in the chemical industry (applicable for, among others, "reforming", "platforming", solidification of fatty acids, liquefication of coal, etc.), and the underlying chemical processes are catalytic hydrogenation, hydrogenolysis, dehydrocyclization, hydrocracking and many others. Another extremely valuable catalytic reaction gives us access to the large atmospheric nitrogen reservoir, namely, the ammonia synthesis from the elements, (the well-known Haber-Bosch process). There are many other catalytic reactions that play a role in daily life. Today, the surface physical chemist is interested, of course, in temperature and pressure conditions at which the respective catalytic reaction is to be carried out, but most all of his questions concern the catalyst material itself. Again, information is required about its structure and chemical composition, before, during, and after the reaction; problems such as maximum turn-over, selectivity, structure stability, possible promoter or inhibitor effects, catalyst poisoning, etc., remain as some of the challenging problems.

For many years university and industry chemists and physicists, researchers as well as theoreticians have been working on the above-mentioned problems, but in our opinion, a still better efficiency of this work could be achieved if there was better communication. These communication problems can and will arise between scientists who were educated in their respective "pure" discipline, e.g solid-state physicists or metallurgists versus inorganic or complex chemists.

At this point it is worthwhile to underline the very important role of *physical chemistry* which can and should act as a *link* between physics and chemistry, and it is especially *surface* physical chemistry where this connecting function is so obvious and easy to establish. Nevertheless, inveterate chemists or physicists still seem to regard the physical chemistry of surfaces from relatively different standpoints, a fact which has led in the past to the two distinguishable sub-disciplines, surface *chemistry* and surface *physics*. Actually, there are not too many differences with respect to methods but with respect to problems: a surface physicist is perhaps more interested in elementary excitation processes of surfaces of a mostly electronic nature under as well-defined and simple conditions as possible. In a sense, the frequently investigated CO molecule interacting with surfaces represents the *most complicated* system for a physicist. The surface chemist, on the other hand, is rather more interested in routes of interaction and reaction, and thermodynamic and kinetic constants that allow a description or better prediction of the chemical behavior of the systems of interest. In terms of the above-mentioned example, the CO molecule and its interaction with surfaces then would be the *simplest* system a *chemist* would regard. So, predominantly, the problem is that of a 'common language'.

However, as far as the experimental methods or even the scientific procedure for addressing a specific problem are concerned, there are now almost no differences between surface physicists and chemists: both groups make extensive use of modern experimental tools, such as various types of electron spectroscopies (e.g., UV and x-ray photoelectron spectroscopy, Auger electron and vibrational loss spectroscopy), thermal desorption spectroscopy, mass spectrometry, gas chromatography, ion and x-ray spectroscopy, scanning tunneling and electron microscopy, etc., and both sides working experimentally are supported by theoretical groups (quantum chemistry, band structure, and cluster calculations). In this book, it will be attempted to consider specific examples that are believed to be relevant for *both* groups, whereby it will be tacitly assumed that the most basic principles of the physical operation of the methods are familiar to both physicists and chemists, since space limitations do not allow to consider many instrumental details here. However, literature references will be given for the benefit of the interested reader.

It is quite revealing to follow the most recent developments in surface physical chemistry. The past three decades saw tempestuous activity in the area of analysis of *static* surface properties, with the determination of *clean* surface structure, binding energies of adsorbates, and surface vibrational frequencies being in the forefront. This has changed somewhat in recent years and now there is an increasing number of studies concerned with surface *dynamic* processes, e.g., atom and molecule scattering behavior of surfaces, time-resolved spectroscopy, determination and calculation of particle trajectories in interaction with solid surfaces. Usually, these latter problems require much larger experimental (and theoretical) efforts, but they must undoubtedly be the final goal towards which surface chemists must move in order to understand the principles of any surface reaction. At present, however, the research in this field is at its beginning; the systems being investigated are still very simple and are far from practical relevance. Despite the importance of surface reaction dynamics there are still great areas of static and equilibrium surface properties yet undiscovered. For example, at present many surface scientists hold to the problem of adsorbate-induced changes of surface structure (relaxation and reconstruction phenomena), often in conjunction with a study of bulk diffusion, incorporation or permeation of the adsorbing gas. The structural changes occurring under catalytic conditions are believed to be crucially important with regard to catalytic activity and selectivity. Quite often, sintering processes occur that reduce surface activity, and it is not always known how these effects depend on temperature and gas pressure or adsorbate coverage. Furthermore, the role of foreign (impurity) atoms must be considered in this context, since there are quite a lot of examples where a certain concentration of impurity is necessary for a desired reactivity. There are many other relevant questions connected with equilibrium surface chemistry and physics, and this is why we shall give this field the highest priority in the context of this book and shall not so much delve into the problems of surface dynamics. In a sense, this work is more devoted to chemists who want to inform themselves about the usefulness of the surface science approach to catalytic chemistry. It is not primarily written for solid state physicists who are interested in certain special surface properties from a more academic point of view, but it may (hopefully) be useful for those of them who are interested in adsorption phenomena and surface chemical reactions, i.e., physical chemistry of heterogeneous catalysis.

1.2 The "Pressure Gap"

If we consider a simple heterogeneous surface reaction, for example, the oxidation of carbon monoxide, it is obvious that the simple net equation

$$CO + \tfrac{1}{2} O_2 \xrightarrow[\text{Pd-catalyst}]{200°C} CO_2$$

does not tell us much about the mechanism of this reaction. Apparently, the role of the metal surface consists of a substantial reduction of the activation barrier in order for this reaction to take place; for the homogeneous gas (chain) reaction, this has been determined to be about 220 kJ/mol [1], with a pronounced temperature dependence. With a Pt(111) surface used as a model catalyst the barrier reduces to only ~100 kJ/mol [2]. In the gas phase reaction, the rate-determining step is the rehybridization and subsequent dissociation of the oxygen molecule, and the beneficial role of the metal surface could be that it provides a much easier path for the dissociation to take place. Once individual oxygen atoms are formed they can easily be attached to a neighboring CO molecule to form CO_2. Without anticipating Chapter 5 of this book, which is devoted to elementary surface reactions, we may state here that a bimolecular surface reaction will proceed as follows: both gaseous reactant molecules become, in a first step, trapped in a (more or less weak) potential well of the surface, from which they enter a chemisorbed state, that is, a much stronger interaction potential. Accordingly, they cannot readily leave the surface again. Surface scientists refer to this process as sticking and adsorption. In the chemisorbed state, molecules frequently undergo dissociation, especially dihydrogen, dioxygen or dinitrogen resulting in the presence of single reactive atoms. Once adsorbed atoms of different kinds collide with each other on the surface the actual *reaction* step can occur, provided certain energy and spatial requirements are fulfilled, and as a result, a product molecule will be formed. Apparently, this reaction step is accompanied by surface diffusion, migration or hopping of individual particles that are trapped in the chemisorbed state. The final step in the reaction sequence then would be the evaporation of the product molecule from the surface back into the gas phase, (a process referred to as *desorption*) where it can then be chemically separated and stored. Again, the reaction sequence consists of the processes: *trapping* and *sticking*, *adsorption* (chemisorption), *reaction*, and product *desorption*. All these processes will be covered in this volume.

The question arises as to how these single steps can be disentangled, and this admittedly simple question leads to a philosophical discussion whenever surface scientists, in particular surface chemists, are asked about the way they approach the surface problems. Up until recently there were (and still are) two main factions – let us call them the "purists" and the "practitioners", and we find, not surprisingly, many more physicists among the first, and more chemists among the second party. The "pure" approach consists of an extensive *modelling* of the reaction in order to keep the parameters as simple as possible. Therefore, reactions are carried out at extremely low pressures (in the ultra-high vacuum (UHV) range, at pressures less than 10^{-7} mbar), surface structure is investigated at even smaller pressures, and small area (~1 cm^2) single-crystal substrates are most widely used with surfaces that are as ideal as possible, i.e., they exhibit a very well-defined geometrical array of atoms onto which the few molecules of the gas phase can adsorb. The *chemical* composition of the substrate is usually also well-defined; mostly extremely high-purity (99.999%) metal crystals are used. In some cases alloy crystals or bimetallic films are also prepared in a well-defined manner by epitaxial growth and then subjected to the aforementioned adsorption studies. Another advantage is that the common UHV instruments also offer the possibility, by using liquid nitrogen cooling devices, to lower the sample temperature to around 100 K. This, in turn, also allows to adjust high adsorbate coverages because the (thermally activated) desorption process is sufficiently slowed at these low temperatures. This holds, too, for activated chemical surface reactions, which can then be studied on a reasonable time scale.

All in all, the UHV approach has many advantages in that not only well-defined structural conditions are provided, but also the whole variety of optical and particle impact (mostly electron) spectroscopies can easily be applied. In contrast, however, there is a serious disadvantage with the UHV model approach, and this is why many practical chemists still have objections to this more "physical" treatment of the catalytic problem. They argue that the UHV model is too far removed from reality, since chemical reactions are typically carried out under quite different conditions, namely, at atmospheric (or even higher) pressures and with much less well-defined surfaces, namely, large-area samples consisting of powders, pellets, thin films or polycrystalline sheets. High-purity one-component materials are therefore a rarity and, instead, multi-component catalysts with a whole variety of additives (promoters) are common in industrial technology. Also, low temperatures (T < 300 K) are of no interest in commercial chemical reactions.

This is the main reason why, in industry, chemists tend to study catalytic problems under conditions that resemble the large-scale chemical process, and they deliberately ignore such details as microscopic structure and purity parameters of their sample materials. The apparatus used in these studies often has much in common with the technical reactor (apart from the much smaller size) and similarly, pressure, temperature, stoichiometry, and purity conditions are adjusted to follow the reaction, and the activity and selectivity of the catalyst. The big advantage here is the close relation to real conditions which enables a more direct transfer of the results to the plant fabrication line and has, therefore, certainly much more impact on the direct improvement of the respective large-scale chemical process. As far as experimental methods are concerned, coupled GC-MS (gas chromatography/mass spectrometry) optical (IR, Raman) or direct preparative methods are widely used, whereby the output of such a small-scale reactor can reach appreciable turnovers. Particularly useful is such a microreactor, if it is combined with surface analytical techniques. The microreactor for methanol oxidation over silver catalysts, as described by Benninghoven and coworkers [3], is shown in Fig. 1.1 and can serve as an example.

However, the more "chemical" approach can seriously suffer from the lack of controllable parameters; usually, neither the structural nor the chemical conditions are sufficiently well- defined on a microscopic scale. This imposes many difficulties on any attempt to correlate, for instance, the catalytic activity of the sample with certain elementary catalytic steps such as adsorption, formation of essential intermediates, product desorption, etc.. The high temperatures usually chosen accelerate the reactions to a rate where it becomes difficult to follow individual kinetic steps or to identify short-lived intermediates. Furthermore, and this intensifies the difficulties, most of the established surface spectroscopic tools fail to function at pressures greater than, say, 10^{-4} mbar, because of the limited lifetime of cathode filaments or the reduced mean free path of the incident and/or emitted particles. In addition, at or near atmospheric pressure, mass transport (diffusion) and energy transport (heat exchange) problems take over and blur most of the correlations between macroscopic reactivity and, for example, adsorption/desorption parameters of the surface in question.

The existence of this famous "pressure gap" separating the UHV single-crystal model studies from technical chemical or catalytic investigations was realized since UHV

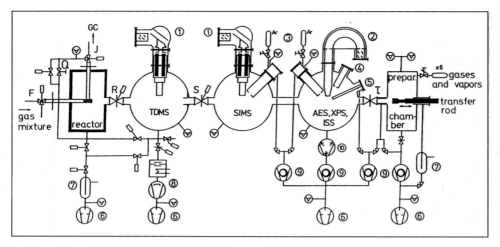

Fig. 1.1. Example of a combined microreactor and UHV surface analytical equipment for in-situ studies of catalytic reactions (oxidation of methanol on Ag). 1 = mass spectrometer, 2 = electron spectrometer, 3 = ion source, 4 = electron gun, 5 = x-ray tube, 6 = rotary pump, 7 = LN$_2$ baffle, 8 = oil diffusion pump with LN$_2$ baffle, 9 = turbomolecular pump, 10 = Ti sublimator with LN$_2$ baffle. After Ganschow et al. [3]

studies were used to explore the elementary steps of catalytic reactions, i.e., since the early 1950s, and it has since inspired many groups to consider means to bridge this gap. A good comprehensive report on the problem of utilizing UHV model studies to elucidate the mechanisms of catalytic reactions was given several years ago by Bonzel [4]. In his work, he presented a matrix (Fig. 1.2) that documents a correlation between reactant pressure and structural "complexity" or degree of dispersion of the catalyst material, and which confirms the existence of the above-mentioned pressure gap. With the aid of some selected examples, Bonzel proved that the reaction mechanism of, for example, the CO oxidation reaction on platinum is the same at low and high pressures, a result which has also been verified for Pd and Rh [5]. Similar considerations have been devoted to the ammonia synthesis reaction, where Ertl and his group have excessively worked on the single-crystal approach [6-10], while recently Nørskov and Stoltze have established a theoretical basis [11, 12]. As far as hydrogenation of CO and CO$_2$ is concerned (the well-known Fischer-Tropsch synthesis reaction), again, Bonzel and Krebs [13] have scrutinized the UHV-model approach and have given examples of how to overcome the pressure gap. The essence of all these considerations is that, although the UHV model study is deemed extremely useful, the ultimate goal is to combine UHV studies with reaction experiments at atmospheric or even higher pressures, and a whole number of apparatus suggestions have been made accordingly. We can only list a few of them here; besides the already cited microreactor designed by Ganschow et al. [3], we refer to the combined UHV-high pressure cells put forward by Blakey et al. [14], Krebs et al. [15], Goodman et al. [16] or Kolb et al. [17]. A common characteristics is that the sample can be transferred from the UHV chamber to the high pressure cell by means of a vacuum-tight sample manipulator. A very intriguing solution was proposed by the group of Somorjai [14], which is shown in Fig. 1.3. A stainless-steel welded bellows separates, in the high pressure mode, the UHV vessel (by means of a gold O-ring seal) from a small high-pressure reaction chamber, which is then connected to a flowmeter and gas chromatograph for further "high pressure" studies.

6

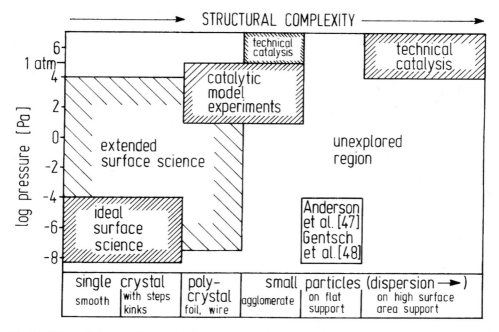

Fig.1.2. Schematic diagram illustrating the pressure gap between surface physical chemistry and practical technical catalysis. "Structural complexity" comprises the surface morphology of the catalytic materials and ranges from small single-crystalline areas (~1 cm^2) on the left-hand side to technical catalysts with areas of several hundred m^2 per gram. After Bonzel [4]

Another solution, which was put forth in our own laboratory in order to combine UHV model studies and atmospheric-pressure electrochemistry [18], uses two of the aforementioned sample manipulators, one long z-travel device with a fork at one end mounted on the UHV chamber and, separated by a valve, another small manipulator on the "high-pressure" reactor which can pull-off the sample from the head of the UHV manipulator and, after a valve between the UHV and the high-pressure cell is closed, allows further studies under atmospheric conditions. A schematic drawing is given in Fig. 1.4. Today, there also exists a variety of commercial solutions, sample transfer rods that allow sample heating and cooling: the reader is referred to the respective vacuum manufacturers. In a systematic way, then the influence of reactant pressure, sample structure, temperature, and chemical purity can be studied. To date, there are many results available [19] that justify the use of model single crystals, low pressures, and temperatures for gaining access to the catalyticly important primary reaction steps.

In effect, we are looking for a means that will enable us to follow a simple chemical surface reaction in all its details, to derive sufficient kinetic and energetic parameters to develop a reaction *mechanism*, and this then is expected to hold regardless of pressure and temperature conditions, as long as the respective reaction is thermodynamically favored.

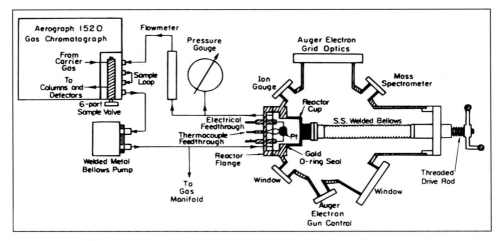

Fig. 1.3. High-pressure cell built into an UHV chamber for surface-reaction studies. A flow loop suitable for atmospheric pressure measurements is indicated. After Blakely et al. [14]. Reproduced with permission.

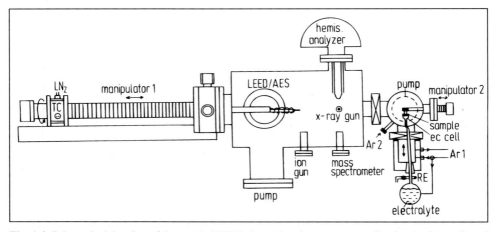

Fig. 1.4. Schematical drawing of the coupled UHV-electrochemistry apparatus allowing in-situ studies of electrochemical processes at surfaces [18].

1.3 Previous and Current Work in Surface Physical Chemistry

At the end of this introductory chapter there remains to present a list of references to other (and, in many cases, more sophisticated) reports on the same subject. Again, the vast literature can be subdivided into physical and chemical approaches. Among the first group, we find books by Prutton [20], Ertl and Küppers [21], Clark [22], Boudart and Djega-Mariadassou [23], Zangwill [24] and various series such as "The Chemical Physics of Solid Surfaces" [25] and "Catalysis" [26] to cite only a few and arbitrarily selected examples, not to mention the various journals devoted to the surface physical aspects of catalysis. There is one book by Roberts and McKee [27] that deserves special attention; it represents a particularly successful and competent attempt to display the whole spectrum of surface physical chemistry, however, with emphasis on metals. On the chemistry side, we recommend books by Gasser [28], Somorjai [19], Hiemenz [29], Adamson [30], Bond [31], and Wedler [32], as well as the series "Advances in Catalysis and Related Subjects" [33] and "Catalysis Review Science and Engineering" [34]. One ought to bear in mind that this is really only a tiny selection of the wealth of general literature pertinent to the subject "Surface Science and Catalysis".

Besides these monographs there is a large number of current journals devoted to surfaces and catalysis. Again, we can only give a selection here. Well established is the journal "Surface Science" [35], along with related periodicals such as "Applied Surface Science"[36] or the review series "Surface Science Reports" [37]. Other relevant journals are "Journal of Catalysis" [38], "Journal of Molecular Catalysis" [39], "Journal of Colloid and Interface Analysis" [40], "Applied Catalysis" [41], the review journal "Progress in Surface Science" [42], the relatively new journal "Langmuir" [43] and, of course, the "surface"sections of the more general physical-chemical journals such as "The Journal of Chemical Physics" [44], "The Journal of Physical Chemistry" [45] or the "Journal of Vacuum Science and Technology" [46].

References

1. Sulzmann KGP, Myers BF, Bartle ER (1965) CO Oxidation. I. Induction Period Preceding CO_2 Formation in Shock-Heated $CO-O_2$-Ar Mixtures. J Chem Phys 42:3969–3979
2. Campbell CT, Ertl G, Kuipers H, Segner J (1980) A Molecular Beam Study of the Catalytic Oxidation of CO on a Pt(111) Surface. J Chem Phys 73: 5862–5873
3. Ganschow O. Jede R, An LD, Manske E, Neelsen J, Wiedmann L, Benninghoven A (1983) A Combined Instrument for the Investigation of Catalytic Reactions by means of Gas Chromatography, Secondary Ion and Gas Phase Mass Spectrometry, Auger and Photoelectron Spectroscopy, and Ion Scattering Spectroscopy. J Vac Sci Technol A1:1491–1506
4. Bonzel HP (1977) The Role of Surface Science Experiments in Understanding Heterogeneous Catalysis. Surface Sci 68:236–258
5. Oh SH, Fisher GB, Carpenter JE, Goodman DW (1986) Comparative Kinetic Studies of $CO-O_2$ and CO-NO Reactions over Single Crystal and Supported Rhodium Catalysts. J Catal 100:360–376
6. Ertl G (1982) Critical Reviews in Solid State and Materials Science. CRC Press, Boca Raton, p 349
7. Ertl G (1980) Surface Science and Catalysis – Studies on the Mechanism of Ammonia Synthesis: The PH Emmett Award Address. Catal Rev Sci Eng 21:101
8. Ertl G (1980) Surface Science and Catalysis on Metals. In: Prins R, Schuit GCA, (eds) Chemistry and Chemical Engineering of Catalytic Processes. Sijthoff and Noordhoff, Germantown, MD, p 271ff
9. Ertl G (1983) Kinetics of Chemical Processes on Well-defined Surfaces. In: Anderson JR, Boudart M (eds .) Catalysis, Science and Technology, vol. 4, Springer, Berlin, p 257ff
10. Schloegl R, Schoonmaker RC, Muhler M, Ertl G (1988) Bridging the "Material Gap" between Single Crystal Studies and Real Catalysis. Catal Lett 1:237–246
11. Nørskov JK, Stoltze P (1987) Theoretical Aspects of Surface Reactions. Surface Sci 189/190:91–105
12. Stoltze P, Nørskov JK (1985) Bridging the "Pressure Gap" between Ultrahigh Vacuum Surface Physics and High-pressure Catalysis. Phys Rev Lett 55:2502–2505
13. Bonzel HP, Krebs HJ (1982) Surface Science Approach to Heterogeneous Catalysis: CO Hydrogenation on Transition Metals. Surface Sci 117:639–658

14. Blakely DW, Kozak E, Sexton BA, Somorjai GA (1976) New Instrumentation and Techniques to Monitor Chemical Surface Reactions on Single Crystals over a Wide Pressure Range (10^{-8}–10^5 Torr) in the Same Apparatus. J Vac Sci Technol 13:1091
15. Krebs HJ, Bonzel HP, Gafner G (1979) A Model Study of the Hydrogenation of CO over Polycrystalline Iron. Surface Sci 88:269–283
16. Goodman DW, Kelley RD, Madey TE, Yates. JT Jr. (1980) Kinetics of the Hydrogenation of CO over a Single Crystal Nickel Catalyst. J Catal 63:226–234
17. Kolb DM (1987) UHV Techniques in the Study of Electrode Surfaces, Z Phys Chemie NF 154:179–199
18. Solomun T, Neumann A, Baumgärtel H, Christmann K (to be published)
19. Somorjai GA (1981) Chemistry in Two Dimensions: Surfaces. Cornell University Press, Ithaca, NY
20. Prutton M (1983) Surface Physics. 2nd edn. Oxford University Press
21. Ertl G, Küppers J (1985) Low Energy Electrons and Surface Chemistry, 2nd edn. Verlag Chemie, Weinheim
22. Clark JA (1970) The Theory of Adsorption and Catalysis. Academic Press, New York
23. Boudart M, Djega-Mariadassou G (1984) Kinetics of Heterogeneous Catalytic Reactions. Princeton University Press, Princeton, NJ
24. Zangwill A (1988) Surface Physics. Cambridge University Press, Cambridge
25. King DA, Woodruff DP, (eds) (1981–1988) The Chemical Physics of Solid Surfaces, vol 1–5. Elsevier, Amsterdam
26. Anderson JR, Boudart M, (eds) (1981–1989) Heterogeneous Catalysis, vol 1–9. Springer, Berlin
27. Roberts MW, McKee CS (1978) Chemistry of the Metal-Gas Interface. Clarendon Press, Oxford
28. Gasser RPH (1985) An Introduction to Chemisorption and Catalysis by Metals. Clarendon Press, Oxford
29. Hiemenz PC (1977) Principles of Colloid and Surface Chemistry, 1st edn. Dekker, New York
30. Adamson AW (1982) Physical Chemistry of Surfaces, 4th edn. Wiley, New York
31. Bond GC (1987) Heterogeneous Catalysis, Principles and Applications, 2nd edn. Oxford Science Publishers, Clarendon Press Oxford
32. Wedler G (1970) Adsorption. Verlag Chemie Weinheim
33. Advances in Catalysis and Related Subjects, vol 1–37 (1948–1990) Academic Press, New York
34. Catalysis Reviews Science and Engineering, vol 1–32 (1968–1990) Dekker, New York
35. Surface Science, vol 1–232 (1964–1990) North-Holland, Amsterdam
36. Applied Surface Science, vol 1–44 (1977–1990) North-Holland, Amsterdam
37. Surface Science Reports, vol 1–11 (1982–1990) North-Holland, Amsterdam
38. Journal of Catalysis, vol 1–137 (1962–1990) Academic Press, New York
39. Journal of Molecular Catalysis, vol 1–60 (1975/76–1990) Elsevier Sequoia, Lausanne
40. Journal of Colloid and Interface Science, vol 1–124 (1907–1990) Academic Press, London
41. Applied Catalysis, vol 1–12 (1978–1990) Elsevier, Amsterdam
42. Progress in Surface Science, vol 1–32 (1971–1990) Pergamon Press, New York
43. Langmuir, vol 1–6 (1985–1990) American Chemical Society, Washington DC
44. The Journal of Chemical Physics, vol 1–92 (1933–1990) The American Institute of Physics, New York
45. The Journal of Physical Chemistry, vol 1–94 (1897–1990) The American Chemical Society, Washington DC
46. The Journal of Vacuum Science and Technology, vol 1–21 (1964–1981) vol A1–A8 (1982–1990); The American Institute of Physics, New York (first and second series).
47. Anderson JR, MacDonald RJ (1970) The Preparation and Use of Ultrathin Metal Films as Model Systems for Highly Dispersed Supported Catalysts. J Catal 19:227–231
48. Gentsch H, Härtel V, Köpp M (1971) Heterogene Katalyse mit Ni- und Pd-Atomen. Methanbildung aus Kohlenstoff und Wasserstoff. Ber Bunsenges Phys Chem 75:1086–1092

2 Macroscopic Treatment of Surface Phenomena: Thermodynamics and Kinetics of Surfaces

To begin with, let us first define the term "surface". We mean by it, simply the termination of the bulk state, that is to say, the region of a solid or liquid phase where the equations based on three-dimensionality are no longer sufficient to describe the complete physical state of the system. This definition implies that a surface is not necessarily confined to the topmost layer of atoms of a liquid or a crystal, but may consist of several such layers extending into the bulk, i.e., that region at or near the surface where the symmetry of the bulk is perturbed so as to give rise to altered interaction forces. This asymmetry is illustrated in Fig. 2.1 and is actually responsible for the peculiar behavior of surfaces and interfaces that lead to phenomena such as surface tension, capillary pressure or enhanced chemical reactivity of surfaces in general.

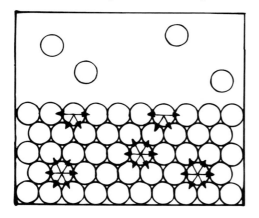

Fig. 2.1. Sketch of the symmetry and asymmetry of interaction forces of particles in the interior and at the surface of a solid or liquid phase. Compared to the condensed state, the interaction forces are practically negligible in the gaseous state.

Surface phenomena can be treated macroscopically by chemical thermodynamics. We recall that in thermodynamical treatment, the concept of atoms or molecules need never be used, because all statements and laws can simply be derived on the basis of macroscopically observable and measurable quantities such as pressure, volume, surface area, temperature or chemical composition. Accordingly, only relatively basic experimental equipment is required to determine surface properties such as surface tension, contact angle or capillary pressure, although extreme care has to be taken with regard to immaculate conditions, as will be pointed out later. This is the reason why, historically, the thermodynamical concept was pursued first, and that we can say that it was, and still is a very successful concept. Of course, since any microscopic description is principally not possible surface thermodynamics cannot provide information about atomic or molecular structure, which may be regarded as an inherent weakness, but it is nevertheless a very simple and powerful tool as far as energetics and entropy are concerned. One should remember that a physical system can only then be regarded as being completely understood when *both* the thermodynamic *and* the atomistic view lead to the same picture. In this chapter, we do by no means intend to give a complete outline of surface thermodynamics; we instead recommend the many excellent textbooks and review articles [1–11].

Rather, we want to present some thermodynamic relations that have proven useful in regards to a combining with the microscopic concept of surfaces and interfaces. This particularly concerns the energetic situation which is, as mentioned before, relatively easy to describe by thermodynamics, but difficult to calculate from microscopic (atomistic) quantum chemical theories. As will be shown below, one important ingredient is the evaluation of the so-called *adsorption isotherm* which can be regarded as the basis for any determination of heats and entropies of adsorption. The heat of adsorption (often also referred to as "adsorption energy") is a decisive quantity if the question is addressed as to whether or not adsorption will occur at all, or to what extent this process will take place at a given pressure or temperature. Adsorption is the enrichment of one or more components at the phase boundary which separates two different phases. Let us first recall that a surface or an interface will principally occur in the following two-phase systems: Solid – solid, solid – liquid, solid – gas, liquid – liquid, and liquid – gas. Note that the thermodynamical treatment does not *a priori* distinguish between the solid or liquid state, it only differentiates between condensed and gaseous phases. Therefore, we shall use the term "surface" in a general sense.

2.1 The Fundamental Equations of Thermodynamics

In ordinary thermodynamics, the internal energy U depends on three independent variables which are chosen to be the total entropy S, the total volume V and the number of moles of each component present in the system. The differential internal energy is then expressed by the Gibbs fundamental equations:

$$dU = TdS - PdV + \sum \mu_i dn_i \qquad 2.1$$

or

$$dG = -SdT + VdP + \sum \mu_i dn_i , \qquad 2.2$$

where T = temperature, P = pressure, V = volume, S = entropy, G = Gibbs energy, n = number of moles of component i and μ_i = chemical potential of component i $[= (\partial G/\partial n_i)_{S,V,n_{j\neq i}}, (\partial G/\partial n_i)_{P,T,n_{j\neq i}}]$, from which U and G, respectively, can be obtained by integration.

Turning to two-dimensional thermodynamics, we consider a two-component condensed phase of n_A moles of a nonvolatile component and n_s moles of a volatile component that is in equilibrium with the gas phase. (The subscript s is chosen to illustrate that this component is able to *adsorb* at the surface). We can now write for the differential internal energy dU of the condensed phase

$$dU = TdS - PdV + \mu_A dn_A + \mu_s dn_s , \qquad 2.3$$

and

$$dG = -SdT + VdP + \mu_A dn_A + \mu_s dn_s . \qquad 2.4$$

These equations mean that any change in energy of the total system cannot only be provided by changes of pressure and temperature, but also by the number of moles of the nonvolatile and the volatile component via their chemical potential. This equation can in principle be used to describe various different thermodynamic systems, for example, hydrogen – titanium, krypton – charcoal, or water – carbon tetrachloride. If the nonvolatile component is a solid it is tacitly understood that its surface area α is proportional to the volume and that a change in surface area $d\alpha$ can always be expressed as the corresponding change in the number of moles of the solid, dn_A. (We recall that these considerations are the basis of the physical phenomenon called "adsorption", and in adsorption terminology, we refer to the condensed phase onto which adsorption occurs as the *adsorbent*, whereas the adsorbing molecules or atoms is called *adsorpt*. The frequently used term *adsorbate* refers to the adsorpt particles which are enriched at the surface).

For the *pure* solid substance, we can again formulate

$$dU_A^0 = TdS_A^0 - PdV_A^0 + \mu_A^0 dn_A . \tag{2.5}$$

For the respective quantities in the two-component system, we define $U_s \equiv U - U_A^0$, $V_s \equiv V - V_A^0$, $S_s \equiv S - S_A^0$, and $\Phi = \mu_A^0 - \mu_A$, where, for example, U_s is just the difference between the total internal energy of the condensed phase U and the energy that n_A moles of pure nonvolatile substance have.

By subtracting the energy part of the pure substance from the total energy of the condensed phase, we obtain the energy part of the adsorbate:

$$dU_s = TdS_s - PdV_s - \Phi dn_A + \mu_s dn_s . \tag{2.6}$$

It is assumed that the n_A moles of the nonvolatile component are inert. The symbol Φ is the difference $\mu_A^0 - \mu_A$, in which μ_A^0 stands for the chemical potential of pure adsorbent (clean surface!) and μ_A is the chemical potential of a pure adsorbent whose surface is covered with a layer of adsorbate. Remembering that μ is defined as the partial derivation of internal energy or Gibbs energy with respect to the number of moles:

$$\left(\frac{\partial U_A^0}{\partial n_A} \right)_{S_A^0, V_A^0} = \mu_A^0 = \left(\frac{\partial G_A^0}{\partial n_A} \right)_{P_A^0, T} , \tag{2.7}$$

and

$$\left(\frac{\partial U}{\partial n_A} \right)_{S, V, n_s} = \mu_A = \left(\frac{\partial G}{\partial n_A} \right)_{P, T, n_s} ; \tag{2.8}$$

Φ represents just the internal energy change per unit of adsorbent in the surface spreading of adsorbate

$$\Phi = - \left(\frac{\partial U_s}{\partial n_A} \right)_{S_s, V_s, n_s} . \tag{2.9}$$

As pointed out before, the number of moles of inert adsorbent n_A is proportional to the surface area α (proportionality factor f) and, hence:

$$\Phi \cdot dn_A = f \cdot \Phi \cdot d\alpha , \tag{2.10}$$

with the definition

$$f \cdot \Phi \equiv \varphi = - \left(\frac{\partial U_s}{\partial \alpha} \right)_{S_s, V_s, n_s} . \tag{2.11}$$

We call φ the "surface tension", which is the difference between the surface tension of the *clean* adsorbent γ_0 and that of the surface covered with adsorbate γ. The expression $\varphi \, d\alpha$ then represents the two-dimensional analogon to the PdV work, that is to say, the work required to create or annihilate surface area. φ is often referred to as "spreading" or surface pressure, its dimension is actually that of a two-dimensional pressure.

Using the definitions, we arrive at

$$dU_s = TdS_s - PdV_s - \varphi d\alpha + \mu_s dn_s \;,$$ 2.12

which describes the differential internal energy of a one-component system of n_s moles of adsorbate on an *inert* substrate surface.

The assumption of a truly inert adsorbent implies that a real separation between the thermodynamic properties of the solid component (subscript A) and those of the adsorbate (subscript s) is possible. This is verified as long as the adsorbate is only weakly interacting with the surface. Equation 2.12 therefore holds well for noble gases interacting with graphite or charcoal surfaces, that representing an example of physisorption forces with energies well below 30 kJ/mol.

In case of a non-inert adsorbent (transition metal surfaces must be ranked among this group) the situation can no longer be described so easily by means of Eq. 2.12. Here, the thermodynamic quantities U, V, S of the fundamental Eqs. 2.3 and 2.4 refer to the complete condensed phase, which means that a differential internal energy change dU is composed of contributions from both the adsorbent and the adsorbate.

For the moment, however, we shall still be concerned with the properties of an inert adsorbent system, and Eq. 2.12 allows us to immediately set up the fundamental equations for adsorption thermodynamics of a one-component system of n_s moles of adsorbed gas. Using the well known interrelations between energy U_s, enthalpy H_s, Helmholtz free energy F_s, and Gibbs energy G_s, we arrive, after integration (all intensive variables are kept constant) at the four equations:

$$U_s(S_s, V_s, \alpha, n_s) = TS_s - PVF_s - \varphi\alpha + \mu_s n_s \;;$$ 2.13

$$H_s(S_s, P, \alpha, n_s) = TS_s - \varphi\alpha + \mu_s n_s \;;$$ 2.14

$$F_s(T, V_s, \alpha, n_s) = -PV_s - \varphi\alpha + \mu_s n_s \;;$$ 2.15

$$G_s(T, P, \alpha, n_s) = -\varphi\alpha + \mu_s n_s \;.$$ 2.16

Apparently, *four* independent variables are necessary to completely describe each thermodynamic energy function, one more than in ordinary three-dimensional thermodynamics, namely just the term containing α and φ, respectively.

2.2 The Adsorption Energy

From the set of four equations (Eqs. 2.13 – 2.16), let us take Eq. 2.13 which we write in differential form:

$$dU_s = TdS_s + S_s dT - PdV_s - V_s dP - \varphi d\alpha - \alpha d\varphi + \mu_s dn_s + n_s d\mu_s \;.$$ 2.17

This can be compared with Eq. 2.12, and we obtain

$$n_s d\mu_s = -S_s dT + V_s dP + \alpha d\varphi \;.$$ 2.18

Upon introducing molar quantities s_s and v_s and rearranging this writes

$$d\mu_s = -s_s dT + v_s dP + \frac{\alpha}{n_s} d\varphi \;.$$ 2.19

Let us now consider the equilibrium condition whereafter the adsorbate phase (subscript s) is in equilibrium with the gas phase (subscript g). Equilibrium means that the chemical potentials of adsorbate and gas phase are equal and remain equal, hence:

14

$$d\mu_g = d\mu_s \, .$$ (2.20)

The gas phase obeys the well-known laws of three-dimensional thermodynamics, i.e.,

$$d\mu_g = -s_g dT + v_g dP \, ,$$ (2.21)

so that

$$-s_g dT + v_g dP = -s_s dT + v_s dP + \frac{\alpha}{n_s} d\varphi \, ,$$ (2.22)

which rearranges, for constant surface pressure ($\varphi =$ const.) to

$$\left(\frac{\partial P}{\partial T} \right)_\varphi = \frac{s_g - s_s}{v_g - v_s} \, .$$ (2.23)

This is the two-dimensional analogon to the well-known Clausius-Clapeyron equation. The common approximation, $v_g \gg v_s$ and the (justified) assumption of ideal gas behavior, yields

$$\left(\frac{\partial \ln P}{\partial T} \right)_\varphi = \frac{s_g - s_s}{RT} = \frac{h_g - h_s}{RT^2} = \frac{\Delta h}{RT^2} \, .$$ (2.24)

This differential equation considers the temperature dependence of the equilibrium gas pressure at *constant surface pressure* φ; h_g and h_s denote the molar enthalpies of the gas and the adsobate phase, respectively.

The slope of a corresponding logarithmic plot of the equilibrium pressure versus reciprocal temperature yields the (equilibrium) adsorption enthalpy Δh which is released upon adsorption of one mol gas. In principle, one can determine Δh theoretically if the molar entropy of adsorbed gas is accessible: this can be accomplished by means of statistical mechanics, provided that the state and configuration of the adsorbate is known.

There is, however, one practical shortcoming of Eq. 2.24 that is, the surface pressure φ is difficult to determine in many adsorbate systems. It is often much easier to measure the number of adsorbed particles, that means, the number of moles of adsorbed gas n_s, in relation to the number of moles of adsorbent n_A. This ratio is called coverage Γ and is defined as

$$\Gamma = \frac{n_s}{n_A} \, .$$ (2.25)

The following procedure is, in close agreement with the foregoing treatment, again based on the existence of the phase equilibrium: gas phase – adsorbate phase.

The chemical potential of the adsorbate, μ_s can be written as

$$\mu_s = \left(\frac{\partial G}{\partial n_s} \right)_{P,T,n_A} \, .$$ (2.26)

Differentiation with respect to temperature, for constant number of moles ($d\Gamma = 0$), yields, according to

$$\left(\frac{\partial \mu_s}{\partial T} \right)_{P,\Gamma} = \left(\frac{\partial^2 G}{\partial n_s \partial T} \right)_{P,T,n_A} = -\left(\frac{\partial S}{\partial n_s} \right)_{P,T,n_A} \, ,$$ (2.27)

the partial molar entropy of adsorbate $\bar{s}_s$:

15

$$\bar{s}_s = \left(\frac{\partial S}{\partial n_s}\right)_{P,T,n_A}.$$ 2.28

Similarly, the partial molar volume of adsorbate $\bar{v}_s$ can be derived by

$$\left(\frac{\partial \mu_s}{\partial P}\right)_{T,\Gamma} = \left(\frac{\partial^2 G}{\partial n_s \partial P}\right)_{T,P,n_a} = \left(\frac{\partial V}{\partial n_s}\right)_{T,P,n_A} = \bar{v}_s.$$ 2.29

A differential change in the chemical potential of the adsorbate is composed of three contributions:

$$d\mu_s = \left(\frac{\partial \mu_s}{\partial T}\right)_{P,\Gamma} dT + \left(\frac{\partial \mu_s}{\partial P}\right)_{T,\Gamma} dP + \left(\frac{\partial \mu_s}{\partial \Gamma}\right)_{P,T} d\Gamma,$$ 2.30

or, by introducing the partial molar quantities

$$d\mu_s = -\bar{s}_s dT + \bar{v}_s dP + \left(\frac{\partial \mu_s}{\partial \Gamma}\right)_{P,T} d\Gamma.$$ 2.31

Again, at equilibrium, $d\mu_s = d\mu_g$ (cf. Eq. 2.20,),

$$-\bar{s}_s dT + \bar{v}_s dP + \left(\frac{\partial \mu_s}{\partial \Gamma}\right)_{P,T} d\Gamma = -s_g dT + v_g dP.$$ 2.32

Constant coverage condition ($d\Gamma = 0$) cancels the Γ term, and one obtains, in close analogy to Eq. 2.23:

$$\left(\frac{\partial P}{\partial T}\right)_\Gamma = \frac{s_g - \bar{s}_s}{v_g - \bar{v}_s}.$$ 2.33

Using the same approximations as before, Eq. 2.33 rearranges to

$$\left(\frac{\partial \ln P}{\partial T}\right)_\Gamma = \frac{s_g - \bar{s}_s}{RT} = \frac{h_g - \bar{h}_s}{RT^2} = \frac{q_{st}}{RT^2};$$ 2.34

$h_g - \bar{h}_s$ denotes the difference between the molar enthalpy of the gas and the partial molar enthalpy of the adsorbate. It is usually called *isosteric* heat of adsorption and can be measured relatively easily in an experiment, since the constant coverage condition is not too difficult to adjust. "Isosteric" means constant coverage. However, in contrast to the (equilibrium) heat of adsorption (cf. Eq. 2.24), it is not possible to calculate the partial molar entropy of the adsorbate, $\bar{s}_s$, by means of statistical mechanics. Consequently, any experimental values of q_{st} cannot be interpreted as easily as those of Δh.

Nevertheless, along with additional microscopic information (how this is provided will be subject of Chapter 3) the isosteric heat of adsorption and its coverage dependence represent an extremely valuable energetic quantity that provides relevant details about the strength of gas-solid interaction forces. On the other hand, we remark that it is not trivial to establish a relationship between bond strength of an adsorbed particle to the surface and the enthalpy measured macroscopically. There exists a variety of considerations on this subject and some empirical relationships have been derived [12,13].

It should be added here that Eq. 2.34 holds for inert as well as for reactive adsorbent – adsorbate systems, and q_{st} values can be determined whether or not the adsorbate induces alterations in the adsorbent surface. However, in case of adsorbate-induced perturbations of the substrate the interpretation of q_{st} becomes difficult, since the heat of adsorption is distributed between both phases in an unknown manner.

The isoteric heat of adsorption, q_{st}, is derived from the differential molar quantities $\bar{s}_s$ and $\bar{v}_s$ at constant T, P and surface area α, whereby $\bar{s}_s$ and $\bar{v}_s$ usually vary with n_s. Therefore, q_{st} is a so-called differential heat of adsorption in contrast to the equilibrium heat of adsorption, Δh (cf., Eq. 2.24)), which is an *integral* heat of adsorption. Note that in its derivation (cf., Eqs. 2.21, 2.22) integral molar quantities are used that are obtained from the partial differential functions by integration with all intensive variables (T, P, φ) held constant, for example,

$$\left(\frac{\partial V_s}{\partial n_s}\right)_{T,P,\varphi} \Rightarrow \frac{V_s}{n_s} \equiv v_s .$$
2.35

Both the equilibrium and the isoteric heat of adsorption are *isothermal* heats since they have been derived for isothermal and isobaric conditions.

From differential heats of adsorption, one can obtain the integral heats by integration, according to

$$Q_{integr.} = \int_0^{n_s} q_{st} dn_s .$$
2.36

Integral heats apply to processes in which n_s moles of adsorbate are transferred in a one-step process from the gas phase to the adsorbent, starting from a bare surface.

2.3 The Measurement of the Isosteric Heat

The question arises of how, for example, the isoteric heat of adsorption, q_{st}, can be measured experimentally. Most conveniently, the equilibrium pressure P at different temperatures T is determined, which leads to the same surface coverage. Actually, this can be achieved by measurements of the so-called adsorption isobars

$$n_s = n_s(T)_P$$
2.37

or adsorption isotherms

$$n_s = n_s(P)_T ,$$
2.38

which will be dealt with in a moment. The basis is the integration of Eq. 2.34, which can be rewritten as

$$\frac{dP}{P} = \frac{q_{st}}{RT^2} dT .$$
2.39

q_{st} is assumed to be temperature-independent, which is usually true for a small temperature interval. This then leads to

$$\ln \frac{P_1}{P_2} = -\frac{q_{st}}{R} \left(\frac{1}{T_1} - \frac{1}{T_2}\right) ,$$
2.40

for two pairs of pressures/temperatures that produce the same surface coverage. For true equilibrium conditions, a straight line with negative slope should be obtained from the corresponding plot, which in turn yields q_{st}. Repeating this procedure for various coverages allows the coverage dependence of the heat of adsorption to be determined. This can be demonstrated for a whole variety of adsorption systems that comprise weakly interacting species such as adsorption of ethyl chloride on charcoal [14] or xenon on single crystalline nickel [15] or palladium [16], as well as more strongly interacting systems such

17

as carbon monoxide on single crystalline palladium [17] or hydrogen on nickel [18]. To give an example, the adsorption of xenon on a Ni(100) surface (squared geometry, cf., Chapter 3) was investigated by Christmann and Demuth [15]. In this experiment single crystal surfaces with an overall surface area α of ca. 1 cm^2 were used, and even at saturation (full coverage, Γ_{max}) less than 10^{15} Xe atoms were adsorbed, i.e., only the incredibly small amount of 10^{-9} moles. This brings us to a well-known difficulty in experimental adsorption thermodynamics, namely, the accurate determination of $\Gamma = n_s/n_A$. We would like to introduce at this point a more common symbol for the surface coverage, namely, Θ, which is not expressed in moles, but actually relates particle numbers to each other. Unfortunately, Θ is often used in two slightly different definitions: i) the number of adsorbed particles per unit area, σ_s, is related to the maximum possible number of adsorbed particles, $\sigma_{s,max}$, according to

$$\Theta_\sigma = \frac{\sigma_s}{\sigma_{s,max}} \, ,$$

2.41

which means that always $0 \leq \Theta \leq 1.0$. ii) Frequently, however, the number of adsorbed particles σ_s, is considered relative to the maximum number of surface atoms per cm^2 of the adsorbent, N_A:

$$\Theta = \frac{\sigma_s}{N_A} \, .$$

2.42

(The symbol N_A must not be confused here with the Avogadro-number N_L which is often also denoted as N_A but has, of course, a completely different meaning). In this case, Θ can exceed the value of 1 if more than one adsorbate particle couples to a surface atom.

While a *direct* measurement of n_s or Θ, respectively, is extremely difficult (a direct gravimetric determination using a microbalance has been reported in few cases only), it is comparatively simple to monitor a physical quantity which is unambiguously correlated with Θ. This can be the optical absorption coefficient (e.g., in ellipsometry), the peak intensity of an adsorbate-induced energy level (e.g. in an ESCA or UV photoemission experiment) or the change of the work function of the metal substrate, $\Delta\Phi$, caused by the adsorbed particles. Since work functions can easily be measured with high accuracy, adsorbate-induced work function changes have long been used as a monitor of the adsorbed amount, among others by Mignolet [19] and later by Tracy and Palmberg in their CO/Pd(100) adsorption experiments [17]. The physical basis here is the validity of the Helmholtz equation which simply considers a homogeneous plate capacitor formed by the solid surface and the outer end of the dipole layer with density σ and charge $\Delta\Phi$:

$$\Delta\Phi = 4\pi \mu_0 \sigma_s f^* \, ,$$

2.43

with

μ_0 = initial dipole moment of the individual adsorption complex,
σ_s = number of adsorbed particles per surface area, and
f^* = $1/(4\pi\varepsilon_0)$ conversion factor from electrostatic cm-g-s units to the SI system. f* carries a dimension [$V\,m\,A^{-1}\,s^{-1}$]. Since in surface thermodynamics cm-g-s units are still frequently being used it is quite helpful to leave the conversion factor f* in the equations.

In all cases where $\Delta\Phi$ is used as a coverage monitor the proportionality between $\Delta\Phi$ and σ_s must be controlled. In doing so, the amount of adsorbed gas has to be determined, which can be done using a technique called thermal desorption spectroscopy (TDS) and which will be considered in greater detail in Sect. 4.4.1: the adsorbate-covered surface is heated with a linear rate $\beta = dT/dt$ and the adsorbate atoms will leave the surface when the thermal energy supplied is sufficient to break the adsorptive bond to the substrate. Then the adsorbed particles leave the surface and return to the gas phase where they can be detected, for example, by a mass spectrometer. These desorption experiments, which have frequently been described in much detail in the literature [20 – 24], lead (in a pumped system) to so-called desorption peaks, the area of which directly reflects the amount of gas adsorbed prior to the desorption process. While this procedure yields, of course, only a correlation between $\Delta\Phi$ and the *relative* coverage, a conversion to absolute numbers can be achieved if the structure of the substrate and the adsorbed layer is known. These and related questions will be the subject of Chapter 4, for thermodynamic purposes it often suffices to know relative quantities.

In the following it will be shown how isosteric heats of adsorption were obtained for the adsorption system xenon on a Ni(100) surface using work function measurements [15]. In Fig. 2.2 we present a series of xenon work function data measured from a Ni(100) surface under isothermic conditions (T_{Ni} = 93.16 K). The measurements were performed in the following manner: the bare surface was, at 93.16 K, exposed to a Xe pressure of 1×10^{-8} Torr (1.33×10^{-8} mbar), and the Xe-induced $\Delta\Phi$ was allowed to adjust to its equilibrium value, which takes approximately 150 s. Then, the pressure was steplike increased to the new equilibrium value of 2×10^{-8} Torr (2.66×10^{-8} mbar). Again, it takes

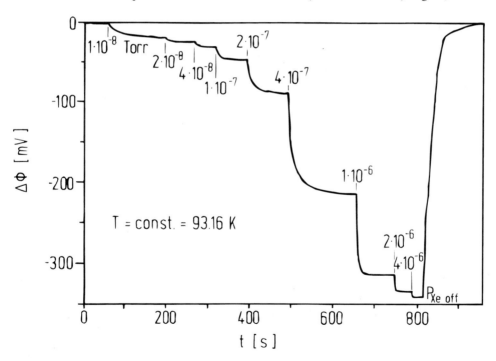

Fig. 2.2. Isothermal (T = 93,16 K) measurement of the Xe-induced work-function change $\Delta\Phi$ (which is proportional to the Xe surface concentration). The experiment was performed in the following manner: certain Xe pressures (indicated in the figure) were step-like adjusted and the work function change of the Ni(100) surface $\Delta\Phi$ was allowed to reach its respective equilibrium value. This takes some time, depending on the kinetics of adsorption and hence on P and T. After Christmann and Demuth [15]

a while after the accompanying $\Delta\Phi$ value is adjusted. In this way it is possible to obtain corresponding P_{Xe} and $\Delta\Phi$ *equilibrium* values *for a given temperature T* (isothermal conditions). At saturation, a $\Delta\Phi$ of some 370 mV is produced which corresponds to a full Xe monolayer ($\Theta = 1$). This layer then contains 5.65×10^{14} Xe atoms/cm^2. A word must be added here as regards the absolute coverage calibration. It is well-known that appreciable mutual depolarization effects occur within layers of ionic or strongly polarized adsorbates (such as xenon), particularly at higher surface concentrations (cf., Sect. 4.4.2). These effects make Eq. 2.43 invalid for $\Theta > 0.5$ and lead frequently to a minimum in a $\Delta\Phi - \Theta_{rel}$ plot when a single monolayer of dipoles is completed. Exactly this calibration was employed in the present case [15].

If we now repeat the isothermal measurements of Fig. 2.2 for different temperatures, the adsorption isotherms of Fig.2.3 are obtained. It can easily be seen that a given Xe coverage can be adjusted by choosing different pairs of temperature T and Xe pressure P: The higher T the higher is the equilibrium pressure P. Figure 2.4 then shows the isosteric

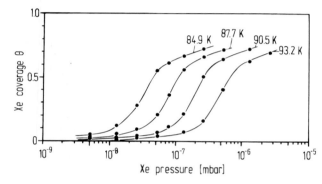

Fig. 2.3. Adsorption isotherms for Xe on Ni(100) as evaluated from the measurements shown in Fig. 2.2 for various temperatures [15].

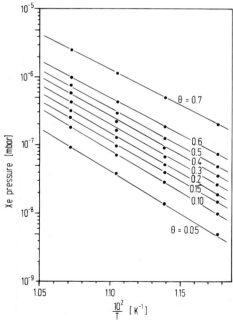

Fig. 2.4. Isosteric plots (ln p_{Xe} vs reciprocal temperature) as derived from the isotherms of Fig. 2.3 for various Xe surface coverages [15].

20

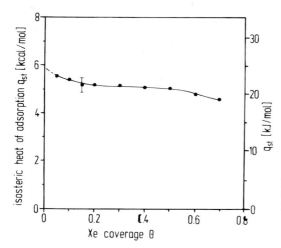

Fig. 2.5. Coverage dependence of the isosteric heat of Xe adsorption on a Ni(100) surface determined from Fig. 2.4 [15].

plot, i.e., in P_{Xe} vs. $1/T$, the so-called isosteres, for different Xe coverages. Evaluation of the slope of these curves finally yields Fig. 2.5, the coverage dependence of the isosteric heat of adsorption, q_{st}, for Xe on Ni(100). Figure 2.5 contains two essential informations, viz., the absolute value of the adsorption energy of Xe on Ni(100) at small coverages, q_{st} = 5 kcal/mol = 21.5 kJ/mol (which may well be interpreted as Xe's *binding energy* to Ni as desorption experiments [15] show), and the coverage dependence of q_{st} which more or less displays the operation of mutual Xe-Xe interactions. It can immediately be seen that q_{st} decreases somewhat with increasing surface concentration of Xe, indicating *repulsive* lateral interactions. (These energetic questions will be dealt with in more detail in Chapter 3).

2.4 The Adsorption Isotherm

When energetic data are interpreted in terms of microscopic interactions we have almost left the thermodynamic concept. However, before doing so in Chapter 3 we first have to evaluate some more thermodynamic and kinetic properties in connection with the adsorption isotherm mentioned in the preceding section. The function $\Theta(P)_T$ relates the adsorbed amount with the equilibrium pressure at a given temperature. Knowledge of this function is, of course, also extremely valuable in practical adsorption technology. Among others, saturation densities of adsorbing and absorbing materials in filter devices can be calculated and predicted. Several analytical expressions have been communicated to describe isotherms [2,5]; here we concentrate on the most important one, the so-called *Langmuir* isotherm, which is both easy to derive theoretically and widely applicable to experimental data, not only in gas adsorption, but also in adsorption from solutions [25]. Most conveniently, its derivation is based on kinetic arguments, i.e., the rate of adsorption is equated with the rate of desorption under equilibrium conditions. In this derivation, three important assumptions are made: First, the adsorption is regarded as being ideally localized, that is to say, one adsorbed particle occupies one adsorption site (immobile adsorption). Second, the adsorption capacity of the surface is completely exhausted as soon as a full monolayer is formed ($\Theta = 1$), and third, the adsorbed particles do not interact whatsoever with their neighbors. This condition implies that all adsorbed particles produce an identical heat of adsorption, Δh_{ad} or q_{st}, which decreases abruptly to zero as soon as the surface is saturated.

The rate of adsorption is proportional to the number of molecules impinging per unit time on a surface with unit area, the so-called particle *flux F*, and to the (dimension-less) efficiency that an impinging particle actually sticks, the so-called sticking probability s_0 (s_0 = number of actually adsorbed particles/number of impinging particles: $0 \le s_0 \le 1$). Furthermore, an eventual activation barrier for adsorption (height ΔE^*_{ad}) must be taken into account. (More details on a microscopic interpretation of the sticking process will be provided in Sect. 3.3). According to kinetic theory, the flux $F = 1/4(N/V)\bar{c}$ ($\bar{c} = \sqrt{8RT/\pi M}$ = mean molecular velocity) is proportional to the gas pressure P, and one can then formulate for the rate of adsorption:

$$r_{ad} = \frac{d\sigma_s}{dt} = \frac{P}{\sqrt{2\pi mkT}} \cdot s_0 \cdot e^{-\frac{\Delta E^*_{ad}}{kT}} \cdot f(\sigma_s) , \qquad 2.44a$$

whereby the function $f(\sigma_s)$ accounts for the increasing loss of empty sites as the adsorption proceeds. This implies that there are different sticking probabilities depending on whether an adsorption site is empty or already occupied (we will further expand on that matter in Sect. 3.3). The rate of adsorption r_{ad} can be expressed in Θ as well to yield, using $\sigma_s = \Theta N_A$,

$$r_{ad}(\Theta) = \frac{d\Theta}{dt} = \frac{P}{\sqrt{2\pi mkT}} \cdot s_0 \cdot e^{-\frac{\Delta E^*_{ad}}{kT}} \cdot f(\Theta) . \qquad 2.44b$$

For atomic or molecular adsorption where one adsorbing particle is consuming just *one* adsorption site, $f(\Theta)$ simply equals $(1 - \Theta)$, for dissociative adsorption where one particle breaks apart into two fragments, each consuming one site, $f(\Theta)$ writes $(1 - \Theta)^2$, etc. In a similar manner, one obtains for the rate of molecular (associative) desorption:

$$r_{des} = -\frac{d\sigma_s}{dt} = k^{(1)}_{des} \cdot \sigma_s , \qquad 2.45a$$

or

$$r_{des}(\Theta) = -\frac{d\Theta}{dt} = k^{(1)}_{des} \cdot \Theta , \qquad 2.45b$$

with

$$k^{(1)}_{des} = \nu^{(1)}_{des} \exp\left(-\frac{\Delta E^*_{des}}{kT}\right) , \qquad 2.46$$

where $\nu^{(1)}_{des}$ denotes the pre-exponential or frequency factor and ΔE^*_{des} the activation energy for desorption (first-order process). For dissociative adsorption with the recombination of fragments prior to desorption being rate-limiting we have

$$r_{des} = -\frac{d\sigma_s}{dt} = k^{(2)}_{des} \cdot \sigma_s^2 , \qquad 2.47a$$

or

$$r_{des}(\Theta) = -\frac{d\Theta}{dt} = k^{(2)}_{des} \cdot \Theta^2 \cdot N_A , \qquad 2.47b$$

and

$$k^{(2)}_{des} = \nu^{(2)}_{des} \exp\left(-\frac{\Delta E^*_{des}}{kT}\right) , \qquad 2.48$$

22

(second-order process). The equilibrium condition now states:

$$|r_{ad}| = |r_{des}| ,$$ 2.49

which yields, for non-dissociative adsorption:

$$\frac{P}{\sqrt{2\pi mkT}} \cdot s_0(1 - \Theta) \cdot e^{-\frac{\Delta E^*_{ad}}{kT}} = \nu^{(1)}_{des} \cdot \Theta \cdot \exp\left(-\frac{\Delta E^*_{des}}{kT}\right) .$$ 2.50

Upon rearranging and introducing a constant b which depends only on temperature,

$$b = \frac{s_0 \exp(-\{\Delta E^*_{ad} - \Delta E^*_{des}\}/kT)}{\nu^{(1)}_{des}\sqrt{2\pi mkT}} ,$$ 2.51

we obtain for Θ the expression

$$\Theta = \frac{b_{(T)}P}{1 + b_{(T)}P} ,$$ 2.52

which is the famous Langmuir isotherm. Combining Eqs. 2.44 and 2.47 yields a similar formula for the dissociative adsorption:

$$\Theta = \frac{\sqrt{b'_{(T)}P}}{1 + \sqrt{b'_{(T)}P}} .$$ 2.53

Figure 2.6 shows a typical example for a Langmuir isotherm in which the coverage is plotted, for two different temperatures $(T_1 < T_2)$, against gas pressure. Apparently, a steep increase in the adsorbed amount is followed by a saturation region, because for $P \gg 1$, $\Theta = 1$ is reached. On the other hand, in the low pressure limit $(bP \ll 1)$ there exists a proportionality between Θ and P for molecular adsorption:

$$\Theta \approx bP ,$$ 2.54

and between Θ and the square root of P for dissociative adsorption:

$$\Theta \approx b'^{1/2} \cdot P^{1/2} ,$$ 2.55

which renders an easy distinction between the possible molecular and dissociative adsorptions.

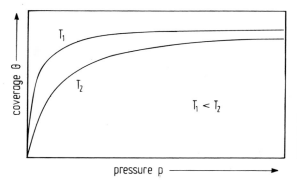

Fig. 2.6. Example of isotherms of the Langmuir type for two different temperatures. Note the typical saturation behavior.

It must be emphasized that the Langmuir model assumptions, particularly the coverage independence of the heat of adsorption, are often far from reality – nevertheless, many adsorption systems display experimental isotherms that resemble Fig. 2.6 and can principally be fitted (by adjusting constant b) to model isotherms of the type described by Eq. 2.52. In this case, however, the interpretation of constant b in terms of molecular constants can be dubious and does not always lead to a significant physical interpretation.

For practical purposes it is often better to develop other isotherms based on more realistic assumptions such as a coverage dependence of the heat of adsorption (i.e., $q_{st} = q_{st}^0(1-a\Theta)$, with a = empirical constant). This leads, for example, to the derivation of the so-called *Temkin* adsorption isotherm [26]. Often, the sorption capacity is not restricted to a single monolayer, rather multilayers can be adsorbed, particularly if weak interactions are involved (physisorption), leading finally to the phenomenon of condensation. In this situation, the well-known BET isotherm describes that behavior quite well which was first introduced by **B**runauer, **E**mmett and **T**eller on the basis of theoretical considerations. More details can be found in the original literature [27]. The BET isotherm has some practical relevance in that it allows to determine the surface area of porous substances such as catalysts [28].

2.5 The Adsorption Entropy

Some remarks shall be devoted to a quantity which has been frequently mentioned in the foregoing sections, namely, the entropy of adsorption S_{ad}. S_{ad} principally contains a wealth of microscopic information about the state of the adsorbate, but it is difficult to extract this clearly from the experimental data. As an example, we shall, in a moment, consider again the system Xe on Ni(100).

The entropy S_s of the adsorbate can be calculated either from the equilibrium heat of adsorption, Δh (cf., Eq. 2.24) (if the surface pressure φ is known) or from the isosteric heat of adsorption, q_{st} (cf., Eq. 2.34), keeping in mind that the entropy is the reversibly exchanged heat divided by temperature T. In addition, the corresponding equilibrium gas pressure must be known for an entropy calculation, as will be shown below. We have to distinguish between the *partial* molar entropy of the adsorbate, $\bar{s}_s$ and the *integral* molar entropy of the adsorbate, S_s. Both are interrelated according to the following considerations:

$$S_s = n_s \cdot s_s , \qquad 2.56$$

(S_s = integral entropy of n_s moles of adsorbate). Differentiation with respect to n_s yields:

$$\left(\frac{\partial S_s}{\partial n_s}\right)_{T,P,\alpha} = \frac{\partial}{\partial n_s}(n_s \cdot s_s) \equiv \bar{s}_s = n_s \left(\frac{\partial s_s}{\partial n_s}\right)_{T,P,\alpha} + s_s . \qquad 2.57$$

This means that the partial molar entropy of the adsorbate contains a contribution of the integral molar entropy of adsorbate, plus a term that depends on n_s, that is, on coverage. Only if $(\partial s_s/\partial n_s)_{T,P,\alpha} = 0$ both entropies are equal. In general, differential experimental entropies can be more readily extracted from experimental isotherms than can integral entropies. The change in Gibbs energy G during isothermal transition of an infinitesimal amount of gas at temperature T and standard pressure $P_0 = 1$ atm into the adsorbed state (equilibrium pressure P) can be written as (g = molar Gibbs energy)

$$\Delta\bar{g} = \bar{g}_s - g_g = (\bar{h}_s - h_g) - T(\bar{s}_s - s_g) , \qquad 2.58$$

where the difference $(\bar{h}_s - h_g)$ equals the isosteric heat of adsorption, q_{st}. Furthermore,

$$\Delta\bar{g} = RT \ln \frac{P}{P_0} \qquad 2.59$$

24

(P = equilibrium pressure, P_0 = standard pressure). For $P_0 = 1$ atm, Eq. 2.59 reduces to $\Delta \bar{g}$ $= RT \ln P$, and we obtain for the partial molar entropy of adsorption

$$\bar{s}_s = s_g - R \ln P - \frac{q_{st}}{T} \ .$$

2.60

From this expression the integral molar entropy can be calculated by integration

$$s_s(n_s) = \frac{1}{n_s} \int_0^{n_s} \bar{s}_s(n_s) dn_s \ .$$

2.61

In order to integrate the expression, the coverage dependence of $\bar{s}_s$ must be known. It is recalled that instead of n_s, the number of moles of adsorbate, the aforementioned coverage Θ can be used as well:

$$s_s(\Theta) = \frac{1}{\theta} \int_0^\Theta \bar{s}_s(\Theta) d\Theta \ .$$

2.62

To summarize, entropies of adsorption (either differential or integral values) can be obtained in a straight-forward manner from experimental heats of adsorption and measurements of the equilibrium pressure P. If the coverage dependence of q_{st} is known, one can also readily deduce the coverage dependence of the entropy of adsorption. The determination of adsorption entropies can be extremely useful if the physical state of an adsorbate is to be determined. For this purpose, of course, some microscopic view is required as far as the configuration of the adsorbed particles, their distribution among the N_A adsorption sites, and their translational, rotational, and vibrational states are concerned. Using statistical mechanics, entropy values can be calculated, provided the partition functions of the various degrees of freedom are separable and, for rotations and translations, the classical approximations hold. Information about the vibrational frequencies of adsorbed molecules can be obtained, e.g. from vibrational spectroscopies such as high-resolution electron energy loss spectroscopy (HREELS) or infrared spectroscopy (cf., Sect. 4.1.5). Furthermore, the different possible arrangements of N_s identical immobile adsorbed particles on N_A equivalent but distinguishable adsorption sites gives rise to a configurational entropy term which, using the approximation of Stirling's formula, can be expressed [29] in integral form

$$s_{s, conf} = -R \left\{ \ln \Theta + \left(\frac{1 - \Theta}{\Theta} \right) \ln (1 - \Theta) \right\} ,$$

2.63

and in differential form (differential molar configurational entropy):

$$\bar{s}_{s, conf} = -R \ln \left(\frac{\Theta}{1 - \Theta} \right) \ .$$

2.64

At $\Theta = 1/2$, $\bar{s}_{s, conf}$ becomes zero. For dissociative adsorption, Eq. 2.63 must be multiplied by a factor 2. Note that the configurational entropy is independent of temperature and hinges only on the number of configurations for given N_s with respect to the total number of adsorption sites provided by the surface lattice.

From isosteric heat measurements, Christmann and Demuth have calculated the coverage dependence of the differential entropy of adsorption at $T = 90$ K for Xe adsorbed on a Ni (100) surface [15]. Using an s_g value for gaseous Xe at 90 K of 34.5 calmol^{-1}K^{-1}, according to Eq. 2.60 $\bar{s}_s$ can also be evaluated. The result is shown in Fig. 2.7 and may be compared with the coverage dependence of q_{st} for the same adsorption system (cf., Fig.

2.5). The adsorption entropy $\bar{s}_s - s_g = \Delta s_{ad}$ exhibits an initial increase by about 1.5 cal-mol^{-1}K^{-1}, thereafter it decreases, until at $\Theta_{Xe} = 0.3$ a value of 21.5 calmol^{-1}K^{-1} is reached. Then, Δs_{ad} again increases to values around 26 calmol^{-1}K^{-1} at $\Theta_{Xe} = 0.7$. Let us discuss the information that we can extract from this behavior. In extreme cases, the Xe configuration can either be completely localized (a very likely situation at low temperatures) or completely mobile (delocalized). The latter case can be assumed to be real at elevated temperatures where the thermal energy content kT is comparable to or larger than the lateral variation of the periodic adsorption potential. For Xe on Ni, this latter condition certainly holds at temperatures of ~90 K. A monoatomic gas in this state has two entropy contributions, namely, the two-dimensional translation parallel to the surface, and the vibration perpendicular to the surface. With the assumption of a dilute perfect two-dimensional gas the coverage dependence of the translational part of the entropy can be calculated via [29]

$$s_{tr,2D} = R \left(\frac{e^2 2\pi M R T b}{h^2 N_L^2} - \ln \Theta \right) , \qquad 2.65$$

where M = molar mass of Xe = 131.32 gmol^{-1}, b = area occupied by one Xe atom at saturation = 1.77×10^{-15} cm^{-2}, T = equilibrium temperature = 90 K and N_L = Avogadro's constant = 6.023×10^{23} atoms mol^{-1}.

The curve resulting from Eq. 2.65 is shown as curve b in Fig. 2.7. A comparison with the experimental curve a yields a downshift of ~2 calmol^{-1}K^{-1} of the theoretical curve. The consideration of the actual size of the adsorbed Xe atoms within the approximation of the so-called Volmer gas [29] would lead to an even lower theoretical translational entropy curve (curve c in Fig 2.7). If we fully attribute the deviation to vibrational effects (the configurational part s_{conf} is quite small) we have to account for a ~4 calmol^{-1}K^{-1} downshift. Assuming a ground state vibrational energy of ~2.5 meV [30] for adsorbed Xe yields indeed about 4 calmol^{-1}K^{-1} for this vibrational contribution, and a satisfactory agreement between theory and experiment results for the coverage range $0.1 < \Theta_{Xe} < 0.4$.

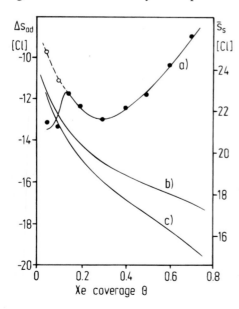

Fig. 2.7. Entropy of adsorption Δs_{ad} (left-hand scale unit: Clausius (Cl) = cal K^{-1} mol^{-1}) and partial molar entropy of adsorption $\bar{s}_s$ (right-hand scale) for Xe on Ni(100) in the low and medium coverage range, as calculated from the data of Fig. 2.6. Curve a) = experiment; open circles are data points which would result if there were a constant isosteric heat observed also at small coverages. Curve b) indicates calculated adsorption entropies for a dilute perfect two-dimensional gas, and curve c) that of a so-called Volmer gas which takes into account the size of the adsorbed atoms.

Yet, there are apparently considerable deviations at small and large coverages. At small coverages the influence of surface defects on the configuration of the adparticles will certainly become noticeable, at larger coverages vibrational coupling effects leading to Θ-dependent vibrational frequencies cannot be ruled out which may influence s_{vib} and hence Δs_{ad}. Besides Xe/Ni(100) there are few other reports on entropy determinations and interpretations for single crystal surfaces/adsorbate systems reported in the literature, for example, Xe on Pd(100) [31] and carbon monoxide and hydrogen on Pd(100) [32, 32a]. Interestingly, also with these latter studies the coverage dependences of the experimental adsorption entropies could, for larger coverages, not be reconciled with the aforementioned simple theoretical predictions. Nevertheless, the data confirmed in all cases the information about the mobility of the adsorbate obtained from microscopic structure-sensitive tools (cf., Chapter 4) which underlines the utility of an entropy determination. Much more entropy data are available for polycrystalline substrate materials (charcoal, Ni films, etc.); a list of references can be found elsewhere [33].

2.6 Surface Kinetics

We have seen, in the preceding section, that a measurement of adsorption entropies can provide valuable information on the microscopic state of adsorbed particles. Besides thermodynamic, also kinetic experiments are capable of giving this information (besides much more), in particular, the already mentioned thermal desorption measurements. Not only do they give access to activation energies for desorption (this issue will be dealt with in Chapter 4), but also to pre-exponential factors (cf., Eqs 2.46 and 2.48) which can be interpreted in terms of transition state theory. Within the framework of statistical mechanics, one can then examine the activation entropy with respect to the molecular structure of the transition state complex. There is a wealth of literature on that subject to which we refer the reader for further specific details [34-38], in particular, for the absolute rate theory treatment of the *adsorption* reaction which must be omitted here for the sake of brevity. The principal considerations are very similar to the *desorption* which we shall deal with in the following. The desorption process can be viewed as a chemical reaction in which the reactants are the adsorbed particles (the adsorbate) which transform to gaseous product molecules via an intermediate state, the so-called activated complex. The situation is illustrated by means of Fig. 2.8. If we first consider the desorption of atoms or intact molecules (for example, Xe or carbon monoxide, respectively) one can write:

$$\underset{\text{reactants}}{CO_{ad}} \rightleftharpoons \underset{\substack{\text{activated}\\\text{complex}}}{\{CO\}_{ad}^{\ddagger}} \longrightarrow \underset{\text{products}}{CO_g} \ .$$

We further assume that the rate-determining step of the overall desorption reaction is the formation of the activated complex and that statistical equilibrium exists between adsorbed molecules and activated complex. These complexes are thought of as vibrating against the surface with frequency v, which is the frequency of decomposition of the complexes. In contrast to the adsorption reaction, surface sites do not play a role, and one can write for the equilibrium constant

$$K = \frac{[\{CO\}_{ad}^{\ddagger}]}{[CO_{ad}]} \qquad\qquad 2.66$$

(the brackets denote surface concentrations).

On the other hand, K can be expressed by the molecular partition functions of the respective species q_i to give

$$K = \frac{[\{CO\}^{\ddagger}_{ad}]}{[CO_{ad}]} = \frac{q^{\ddagger}_{CO}}{q_{ad}} .$$ 2.67

Separating the zero-point energies from the partition functions, and from $q^{\ddagger}_{CO}$, the term $kT/h\nu$ due to the vibration perpendicular to the surface, we obtain

$$\frac{[\{CO\}^{\ddagger}_{ad}]}{[CO_{ad}]} = \frac{kT}{h\nu} \cdot \frac{q^{\ddagger}_{-1}}{q_{ad}} \cdot \exp\left(\frac{-\Delta E^*_{des}}{kT}\right) ,$$ 2.68

where ΔE^*_{des} is the activation energy for desorption at absolute zero and $q^{\ddagger}_{-1}$ denotes the complete partition function of the activated complex minus one vibrational degree of freedom.

The rate of desorption can now be expressed as

$$r_{des} = \nu[\{CO\}^{\ddagger}_{ad}] = [CO_{ad}] \cdot \frac{kT}{h} \cdot \frac{q^{\ddagger}_{-1}}{q_{ad}} \cdot \exp\left(\frac{-\Delta E^*_{des}}{kT}\right) .$$ 2.69

The CO surface concentration $[CO_{ad}]$ is actually equal to σ_s (cf., Eq. 2.42) and can be replaced by a coverage-dependent function $f(\Theta)$ times the total number of adsorption sites per unit area, N_A, via

$$[CO_{ad}] = \sigma_s = N_A \cdot f(\Theta) .$$ 2.70

In case of associative desorption $f(\Theta)$ simply equals Θ. We then have for the desorption rate, formulated as $-d\Theta/dt$, the change of coverage with time:

$$r_{des}(\Theta) = -\frac{d\Theta}{dt} = \frac{kT}{h} \cdot \frac{q^{\ddagger}_{-1}}{q_{ad}} \cdot \exp\left(\frac{-\Delta E^*_{des}}{kT}\right) \cdot \Theta .$$ 2.71

This rate expression has to be compared with the phenomenological Wigner-Polanyi-equation (Eqs. 2.45 and 2.46), which describes desorption from a uniform surface:

$$r_{des}(\Theta) = -\frac{d\Theta}{dt} = \nu^{(1)}_{des} \cdot \exp\left(\frac{-\Delta E^*_{des}}{kT}\right) \cdot \Theta .$$ 2.72

The combination of Eqs. 2.71 and 2.72 enables us to interpret the pre-exponential factor ν_{des} in terms of statistical theory:

$$\nu^{(1)}_{des} \approx \frac{kT}{h} \cdot \frac{q^{\ddagger}_{-1}}{q_{ad}} .$$ 2.73

For the case of dissociative adsorption and associative desorption as it is observed, for example, in hydrogen desorption, similar considerations as above lead to the equation

$$r_{des}(\Theta) = -\frac{d\Theta}{dt} = \frac{kT}{h} \cdot \frac{q^{\ddagger}_{-1}}{q^2_{ad}} \cdot \exp\left(\frac{-\Delta E^*_{des}}{kT}\right) \cdot N_A \cdot \Theta^2 ,$$ 2.74

which can again be compared with the Wigner-Polanyi-equation (cf., Eq. 2.47b) to yield:

$$\nu^{(2)}_{des} \approx \frac{kT}{h} \cdot \frac{q^{\ddagger}_{-1}}{q^2_{ad}} \cdot \frac{1}{N_A} .$$ 2.75

Discussing the molecular desorption first, Eq. 2.73 suggests that $v_{des}^{(1)}$ equals $kT/h \cong 6 \times 10^{12}\,\text{s}^{-1}$ at $T = 300$ K if the activated complex and the adsorbed species have the same partition function, i.e., possess the same degrees of translational, rotational and vibrational freedom. However, if the transition state complex is less strongly bound to the surface (i.e., represents a molecule "ready" for desorption) it may have higher degrees of freedom so that $q_{-1}^{\ddagger} > q_{ad}$. For example, a weakly held activated complex may be delocalized parallel to the surface and therefore has gained translational freedom which can lead to an increase of $q_{-1}^{\ddagger}$ by a factor of 10^3. In this case, relatively high pre-exponentials arise up to $10^{-16}\,\text{s}^{-1}$ (as they have been reported by Pfnür et al. [39] for the system Ru(0001)/CO).

If we deal with dissociative adsorption in which recombination of individual atoms is the rate-determining step during desorption, for example, in the reaction

$$H_{ad} + H_{ad} \leftrightharpoons \{H_2\}_{ad}^{\ddagger} \longrightarrow H_{2g} \,, \qquad\qquad 2.76$$

we must consider various different situations: First, one can again envisage that the adsorbed particles are less mobile than the activated complex, owing to their higher binding energy to the surface, and $q_{-1}^{\ddagger}$ considerably exceeds q_{ad}. Again, this results in fairly large pre-exponential factors that may amount up to 100 $\text{cm}^2\text{s}^{-1}\text{particle}^{-1}$ at 300 K (note that in second-order reactions $v_{des}^{(2)}$ has a different dimension). Because both the adsorbed atoms as well as the activated complex are localized, $q_{-1}^{\ddagger}$ is approximately equal to q_{ad}^2, and hence $v_{des}^{(2)} \approx (kT/h) \times 1/N_A$ which leads to values of about $6 \cdot 10^{-3}\text{cm}^2\text{s}^{-1}\text{particle}^{-1}$ at room temperature.

A third situation may be given when the adsorbate is less localized than the activated complex (e.g., in sterically demanding reactions with strained transition complexes). In this case, $q_{-1}^{\ddagger} / q_A^2$ may be considerably smaller than unity thus leading to a very low pre-exponential factor.

We should add that application of TST may fail in some cases, such as in ordinary three-dimensional reaction kinetics. Sometimes it helps to introduce a so-called transmission coefficient $\varkappa$ to Eq. 2.69 which accounts for the deviations. In desorption experiments this would show up in a much slower rate than anticipated from the simple Wigner-Polanyi-equation with a normal v-value.

A microscopic interpretation of v is given in Sect. 3.3 together with a listing of various experimentally determined v-values. So far, we have concentrated on the characterization and interpretation of the pre-exponential factor. There remains to provide a thermodynamic understanding of the activation energy term, ΔE_{des}^*. For this purpose, we refer to Fig. 2.8, which displays the potential energy of a general reaction system as a function of the reaction coordinate.

The basis here is provided by van't Hoff's approach, and we define the standard internal energy change for the overall reaction, ΔU_r^0. Between the initial and final stage (and vice versa) the energy usually passes a maximum (the activated complex). For reaction from left to right, a barrier of height ΔE_1^*, and for the reverse reaction from right to left a barrier ΔE_2^*, has to be surmounted. These energies and ΔU_r^0 are interrelated by

$$\Delta U_r^0 = \Delta E_2^* - \Delta E_1^* \qquad\qquad 2.77$$

Assuming equilibrium between products and reactants (equilibrium constant K), one may write for the temperature dependence of K (van't Hoff's reaction isochore)

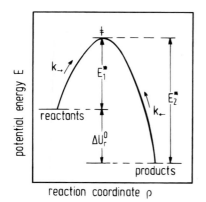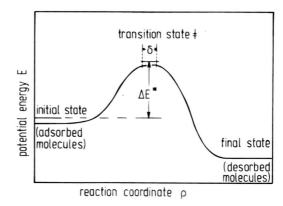

Fig. 2.8. Schematic potential-energy situation (potential energy E vs reaction coordinate ρ) for a surface reaction in general (left-hand side) and for the special case of a desorption reaction (right-hand side), using the view of transition state theory (TST).

$$\left(\frac{\partial \ln K}{\partial T}\right)_v = \frac{\Delta U_r^0}{RT^2} , \qquad\qquad 2.78$$

and with $K = k_\rightarrow / k_\leftarrow$, one obtains

$$\frac{\partial \ln k_\rightarrow}{\partial T} - \frac{\partial \ln k_\leftarrow}{\partial T} = \frac{\Delta E_2^*}{RT^2} - \frac{\Delta E_1^*}{RT^2} , \qquad\qquad 2.79$$

with

$$\frac{\partial \ln k_\rightarrow}{\partial T} = \frac{\Delta E_1^*}{RT^2} , \qquad\qquad 2.80a$$

and

$$\frac{\partial \ln k_\leftarrow}{\partial T} = \frac{\Delta E_2^*}{RT^2} . \qquad\qquad 2.80b$$

Assuming the activation energies independent of temperature, integration yields

$$k_\rightarrow = \text{const}(T) \cdot \exp\left(-\frac{\Delta E_1^*}{RT}\right) , \qquad\qquad 2.81a$$

and

$$k_\leftarrow = \text{const}'(T) \cdot \exp\left(-\frac{\Delta E_2^*}{RT}\right) . \qquad\qquad 2.81b$$

The above relations not only hold for an adsorption-desorption reaction, but also hold for any kind of surface reaction. Instead of using isochoric conditons, one can, of course, also assume an isobaric situation, and the van't Hoff reaction isobar

$$\left(\frac{\partial \ln K}{\partial T}\right)_p = \frac{\Delta H_r^0}{RT^2} , \qquad\qquad 2.82$$

containing the standard enthalpy of reaction, ΔH_r^0, serves as a starting point to derive the relations between rate and activation enthalpies.

We may now pursue the thermodynamic concept by applying equilibrium thermodynamics to the equilibrium constant K of Eq. 2.67. Remembering the relation $\Delta G = -RT \ln K$, we may write for the rate of desorption:

$$r_{des}(\Theta) = -\frac{d\Theta}{dt} = \frac{kT}{h} \cdot N_A \cdot \exp\left(-\frac{\Delta G^{\ddagger}}{RT}\right) , \qquad 2.83$$

where $\Delta G^{\ddagger}$ denotes the change in Gibbs energy associated with the formation of the activated complex. $\Delta G^{\ddagger}$ can be correlated with the energy of desorption and the entropy of the activated complex $\Delta S^{\ddagger}$ according to

$$\Delta G^{\ddagger} = \Delta H^{\ddagger} - T \cdot \Delta S^{\ddagger} , \qquad 2.84$$

and

$$\Delta H^{\ddagger} = \Delta E^*_{des} + \Delta (pV)^{\ddagger} , \qquad 2.85$$

where $\Delta H^{\ddagger}$ denotes the enthalpy of formation of the activated complex, and $V^{\ddagger}$ the corresponding "activation volume". For condensed phases (which we deal with here) the term $\Delta(pV)^{\ddagger}$ vanishes, that is, $\Delta H^{\ddagger}$ equals the activation energy ΔE^*_{des}. We may then write for Eq. 2.83

$$r_{des}(\Theta) = \frac{kT}{h} \cdot \Theta \cdot \exp\left(-\frac{\Delta E^*_{des}}{RT}\right) \cdot \exp\left(\frac{\Delta S^{\ddagger}}{R}\right) . \qquad 2.86$$

A comparison with the phenomenological Wigner-Polanyi-equation immediately yields

$$\nu_{des}^{(1)} \approx \frac{kT}{h} \cdot \exp\left(\frac{\Delta S^{\ddagger}}{R}\right) , \qquad 2.87$$

which clearly shows that the activation entropy is implicit in the frequency factor. Equation 2.87 can be rearranged to yield

$$\Delta S^{\ddagger} \approx R \cdot \ln \frac{\nu_{des} h}{kT} . \qquad 2.88$$

The upper limit of $\Delta S^{\ddagger}$ would be reached if desorption occurs from a completely immobile layer into a completely mobile transition state and would thus correspond to the entropy of a two-dimensional gas. As in three-dimensional reaction kinetics, also negative activation entropies are possible, for example, if, in the transition state, a complicated configuration is formed from a delocalized adsorbate.

It is evident at this point that the physical body of the activation entropy calls for a microscopic interpretation, which we had already announced in the discussion of the pre-exponential factor. While the magnitude and the sign of $\Delta S^{\ddagger}$ allow only fairly indirect conclusions about transient or intermediate stages of a reaction, there are more and more modern spectroscopic tools available which render a more direct physical characterization of short-living reaction intermediates possible. Among others, this will be addressed in the subsequent chapters.

References

1. Clark A (1970) The Theory of Adsorption and Catalysis. Academic Press, New York
2. Hiemenz PC (1977) Principles of Colloid and Surface Chemistry, 1st edn. Dekker, New York
3. Bikerman J (1970) Physical Surfaces. Academic Press, New York
4. Jaycock MJ, Parfitt GD (1981) Chemistry of Interfaces. Ellis Horwood Ltd., Chichester
5. Wedler G (1979) Chemisorption. Butterworths, London
6. Young DM, Crowell AD (1962) Physical Adsorption of Gases. Butterworths, London
7. Dash JG (1975) Films on Solid Surfaces. Academic Press, New York
8. Hayward DO, Trapnell BMW (1964) Chemisorption. 2nd edn. Butterworths, London
9. Černy S (1983) Energy and Entropy of Adsorption. In: King DA, Woodruff DP, (eds) The Chemical Physics of Solid Surfaces and Heterogeneous Catalysis, vol 2. Elsevier, Amsterdam, pp. 1–57
10. Miller AR (1949) The Adsorption of Gases on Solids. Cambridge University Press, London
11. Steele WA (1974) The Interaction of Gases with Solid Surfaces. Pergamon Press, Oxford
12. Roberts MW (1960) Heats of Chemisorption of Simple Diatomic Molecules on Metals. Nature 188:1021–1021
13. Tanaka K, Takamura K (1963) A General Rule in Chemisorption of Gases on Metals. J Catal 2:366–370
14. Goldmann F, Polanyi M (1928) Adsorption von Dämpfen an Kohle und die Wärmeausdehnung der Benetzungsschicht. Z Phys Chem 132:321–370
15. Christmann K, Demuth JE (1982) Interaction of Inert Gases with a Nickel (100) Surface. I. Adsorption of Xenon. Surface Sci 120:291–318
16. Palmberg PW (1971) Physical Adsorption of Xenon on Pd(100). Surface Sci 25:598–608
17. Tracy JC, Palmberg PW (1969) Structural Influences on Adsorbate Binding Energy. I. Carbon Monoxide on (100) Palladium. J Chem Phys 51:4852–4862
18. Christmann K, Schober O, Ertl G, Neumann M (1974) Adsorption of Hydrogen on Nickel Single Crystal Surfaces. J Chem Phys 60:4528–4540
19. Mignolet JCP (1950) Studies in Contact Potentials. II. Vibrating Cells for the Vibrating Condenser Method. Disc Faraday Soc 8:326–329
20. Redhead PA (1962) Thermal Desorption of Gases. Vacuum 12:203–211
21. King DA (1975) Thermal Desorption from Metal Surfaces: A Review. Surface Sci 47:384–402
22. Menzel D (1975) Desorption Phenomena. In: Gomer R, (ed) Interactions on Metal Surfaces, Topics in Applied Physics, vol. 4, Springer, Berlin New York, pp. 101–142
23. Peterman LA (1972) Thermal Desorption Kinetics of Chemisorbed Gases. In: Progress in Surface Science, vol. 3. Pergamon Press, Oxford, pp. 2–61
24. Ehrlich G (1963) Modern Methods in Surface Kinetics. Flash Desorption, Field Emission Microscopy, and Ultrahigh Vacuum Techniques. Adv Cat Rel Subj 14:256–427
25. Langmuir I (1918) The Adsorption of Gases on Plane Surfaces of Glass, Mica and Platinum. J Am Chem Soc 40:1361
26. Temkin MJ, Pyzhev V (1940) Kinetics of Ammonia Synthesis on Promoted Iron Catalysts. Acta physicochim USSR 12:327–356
27. Brunauer S, Emmett PH, Teller E (1938) Adsorption of Gases in Multimolecular Layers. J Am Chem Soc 60:309–319
28. Hiemenz PC (1977) Principles of Colloid and Surface Chemistry, 1st edn. Dekker, New York, pp. 322ff.
29. Clark A (1970) The Theory of Adsorption and Catalysis. Academic Press, New York, p. 42
30. Gadzuk JW, Holloway S, Mariani C, Horn K (1982) Temperaturabhängige Linienbreiten im Photoelektronenspektrum von physisorbiertem Xenon. Verh Dtsch Phys Ges 5:935
31. Miranda R, Daiser S, Wandelt K, Ertl G (1983) Thermodynamics of Xenon Adsorption on Pd(S)[8(100)x(110)]: From Steps to Multilayers. Surface Sci 131:61–91

32. Behm RJ, Christmann K., Ertl G, van Hove MA (1980) Adsorption of CO on Pd(100). J Chem Phys 73:2984–2995;

32a.Behm RJ, Christmann K, Ertl G (1980) Adsorption of Hydrogen on Pd(100). Surface Sci 99:320–340

33. Everett HD (1957) Some Developments in the Study of Physical Adsorption. Proc Chem Soc 37:38–53

34. Hill TL (1960) Introduction in Statistical Thermodynamics. Addison-Wesley, Reading, MA

35. Laidler KJ (1987) Chemical Kinetics, 3rd ed. Harper and Row, New York

36. Johnston HS (1966) Gas Phase Reaction Rate Theory. Ronald Press, New York

37. Glasstone S, Laidler KJ, Eyring H (1941) The Theory of Rate Processes. McGraw-Hill, New York, p. 347

38. Morris MA, Bowker M, King DA (1984) Kinetics of Adsorption, Desorption, and Diffusion at Metal Surfaces, In: Bamford CH, Tipper CFH, Compton RG, (eds) "Simple Processes at the Gas-Solid-Interface", Comprehensive Chemical Kinetics, vol. 19. Elsevier, Amsterdam, pp. 1–179

39. Pfnür H, Feulner P, Engelhardt HA, Menzel D (1978) An example of "fast" desorption: Anomalously high pre-exponentials for CO desorption from Ru(001). Chem Phys Lett 59:481–486

3 Microscopic Treatment of Surface Phenomena

When dealing with surface – gas interaction we again recall that the complete thermodynamic system consists, in the simplest case, of three phases, namely, the solid (bulk) phase of the substrate (for example, a metal or alloy crystal), the gas phase (containing one or more individual gases), and a two-dimensional interface at the boundary: gas – solid. We have, in the preceding chapter, stated that with chemically sufficiently active gases and/or at low enough temperatures this boundary face is enriched in one or more constituents of the gas phase, a process which we have called adsorption. We have also seen that well-defined thermodynamic relationships hold for the various phase equilibria. The knowledge of heats and entropies of adsorption can provide some insight in the microscopic structure of the adsorption systems, but much more powerful in this respect is the microscopic approach, which we shall pursue in this chapter. In the first part, we shall describe some of the physical properties of the phases involved, whereby, in the beginning, the clean substrate phase and thereafter the adsorbate phase deserve the greatest attention. For the sake of brevity, we shall not expand too much on the derivation of the fundamental physical laws and relations which can be found in the respective textbooks; instead, we would like to develop a basic understanding of how the macroscopic properties can be deduced from the microscopic (atomistic) behavior of matter.

3.1 The Structure of Surfaces

3.1.1 Clean Surface Structure

Beginning with a short excursus of the surface properties, it has become clear from the introductory chapter that there are two most prominent characteristics of a surface, namely, its *structure* and its chemical *composition*. At this point, we shall definitely leave the liquid surface which is principallly included in most of the thermodynamic formulae presented in Chapter 2, and concentrate exclusively on crystalline solid surfaces. Usually, the structure of a crystal is a direct consequence of its chemical composition, that is to say, chemical elements crystallize in their characteristic lattice, and so do stoichiometric chemical compounds. Closely related to the geometrical structure is the *electronic* structure which describes the energetic states of the electrons in the surface region of the crystal. In the context of this chapter, we shall not be too much concerned with the surface electronic structure, but we should always keep in mind that surface geometry is a direct consequence of the electronic structure, that is, the charge distribution parallel and perpendicular to the surface.

The crystal structure depends on temperature and pressure, whereby it is sufficient for most purposes to consider the temperature influence. Somewhat more complex is the situation if we do not deal with stoichiometric chemical compounds or elements, but with materials that consist of mixtures, precipitates, thin films, metallic glasses, etc.. Often these do not possess a characteristic homogeneous crystal structure on the macroscopic scale. This holds in particular for catalyst materials used in heterogeneous chemical reactions. Catalysts may consist of bimetallic precipitates, of thin films supported onto alumina (Al_2O_3), silica (SiO_2) or titania (TiO_2), or highly dispersed metals such as platinum black. However, even these materials have a regular structure on the microscopic scale which can be equally well described by crystallography. Therefore, in view of their role in heterogeneous catalysis, it appears reasonable to distinguish *microscopic* and *macroscopic* surface structure (or surface morphology). Accordingly, we find it useful to present in brief some basic principles of surface crystallography. We can, more or less, only touch on the most important points here, since there exist many elaborate textbooks on this subject, and we refer to these presentations and monographs [1-7]. For most purposes in surface chemistry it would be useful if the reader is familiar with at least some fundamentals of crystallography such as the existence of crystal symmetry classes, Miller indices, or x-ray diffraction phenomena. In many cases reading of the respective chapters of textbooks of physical chemistry [8-10] provides the necessary information.

Consider now any bulk crystal. Its most prominent feature is its crystal lattice which is composed of many strictly periodically arranged unit cells with identical positions of atoms or molecules inside. A unit cell is spanned by three vectors *a*, *b*, and *c* with well-defined relative lengths and directions. Actually, the regular arrangement of these atoms or groups of atoms is one of the most striking properties of a solid, and one can understand its whole structure if just the structure of a single unit cell and the relations of its repetition in space is understood. The relations between lengths and directions of the unit cell vectors determine the type of crystal lattice and, thus, the habitus of a macroscopic crystal. Any plane in space can then be related to the direction of *a*, *b*, and *c* and it has become very useful to define the position of this plane by the so-called Miller indices, a triplet of figures (hkl) which is obtained in the following manner (Fig. 3.1.). Consider the hatched planes 1, 2, 3. Their intercepts with the coordinate axes x, y, z (chosen parallel to the cell vectors *a*, *b*, *c*) are expressed in fractions or multiples of *a*, *b*, *c*. These three coefficients are then inverted and, if necessary, multiplied by a factor so as to obtain whole numbers which are put in round brackets and are called "Miller indices ". In our example in Fig. 3.1 we have chosen three parallel planes which form the intercepts $1/_4a$ ($1/_2a$, $3/_4a$) with the x-axis, $1/_2b$ ($1/_1b$, $3/_2b$) with the y-axis and $1/_1c$ ($2/_1c$, $3/_1c$) with the z-axis. For all three planes we end up with the index notation (421), in other words $h = 4$, $k = 2$, $l = 1$.

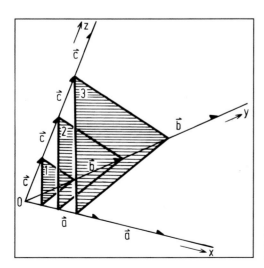

Fig. 3.1. Crystal axes x, y, and z, which are intercepted by crystal planes (hatched areas) to understand the Miller indices. See text for details.

The definition of these indices directly corresponds to the law of rational intercepts of Haüy, and the use of reciprocal intercepts (hkl) to define the position of a crystal face was proposed by W.H. Miller as early as in 1839. If a face is parallel to a, b, or c, the intercept is at infinity ∞, with the respective Miller index becoming zero. If the coordinate axis are cut on the negative sides, the respective indices become negative, too, which is indicated by bars according to $(\bar{h},\bar{k},\bar{l})$. In the hexagonal crystal system, sometimes four Miller indices are chosen (h,k,i,l), because four lattice vectors are used to describe the hexagonal unit cell. The fourth index i is related with the first two indices via:

$$h + k = -i, \tag{3.1}$$

if one prefers to use the three-index notation which is, of course, also possible. A *direction* in a crystal is specified by a line perpendicular to a crystal plane, and the corresponding Miller indices are put in square brackets $[h,k,l]$. The terminating face of a cubic crystal (rock salt, aluminium, nickel, etc.) then has a (100) orientation. The same orientation is obtained if we cut this crystal along a high symmetry plane in [001] direction. The former bulk atoms have now become *surface* atoms and, in a first approximation, still maintain the positions that they had before in the bulk crystal. In our example, we always obtain *squared* arrangements of surface atoms, that is to say, atoms in the two perpendicular surface directions are most densely packed and exhibit identical distances between nearest neighbors. In the same way one can think of cuts along other directions of a cubic crystal, or cuts through crystals of other symmetry (hexagonal, orthorhombic, monoclinic, triclinic etc.), and one will always obtain surfaces in which the atoms are arranged in a certain regular way with a symmetry that can be related to the structure of the bulk crystal. Again, the situation is most obvious or the cubic lattice system. In Fig. 3.2 we show three different cuts through a face-centered cubic (fcc) crystal namely, parallel to the (100), (110), and (111) planes, and we obtain the respective surfaces which are characterized by fourfold, twofold, and threefold symmetry, respectively. Also shown are perspective views of the corresponding surfaces (Fig. 3.3). There are various experimental methods to image the structure of a surface. Some of these methods will be presented in Chapter 4. A particularly elegant probe for surface structure is the Scanning Tunneling Microscopy (STM) [11, 11a]. Using this method, Wintterlin et al. [12] were able to directly observe the hexagonal structure of an aluminium (111) surface. An example is shown in Fig 3.4.

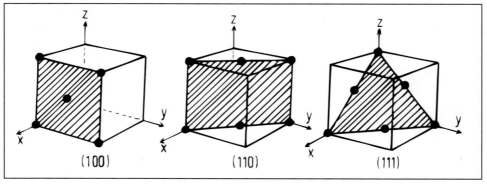

Fig. 3.2. Illustration of the three most important low-index crystal planes of the facecentered cubic (fcc) lattice (100), (110), and (111).

Fig. 3.3. Perspective view of ball models for the surfaces of Fig. 3.2. a) (100) surface, b) (110) surface, c) (111) surface.

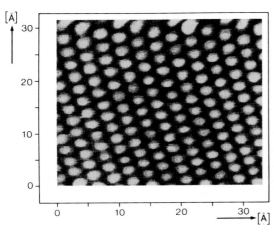

Fig. 3.4. Scanning-tunneling-microscope (STM) image of a (hexagonal) Al(111) surface with atomic resolution. Low levels are shaded dark. The maximum corrugation amplitude is 0.3Å. Tunneling voltage −50 mV, tunneling current $6,3 \times 10^{-9}$ A. After Wintterlin et al. [12]

According to the lower dimensionality of surfaces there exists only a limited number of symmetry operations which can be carried out with *surface* lattices. Actually, one ends up with only five so-called surface Bravais lattices which are reproduced in Fig. 3.5. The surface lattice points can be connected by translation vectors

$$T = ma_1 + na_2 \quad (m,n = \text{integers}),$$ \hfill 3.2

whereby, according to conventions of x-ray crystallography, a_1 and a_2 are chosen so that $|a_1| \leq |a_2|$.

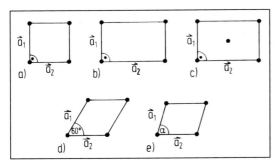

Fig. 3.5. The five surface Bravais lattices a) square, $a_1 = a_2$, $\alpha = 90°$ b) rectangular primitive, $a_1 \neq a_2$, $\alpha = 90°$; c) rectangular centered, $a_1 \neq a_2$, $\alpha = 90°$ d) hexagonal, $a_1 = a_2$, $\alpha = 60°$; e) oblique, $a_1 \neq a_2$; $\alpha \neq 90°$.

From the way we have generated our surface by cleavage of bulk material we expect that the surface periodicity is the same as in the bulk substrate. Relatively often, however, clean surfaces do not exhibit the characteristic periodicity of the bulk crystal. Owing to the asymmetric binding forces in the surface the topmost atoms can be displaced from their normal lattice positions. In the simplest case, the perpendicular distance between the first and second atomic layers is somewhat contracted, by about 1–10% of the nominal bulk layer distance. This phenomenon is called *layer relaxation*. Actually, the displacements and lattice perturbations introduced by the surface are not restricted to the top two layers, but may extend well into the bulk, especially with covalent crystals, for example, semiconductors. We then have the phenomenon of *multilayer* relaxation. Some experimentally determined relaxation parameters are listed in Table 3.1 [13].

Table 3.1. Multilayer Relaxation of various metal surfaces as determined by LEED and other methods [13]

Surface	Δ_{12}	Δ_{23}	Δ_{34}	Reference
Al(111)	+ 0.9			[14]
V(110)	− 0.3			[14]
Cu(100)	− 1.1			[15]
Al(110)	− 8.6	+ 5.0	−1.6	[14, 16]
Al(110)	− 8.5	+ 5.5	+2.2	[17]
V(100)	− 6.7	+ 1.0		[14]
Fe(211)	−10	+ 5		[18]
Fe(310)	−16	+12	−4	[19]
Ni(110)	− 8.4	+ 3.1		[20]
Ni(311)	−15.9	+ 4.1	−1.6	[21]
Cu(110)	−10	+ 1.9		[15]
Cu(110)	− 7.9	+ 2.4		[15]
Cu(110)	− 9.5	+ 2.6		[15]
Cu(110)	− 8.5	+ 2.3	−0.9	[14]
Ag(110)	− 5.7	+ 2.2		[22]
Re(0101)	−17	+ 1		[23]

The (−) sign indicates a layer distance contraction, (+) indicates an expansion; Δ_{12} denotes first, Δ_{23} second and Δ_{34} third layer distance; all values are given in [%] of the unrelaxed distances.

In the case of relaxation, there is no change of the (lateral) surface periodicity, still the surface possesses the crystal structure pertinent to the bulk. This is not so for surfaces which undergo the so-called *surface reconstruction*. This phenomenon is observed with a variety of clean single crystal surface orientations, e.g., (100) faces of W, Pt, Ir or Au, or with Si(111) with its famous 7×7 structure. Driven again by the asymmetry of the chemical binding forces in the surface region the topmost atoms also move laterally, thereby forming new and deviating surface periodicities with respect to the bulk structure. In this case, the surface unit meshes give rise to new diffraction features; they can be described in terms of Wood's nomenclature, which will be explained in a moment. Quite often, surface reconstructions are also induced by chemically active adsorbates; we shall return to this adsorbate-induced reconstruction later. Some examples for reconstructed surfaces are given in Fig. 3.6. There are various types of reconstructions – some of them require substantial mass transfer since the displaced atoms have to move over appreciable distances to new equilibrium positions, others simply consist of a pairing of adjacent atoms and do not require extensive mass transport. Relaxation and reconstruction phenomena have been repeatedly described in the literature [13-30].

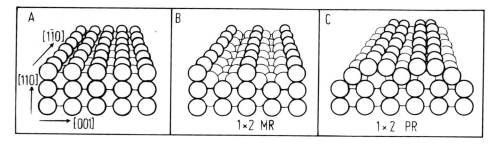

Fig. 3.6. Perspective view of an fcc(110) surface. A) unreconstructed suface, B) missing-row (MR) reconstructed, C) pairing-row (PR) reconstructed.

Other kinds of surfaces which have gained interest in the past are *stepped* and *kinked* surfaces, because their structures can be regarded as a first step towards "real life"surfaces [31]. Regular steps on single crystal surfaces can be produced by making a cut through the crystal at a small inclination angle with respect to a low-index plane. As a result, surfaces are generated (and, surprisingly, could be shown to be thermodynamically relatively stable) that consist of large flat areas (the "terraces") interrupted by small steps of monoatomic or multiple height. Some examples, taken from Somorjai's book [31] are given in Fig. 3.7. Actually, it was Somorjai who first drew attention to stepped surfaces [33] and who developed a nomenclature to assign these surfaces (although stepped surfaces could also be conventionally denoted using Miller indices).

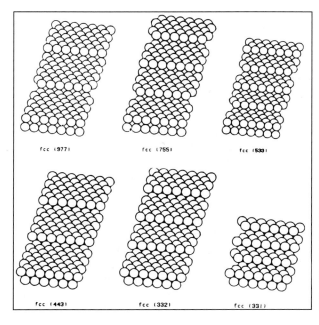

Fig. 3.7. Perspective representation of six stepped fcc surfaces with different step orientations and terrace widths. Reprinted from [31]: G.A. Somorjai *Chemistry in Two Dimensions.* Copyright© 1981 by Cornell University. Used by permission of the publisher, Cornell University Press.

Fig. 3.8. Ball model of a stepped surface of an fcc crystal: [7 (100) × 331] orientation. After [34]

Regarding this nomenclature of stepped and kinked surfaces, there are two notations in use: the "step" notation proposed by Lang et al. [32] and the more generally applicable "microfacet" notation introduced by van Hove and Somorjai [33]. Particularly for high-index planes which are often stepped *and* kinked, the nomenclature is relatively compli-cated, and one cannot easily transform the Miller-index notation to the microfacet nota-tion and vice versa. As an example for a stepped and kinked surface we present, in Fig. 3.8, a (29,3,1) surface which, in the step notation, reads [7(100)×(331)].

Turning to even more complex surface structures, we have to mention so-called faceted surfaces which contain, in sort of a hill-and-valley structure, microfacets of differ-ent orientation, which, however, still exhibit long-range periodicity. Further structural complications are all kinds of lattice defects – grain boundaries, crystal twins, stacking faults, screw dislocations, mosaic structures or vacancies – again, as before the step and kink sites all these disturbances of the ideal surface periodicity represent additional centers of chemical reactivity and they can act as nuclei for chemical attack in that these parts of the surface provide favorite adsorption sites for gas – surface interaction, or entrance channels for gas absorption, bulk diffusion or permeation.

At this point, we have almost left the ideal single crystal surfaces and turned to polycry-stalline materials. The solid bulk material of daily life is usually polycrystalline; metals, for example, are agglomerates of statistically mixed and oriented microcrystallites separ-ated by grain boundaries. Of more chemical interest because of their enhanced reactivity are highly dispersed materials, for instance, dusts and powders which, in the first place, have a much larger surface-to-volume ratio. Particularly important are the so-called sup-ported catalysts which represent active metal-containing chemical compounds precipi-tated from solution onto chemically fairly inert, but porous support materials such as non-metal oxides. As far as the specific structural properties are concerned, we first con-sider the role of this support (γ-Al_2O_3, SiO_2, zeolites, activated carbon, etc.). Its most prominent function is to maximize the surface area of the precipitated "active" phase (Pt or Pd, for example) by its inherent porous structure – surface areas of 100 to 1000 m^2 per gram can be achieved. The actual surface structure of these oxides consists of small single crystalline areas interrupted by more or less deep pores. Sometimes, particularly with zeolites, there exist caverns or cages of a specific size which are capable of trapping certain gas molecules with high selectivity. Chemically, it is of interest that these oxide supports reveal a certain acidity and, therefore, have affinity to hydrogen, oxygen and hydroxo groups, particularly in conjunction with the supported active material, for example, metals. These metals (in some cases, precious metals such as Pt, Ru or Rh) are dispersed over the support surface in an overall relatively small concentration. They can often form local agglomerates, small spheric clusters of regular shape (e.g., octahedral, and icosahedral structures have been found) which expose well-defined small area single-crystal surfaces of various orientation to the reactive gases [35] – again, a hint that the single-crystal approach supported in the introductory chapter rests on solid ground.

The enhanced chemical reactivity of those supported catalyst materials is almost entirely based upon the high degree of dispersion, although in some cases a significant interaction between support and precipitated metal can also occur ("strong-metal-support interaction" SMSI) which leads to peculiar chemical reactivity [36].

Under practical conditions the large surface area must not be reduced by sintering or conglomeration during the reaction (which easily occurs with powders). Therefore, if dispersed material is chosen in heterogeneous catalysis it is often preformed as larger grains, spheres, granules or pellets, which nevertheless exhibit, on a more microscopic scale, high porosity and thus reactive surface. Although this is the topic of "macroscopic" surface structure we give, in Fig. 3.9A, some common forms of examples of coarse catalyst particles taken from the book by Bond [37]. Mechanical requirements here are crushing strength and attrition resistance. Sometimes, it is necessary to have fine particles (for example in fluidized-bed reactors), or it is advantageous (among others, in air-pollution control to have a catalyst of monolithic structure which is made up from a block of α-alumina with fine parallel channels in order to reach high surface area. These channels can have any regular shape; two examples are given in Fig. 3.9B. Of course, it depends largely on the type of reaction, as well as on the chemical reactor, as to which type of catalyst is most advantageous for achieving high turnover numbers and long-term stability in the actual technical process.

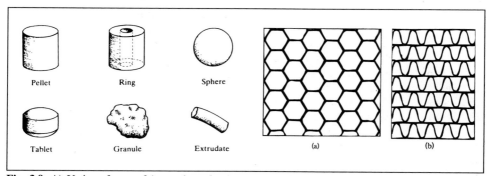

Fig. 3.9. A) Various forms of 'coarse' catalyst particles used in practical catalysis;   B) Typical cross-sections of monolithic support materials: a) honeycomb, b) corrugated. After Bond [37]. Reproduced with permission.

41

Finally, a short comment on the structure of the so-called *bimetallic cluster catalysts* and alloy catalysts in general is worthwhile; a class of catalytically active materials first propagated by the Dutch school of Sachtler [38,38a,39], Ponec [40,41] and others, and then by Sinfelt and his group at EXXON [42-48a]. They are characterized by a unique selectivity in certain hydrocarbon reactions [44,45]. In Sect. 5.4, we will present some more details and an explanation for their chemical behavior. One can easliy understand why these materials have gained such an interest, and accordingly, there exists a vast literature on that subject in which still more details about catalyst structure and function can be found. Here, we must be satisfied with a short description of the structural properties of Cu-Ru bimetallic catalysts, following Sinfelt et al. [46,47]. The Ru and Cu were, in a monometallic form, dispersed onto a silica carrier either by sequential precipitation or coprecipitation from solution. Thereafter, cluster size distributions and shapes were determined by electron microscopy, whereby it is important to mention that Cu and Ru are immiscible in the bulk, although there is evidence of some chemical interaction between the two metals. An example, taken from Prestridge et al. [47], is reproduced in Fig. 3.10. The average diameters of the Ru and Cu-Ru cluster were about 30 Å (sometimes up to 60 Å) with fairly thin layers of Cu deposited onto Ru so as to form "raft-like" aggregates. This particular surface structure promotes again UHV single-crystal model experiments in which Cu (or any other immiscible noble metal) is precipitated onto a Ru substrate and investigated with regard to structural and adsorptive properties [49-52].

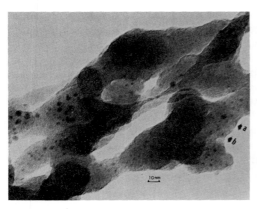

Fig. 3.10. Electron micrographs of silica-supported Ru-Cu catalysts.
a) 1% Ru, 0.63% Cu. The Ru-Cu clusters (dark spots) all have thin raft-like structures. Arrows a and b point to regions at the boundary of the specimen where side views of the clusters are visible.

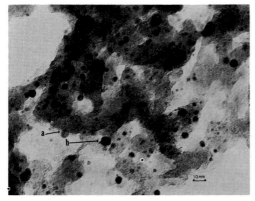

b) 5% Ru, 3.15% Cu. Ru-Cu clusters with a three-dimensional character, in addition to raft-like clusters, are observed as evidenced by differences in contrast with the SiO_2. Arrows a and b indicate clusters where ridges or boundaries of the underlying silica carrier are observable through the clusters. After Prestridge et al. [47]. Reproduced with permission.

Since it is not the intention of this book to present an exhaustive description of all possible microscopic or macroscopic structures of surfaces, we would instead like to turn to the subject of *adsorbate* structure, which is certainly equally important.

3.1.2 The Adsorbate Structure

It is well-known that any additional atom that arrives on a surface (*hkl*) at low enough temperature sticks on that surface and forms a chemical or physical bond to the adjacent surface atom(s). This is the microscopic view of the thermodynamical adsorption process. In terms of surface crystallography, the particle occupies a so-called adsorption site. This site has a defined geometry, which in most cases is closely related to the structure of the surface underneath.

Before we expand on the term and properties of an adsorption site, we present, in brief, an appropriate description of how to assign the crystallographic structure of adsorbate-covered surfaces. Consider a whole ensemble of adsorbed particles which reside on a regular surface and are allowed to interact with each other by repulsive forces. The particles will tend to occupy sites with identical (favorable) local binding geometry which are separated by the largest mutual distances possible. It can be immediately rationalized that surface phases with long-range periodicity in x- and y-direction are formed. Depending on the overall density of the adsorbate, every second, third, fourth etc. surface site is occupied in a very regular manner. This is illustrated in Fig. 3.11 for a cubic (100) (a, b) and a hexagonal (c) surface and for an adsorption site with fourfold (a, b) and twofold (bridge) symmetry (c). Unless we have as many adsorbate particles as adsorption sites (which are, in this example, assumed to equal the number of surface atoms) the adsorbate phase is always more "diluted" than the substrate surface, which means that the unit mesh spanned by the adsorbate lattice vectors b_1 and b_2 (remember that the substrate unit mesh was defined by a_1 and a_2) is larger than that of the substrate. We may write, analogous to Eq. 3.2,

$$T_{ad} = rb_1 + sb_2 \quad (r,s = \text{integers}) \tag{3.3}$$

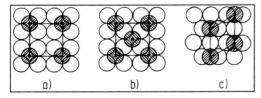

Fig. 3.11. Adsorption sites on a fcc(100) surface a, b) and a hexagonal fcc(111) or hcp(0001) surface c) occupied by the shaded atoms. Sites a) and b) have fourfold, site c) has twofold symmetry.

The surface structures can be classified according to a suggestion made by Wood [53] based on x-ray crystallography. The relation between the adsorbate lattice and the substrate surface lattice is expressed by the ratios of the lengths of the vectors of the unit cell, i.e., $|b_1|/|a_1|$ and $|b_2|/|a_2|$. If the adlattice is rotated by an angle α with respect to the substrate lattice, the value of α (except $\alpha = 0°$) is also indicated. Primitive (p) and centered (c) unit cells are indicated by p and c, respectively. Taken together, we may write for the surface structure of an overlayer of adsorbate species on the $\{h, k, l\}$ plane of a crystal M

$$M\{h, k, l\} - \left(\frac{|b_1|}{|a_1|} \times \frac{|b_2|}{|a_2|} \right) - R \cdot \alpha \ . \tag{3.4}$$

Most of the so-called adsorbate superstructures can be assigned in Wood's nomenclature. There exist, however, more complicated incoherent structures with a lack of common periodicity between substrate and adsorbate which must be assigned in a different way, namely the matrix notation. For details the reader is referred to the original publication by Park and Madden [54].

We take the opportunity and remind the reader of some other important terms of crystallography. We recall that although a crystal is made of elementary material units, its structure is geometrically idealized in that the term "lattice" is introduced, that being only a regular array of points in space. The lattice points can be connected by a regular network of lines in various directions. With two dimensions we have a "surface" lattice (completely analogous to the three-dimensional case), and the lattice area is broken up into many "unit cells" or "unit meshes" each terminated by the aforementioned lattice vectors a_1 and a_2, (or b_1 and b_2, if adsorbate lattices are considered). If each lattice point is replaced by a single identical atom, we obtain a so-called *primitive* lattice, if it is replaced by a whole group of atoms, we deal with a *non-primitive* lattice. We repeat that the artificial term lattice is nothing but an array of points, in the *crystal structure* each such point is replaced by a material unit.

Any ordered surface overlayer can be characterized by surface crystallography (Eq. 3.3). For adsorbate atoms or molecules present on a surface, we must distinguish two cases – the situation, where only a *single* particle is adsorbed, and that where a *whole ensemble* of particles interacts with the surface and with each other. In the first case it is the local geometry of the adsorption complex which deserves interest, whereby the adsorption *complex* consists of the adsorbed particle and all surface atoms that participate in the adsorptive bonding. In the second case *additional* information is required about the long-range correlations between the adparticles (of equal or different kind), that is to say, their long-range order and, of course, about any alterations of the local morphology of the adsorption complex as induced by particle-particle interactions at higher adsorbate concentrations on the surface. In a moment, we shall understand the difference between single and multi-particle phenomena as a consequence of the energetics of a surface.

Let us now first consider the single-particle approach with emphasis on the adsorption site geometry.

As in coordination chemistry, one distinguishes sites with fourfold, threefold, twofold (bridge) or linear (atop) coordination. Each of these sites can be symmetrical or asymmetrical with respect to the surface. If we consider the *symmetrical* sites first and assume a single adsorbed atom, a perpendicular rotational axis connecting the atom with the surface atoms involved in the bonding would have C_{6v}, C_{4v}, C_{3v} or C_{2v} symmetry as illustrated in Fig. 3.12a. For *asymmetric* sites, the same axis would possess only C_s or C_1 symmetry (Fig. 3.12b). These symbols are used in group theory, for more details, monographs of this subject are recommended for further reading [55, 56]. Consider now the adsorption of a molecule consisting of two atoms (CO, for example) for which the same rules apply, if the molecular axis coincides with the rotational axes. The same holds, of course, for all polyatomic *linear* molecules such as CO_2 or C_2H_2. If, however, a triatomic nonlinear (bent) adsorbed molecule such as H_2O is considered there is now a greater variety of possible configurations, all of lower symmetry. Some of the possible geometries are depicted in Fig. 3.13. Particularly in organic chemistry, there are many relatively complicated molecules that still have fairly high symmetry, for example, benzene. In these cases the molecule can take advantage of the surface symmetry – it is expected that benzene would favor adsorption in a flat position into sites with C_{6v} or C_{3v} symmetry which

44

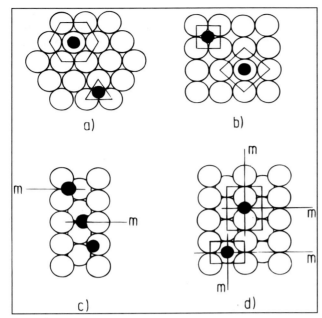

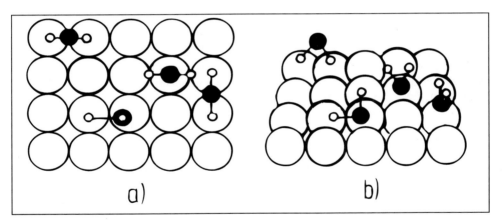

Fig. 3.12. Illustration of the local symmetry of adsorption sites: a) with C_{6v} and C_{3v} symmetry, respectively; b) with C_{4v} symmetry (fourfold hollow and fourfold atop); c) with (quasi) threefold coordination (C_s symmetry with the mirror planes m indicated) and (lower right) with C_1 symmetry (asymmetric site); d) with C_{2v} symmetry (bridge site, hollow site) with the two mirror planes m being indicated.

Fig. 3.13. Schematical representation of different possible configurations of a triatomic-bent molecule (H_2O) on a squared surface: a) top view, b) perspective view.

are provided by hexagonal or trigonal surfaces. Although this naive view is often true, there are also exceptions, particularly if the electron charge distribution in the adsorbed particle is asymmetric and different from the geometry of the molecular skeleton.

In catalysis, the correlation between the shape of a molecule, surface structure, and macroscopic reactivity represents a very important and long discussed problem. One distinguishes between *structure-sensitive* and *structure-insensitive* surface reactions. Special site requirements for adsorption have been discussed in terms of the so-called *ensemble effect* whereafter a molecule can only adsorb with sufficient energy if a certain group of adjacent surface atoms, for example, in a binary alloy AB, consists exclusively of atoms of kind A. As soon as B atoms are mixed in, the adsorption capacity is lost or considerably impaired. The reason why an ensemble is necessary can be at least two-fold. There may be kinetic and/or energetic effects. In the first case the molecule impinging on the surface requires sort of a flat "runway" in order to become adsorbed or dissociated, once it is accommodated on the surface, it can adsorb everywhere by means of the so-called *spill-over* effect. (Spillover designates the mobility of an adsorbed species from one phase onto another where it does not directly (i.e., by impact from the gas phase) adsorb). In the other case, the whole ensemble is necessary to gain sufficient adsorption energy to keep the molecule on the surface. A thorough discussion of the ensemble effect in view of catalysis was given by Sachtler [57,58,58a], but actually, it was Balandin who first pointed to the important role of ensembles in adsorption and cata-lysis when he formulated his famous "multiplet"theory [59,59a].

The aforementioned example of adsorbed benzene is also suited to shed light on another interesting point: On an ideal clean surface there is a well-defined lateral sequence of adsorption sites, according to Eq. 3.2, that is to say, the shortest distance between two identical sites is given by

$$T_1 = a_1 \text{ or } T_2 = a_2, \tag{3.5}$$

depending on the direction of propagation. This, however, cannot mean that all these possible sites are actually occupied by adatoms – as we shall see later there are steric (spa-tial) effects or repulsive electronic interactions between neighboring adatoms which often prevent the simultaneous occupation of adjacent adsorption sites. Adsorption of the rather large benzene (C_6H_6) molecule (5.70 Å diameter [60]) on a Ag(111) surface ($d_{Ag-Ag} = 2.88$ Å) actually excludes adsorption on six adjacent sites. Other examples are the voluminous Xe atom (4.5 Å diameter) on hexagonal graphite (0001) etc. [61] whereby again the six nearest neighbor sites are blocked. This is illustrated by means of Fig. 3.14. On the other hand, there are also particularly small atoms or molecules such as hydrogen or deuterium, and since the H atom has a diameter which is smaller than any other atom, there should exist no such steric limitations with regard to adsorption sites – in other words, it should be possible to occupy each adsorption site by a H atom. Although there are systems where this is observed (so-called (1x1) structures are formed, among others, with Pt(111) [62] or Ru(0001) surfaces [63]), in most of the cases H atoms

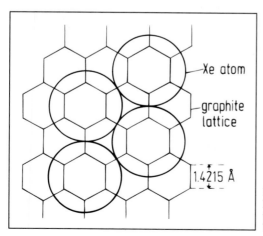

Fig. 3.14. Site occupancy of Xe adsorption on a hexagonal graphite surface (honeycomb struc-ture). One Xe atom effectively blocks six neigh-boring sites. After Morrison and Lander [61].

that are brought closely together repel each other because of quantum-chemical interactions (cf., Sect. 3.2.2), or induce a surface reconstruction, so that even with hydrogen particularly high densities of adsorbed layers on single crystal surfaces are seldom reached. However, on crystallographically rough surfaces, for instance on fcc (110) surfaces, in some cases unusually high adsorbate concentrations could be found; on Rh(110) and Ru(10$\bar{1}$0) hydrogen saturation densities of 2 H atoms/ metal surface atom were reported [64,65]. In Fig. 3.15 a structure model for the system H/Rh(110) is presented [64] which could also be confirmed by LEED [66] (cf., Chapter 4).

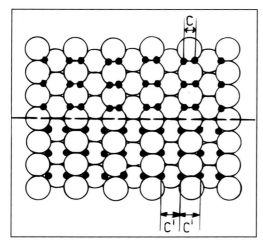

Fig. 3.15. Hydrogen (1 ×1)–2H saturation structure on Rh(110). The small dark circles represent H atoms. In the top part, the H atoms are placed into the threefold sites on both sides of a row of Rh atoms (mutual distance c); in the bottom part, the hydrogen atoms are somewhat displaced laterally to achieve a more homogeneous array (distance c'). After [64].

In the following list we present some examples of important adsorbate structures (Wood's nomenclature) in which the local coordination could be determined by low-energy electron diffraction or other structure-sensitive methods (cf., Chapter 4) (Table 3.2). Where available, also bond lengths and angles are given, more such data are listed in articles by van Hove et al. [67,68].

Table 3.2. Coordination numbers and bond lengths for some resolved adsorbate structures [67]

System	Coverage	Structure	Coordination	Bond length [Å]	Reference
H/Ni(111)	0.5	c(2×2) (honeycomb)	3-fold hollow	1.84 (±0.06)	[70]
H/Ru(0001)	1.0	(1×1)	3-fold hollow	1.91	[71]
O/Ni(100)	0.5	c(2×2)	4-fold hollow	1.92 (±0.02)	[72]
O/Cu(100)	0.5	(√2×2√2)R 45°	(reconstr.) hollow site	?	[73]
N/Fe(100)	0.5	c(2×2)	4-fold hollow	1.81	[74]
CO/Pd(100)	0.5	c(2√2×√2)R 45°	2-fold (bridge)	1.93 (±0.07) (Pd-C distance)	[75]
S/Ni(100)	0.5	c(2×2)	4-fold hollow	2.28	[76]
Se/Ni(100)	0.25	p(2×2)	4-fold hollow	2.34 (±0.07)	[77]
Te/Cu(100)	0.25	p(2×2)	4-fold hollow	2.48 (±0.10)	[78]

It seems appropriate to return to the problem of adsorbate-induced reconstruction of surfaces. We have seen that, driven by minimization of free energy, clean surfaces can and do sometimes lower their surface energy by restructuring. Quite frequently, these processes also occur in the course of gas adsorption. Due to a more or less strong interaction the adsorbate atom(s) remove electron charge from the surrounding metal surface atoms which somewhat weakens their metallic bonds to their neighbors and facilitates lateral and/or vertical displacements of these atoms. One can basically distinguish two types of adsorbate-induced reconstructions, namely, a short-range and a long-range type. In the first case, shifts of surface atoms can only occur in the direct vicinity of an adatom and thus become noticeable only at appreciable adsorbate concentrations when a substrate atom is really surrounded by adsorbed particles. The other type is more induced by long-range changes of the electronic band structure of a surface in that, e.g., electronic surface states are quenched by only a few adsorbate or impurity atoms, which then affects the entire crystal surface and leads to a flip-over to another more stable surface lattice configuration. In reality, both types are frequently observed – examples for local reconstructions are H on Ni(110)-1×2 [79] or Pd(110) [80], for long-range reconstructions H on W(100) [81-84] or K on Ag(110) [85]. Of course, the structural changes can also occur in a reverse direction, that is to say, inherent reconstructions of surfaces can be removed by adsorbed atoms, for the same reasons as mentioned above. An example here is the lifting of the Pt(100)-5×20 reconstruction by adsorbed carbon monoxide [86,86a] or of Ir(110)-1×2 by adsorbed oxygen [87].

In view of practical catalysis the adsorbate-induced restructuring of surfaces is of great importance, since it may provide catalyst surfaces with a greater chemical reactivity under reaction conditions. Particularly at elevated temperatures and gas pressures reconstruction phenomena are believed to play a decisive role in that they may be regarded as a precursor to an actual (reversible) surface compound formation (oxides, nitrides, hydrides, carbides, etc.) thus considerably facilitating certain reaction paths. There exist a great many of studies on adsorbate-induced reconstruction, for more details the reader is referred to the relevant literature [24-30].

3.2 The Energetics of Surfaces

The geometrical structure of clean surfaces is, as pointed out before, just a consequence of the thermodynamic principle of minimization of surface free energy. Under equilibrium conditions, each surface atom will search for its respective site with lowest free energy and find and occupy this site unless it is inaccessible due to diffusion activation energy barriers.

In the microscopic quantum chemical description the appropriate starting point is to set up the Schrödinger equation for the complete system consisting of N_i particles

$$\hat{H}\psi = E\psi \, (r_i, R_i),$$

3.6

where ψ denotes the wave function that depends on the coordinates of all electrons (r_i) and all nuclei (R_i).

The Hamiltonian $\hat{H}$ can be split up into a part for the electrons $\hat{H}_{(el)}$ and for the nuclei $\hat{H}_{(nucl)}$ (V stands for the respective interaction potentials):

$$\hat{H} = \hat{H}_{(el)} + V_{(el-nucl)} + V_{(el-el)} + V_{(nucl-nucl)} + \hat{H}_{(nucl)} \, .$$

3.7

Within the framework of the Born-Oppenheimer approximation the contributions of all electrons and nuclei can be separated:

$$\psi = \psi_{(el)} \cdot \psi_{(nucl)} \cdot \qquad\qquad 3.8$$

and the Schrödinger equation for the electronic part reads

$$\hat{H}_{(el)} \cdot \psi_{(el)} = E_{(el)} \cdot \psi_{(el)} , \qquad\qquad 3.9$$

where $E_{(el)}$ represents the potential $V_{(R_j)}$ of the surface.

These considerations hold for $T = 0$ K, and all the particles would be located at the bottom of the minima of the potential $V_{(R_j)}$, thus the true energy minimum would be reached. For $T > 0$ K, we must (as mentioned before) consider the *surface free energy F* or *Gibbs energy G* which tends to a minimum according to:

$$F = U - TS, \qquad\qquad 3.10a$$

and

$$G = H - TS, \qquad\qquad 3.10b$$

The entropy term is responsible for any structural changes, phase transitions etc., which may occur at higher temperatures. The above equations hold for clean as well as for adsorbate-covered surfaces.

The time-independent Schrödinger equation describes stationary states, whereas the dynamics of surface processes corresponds to changes in the function $\psi_{(nucl)}(t)$ in Eq. 3.7. Even if we neglect those dynamic processes which introduce tremendous additional complexity, we immediately realize that it is hopeless to attempt a solution even of the stationary Schrödinger equation since it refers to a N-dimensional problem for any surface/gas system under consideration.

It is therefore advisable not to start off with a rigorous quantum chemical treatment of the complete surface + adsorbate system, but rather to start by trying to understand the two-particle system adatom-surface atom. Attempts of this kind have been known for a long time and have led, for example, to the Lennard-Jones potential ansatz [88]. At a later stage, the individual potentials can be allowed to interact with each other so as to mimic an extended two-dimensional periodic surface. Details can be found in the respective literature on solid state physics [89].

3.2.1 The Single Particle Interaction, Activated and Non-activated Adsorption

Neglecting for the moment all problems in conjunction with the question how a gas-phase particle is actually trapped in a bound state by a solid surface, we simply consider the interaction potential between this particle and an arbitrarily chosen surface atom M. In the simplest case, the gas particle consists only of a single atom. As soon as this atom is brought so close to the surface that noticeable interaction can occur, that is to say, that overlap between the wave functions of the metal and the gas atom becomes possible, a lowering of the total energy of the combined system can take place, resulting in a bound (adsorbed) state, and any excess energy is released as heat of adsorption. On the other hand, in order to remove the adsorbed particle from the bottom of the potential well back into the gas phase, at least the respective energy has to be supplied to the bound system, in some cases even additional activation barriers must be surmounted. The situation is illustrated by means of Fig. 3.16, which shows the well-known Lennard-Jones potential energy diagram. It can be regarded as a superposition of attractive and repulsive interaction forces between the adsorbing particle and the surface (atom) according to the expression:

$$E_{(z)} = -Az^{-6} + Bz^{-12},$$ 3.11

where A, B = empirical constants, and z = distance of the particle from the surface atom.

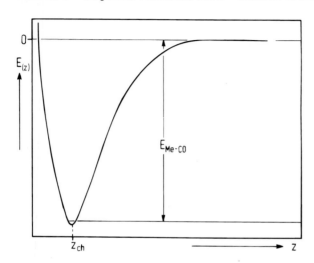

Fig. 3.16. One-dimensional potential energy curve ($E_{(z)}$ vs z) for a molecular adsorbate (CO) approaching a surface along the perpendicular distance coordinate z. At the equilibrium distance z_{ch} the adsorption energy E_{Me-CO} is gained.

The second (positive) term describes the (very short-range) repulsion, and the first (negative) term is the attraction which dominates at somewhat larger atom-surface distances. The energy zero is chosen such that $E_{(z)} = 0$ for $z = \infty$. It is self-evident that deep potential wells indicate strong interaction and vice versa, also the equilibrium position z_0, i.e., the location of the potential minimum, increases with decreasing surface-adatom bond strength. It is also clear that the shape of the interaction potential is entirely determined by the interaction *forces* between an adsorbed particle and the surface, and there are essentially three different types of these forces operating, depending on the system under consideration. In principle, it is the same forces which provide the various types of bonding between isolated atoms (i.e., van-der-Waals, ionic and covalent forces), however, the situation with a surface is much more complex since an adatom is usually not coupled to a single surface atom, but rather to a whole array of atoms. Whereas with distinct atoms the quantum chemical interaction is between discrete atomic orbitals of sharp energies, in the adatom-surface interaction we have delocalized electronic bands on the surface side, and sharp orbitals on the adatom only when it is far away from the surface. As the distance z gets smaller these latter orbitals also broaden and shift, and certain degeneracies are lifted, owing to the reduction of overall symmetry caused by the presence of the surface. This is roughly the physical basis of quantum chemical theories that make use of the band structure model [89,90]. However, in many theoretical treatments of the chemisorption problem the so-called cluster approach is chosen. The cluster consists of a surface molecule (= the adatom/molecule + a small unit of adjacent surface atoms that contributes most to the bonding) which can be calculated quantum-chemically with relatively great precision and with a smaller effort than the corresponding band structure treatments [91].

50

In summary, the interaction forces leading to adsorption can be distinguished with respect to their physical origin. Their strengths determine the depth of the potential well of Fig. 3.16.

We have operative in every system *van-der-Waals forces* which are caused by mutually induced dipole moments in the electronic shells of the adsorbate and the surface atoms. These forces are very weak and are responsible for noble gas or other closed shell particles' (CH_4, H_2) adsorption, as well as condensation of non-polar organic molecules. The corresponding heats of adsorption are quite small, they usually range between 10 and 20 kJ/mol. We refer to this process as *physisorption*. Accordingly, the respective surface must be kept at very low temperatures (below liquid N_2 temperature) in order to reach appreciable surface concentrations of these species. Furthermore, the electronic structure of the surface does not seem to play a major role – van-der-Waals bonding occurs on transition metals, graphite or insulators almost equally well. Of course, very polar surfaces (metal oxides etc.) can induce dipole moments in non-polar adsorbates provided they are easily polarizable, and additional bonding contribution comes into play which reinforces the physisorptive bond. This is the reason why the large and polarizable Xe atom exhibits larger physisorption energies than the small and almost non-polarizable He or Ne atoms. If polar molecules are used instead of non-polar species *dipole-dipole interaction forces* arise which will dominate the adsorptive bonding. Examples are adsorption of water or hydrogen cyanide on metal surfaces, where significantly increased heats of adsorption are observed as compared to noble gases, the more so, if also polar surfaces such as Al_2O_3, SiO_2, TiO_2 etc. are used as adsorbents. The magnitude of these interaction energies ranges between 20 to 50 kJ/mol and even more in some cases.

The forces responsible for *chemisorption* are solely based on quantum mechanical interaction between adatoms and the surface (that is, overlap between the respective wave functions) and generally lead to quite appreciable bonding strengths, comparable to normal chemical bonds. The magnitude of the corresponding heats of adsorption ranges from ca. 80 to 500 kJ/mol; examples are H, CO, O or N interaction with transition metal surfaces (Fe, Ni, W, Pt). Quite often, particularly when electropositive metals are involved, a real chemical compound is formed, leading for instance in the interaction between H, O or Cl with typical sp-electron metals (Na, K, Cs, Ca, etc.), to hydrides, oxides or chlorides. The respective surface compounds can no longer be distinguished from bulk compounds.

A good example of a typical chemisorption case is carbon monoxide on transition metal surfaces. The interaction mechanism can be reasonably well understood in terms of the Blyholder model [97] which considers a surface complex being formed between CO and the metal underneath, due to quantum-chemical interaction between the metal's d-band and the CO molecular orbitals (MOs). In this model, it is assumed that electronic charge flows from the occupied $CO - 5\sigma$ MO to the d-band, and that *backbonding* occurs whereby metal electrons partially populate the antibonding $2\pi^*$-MO of the CO molecule. A strong backdonation therefore has the effect of providing a strong metal – CO bond, but weakens the C–O bond considerably. Hence, the Blyholder model predicts a strong correlation between the CO chemisorption energy and the degree of backbonding, which in turn is favored by a large d-electron concentration close to the metal's Fermi level. This is the reason why transition metals with their high density of d-states at E_F exhibit strong heats of CO adsorption. In the following table we present some selected initial heats of adsorption of CO on various metal single crystals (Table 3.3). The observed heats are between 58 and 160 kJ/mol, whereby the low value refers to Cu, which is *not* a typical d metal.

Table 3.3. Initial heats of adsorption (q_{st}) of carbon monoxide on various metal surfaces

Surface	q_{st} (kJ/mol)	Reference
Ni(111)	111 (± 5)	[92]
Pd(100)	150 (± 5)	[93]
	161 (± 8)	[75]
Pd(111)	142 (± 3)	[94]
Ru(0001)	160 (±10)	[95]
Ru(1010)	157 (±10)	[96]
Cu(100)	58 (±10)	[98]
Ni(100)	125 (± 5)	[99]

For oxygen, heat values are approximately twice as large as for CO on transition metals, due to the strong chemical affinity between oxygen and metals. With hydrogen, heats of adsorption around 80 to 100 kJ/mol are the rule. Table 3.4 displays experimentally determined isosteric heats of hydrogen adsorption at vanishing coverages for a variety of single crystal surfaces.

Table 3.4. Initial heats of adsorption (q_{st}) of hydrogen on various metal surfaces [100]

Surface	q_{st} (kJ/mol)	Reference
Ni(100)	96.3 (±5)	[101]
Ni(110)	90.0 (±5)	[101]
Ni(111)	96.3 (±5)	[101]
Ni(111)	85 (±5)	[102]
Pd(111)	88 (±5)	[104]
Pd(110)	103 (±5)	[103]
Pd(100)	102 (±5)	[104]
Rh(110)	92 (±5)	[64]
Ru(10T0)	80 (±5)	[65]
Co(10T0)	80 (±3)	[105]

There is an important difference between the energetics of CO adsorption and that of dioxygen or dihydrogen. With O_2 and H_2, dissociation of the molecular entity can occur relatively easily and is indeed frequently observed in the course of the chemisorption process on transition metals. Accordingly, one must distinguish between molecular (for which the simple Lennard-Jones potential of Fig. 3.16 is a good description) and dissociative adsorption, whereby the occurrence of dissociation requires certain modifications of the potential energy diagram in that both the potential energy curve of the molecule's interaction with the surface and that of the dissociated atoms must be taken into consideration. Again, a one-dimensional diagram of the Lennard-Jones type can help to understand the situation (Fig. 3.17). The covalent dihydrogen molecule experiences certain (weak) van-der-Waals interaction forces as it approaches the surface. In equilibrium, it could reside in the potential minimum E_p relatively far away from the surface, a state which is entirely determined by van-der-Waals interaction forces. However, if it possesses a little more thermal energy, it can and will reach the potential energy curve which describes the interaction of a single H *atom* with the surface. Because of the unsaturated 1s orbital configuration of an isolated H atom, there occurs a much stronger (chemical) interaction with the surface, leading to the comparatively deep potential energy well E_{ch} with an equilibrium distance, z_{ch}, quite close to the surface. Of course, in order to produce H atoms, a H_2 molecule has to be dissociated prior to any interaction with the surface, which simply requires spending of the dissociation (or molecular bonding) energy of 432 kJ/mol in this case. The energy balance then reads

$$2\,E_{\text{Me-H}} = E_{\text{diss}} + E_{\text{ch,H}}. \qquad 3.12$$

Because *two* H atoms are formed by a single dissociation event, the H-Me bond energy $E_{\text{Me-H}}$ is gained twice. This is one of the "secrets" of why transition metal surfaces are such efficient catalysts in hydrogenation reactions – they easily provide dissociation of strongly bound molecules and convert them to a much more reactive metal: hydrogen (or oxygen and nitrogen) complexes which then can further react to desired species. Equation 3.12 can also be immediately used to deduce chemisorption bond energies from measured heats of adsorption. Remember that for non-dissociative adsorption the chemisorption bond energies simply equal the heat of adsorption, for example

$$E_{\text{Me-CO}} = E_{\text{ch,CO}}. \qquad 3.13$$

Therefore, if a dihydrogen molecule comes into contact with a surface, at the cross-over point P of Fig. 3.17, spontaneous dissociation will take place and the system can reach its energy minimum.

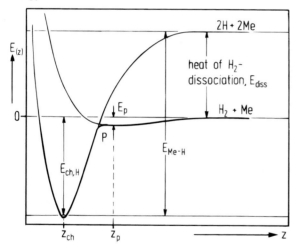

Fig. 3.17. One-dimensional potential energy diagram for the interaction of a homonuclear diatomic molecule (for example, hydrogen) with a surface. The subscripts p and ch mean "physisorption" and "chemisorption", respectively. Indicated are two interaction curves, one of a H atom formed by predissociation (shifted by E_{diss} to positive energies and leading to a deep well of depth $E_{\text{ch,H}} + E_{\text{diss}} = 2E_{\text{Me-H}}$ at z_{ch}), the other describing the (van-der-Waals-like) molecular interaction giving rise to a shallow minimum (E_p) at z_p. The actually observed potential energy curve is a superposition of the individual functions (bold line). The crossing-over point P is located below the energy-zero line, which implies spontaneous dissociation of the diatomic molecule at P.

Actually however, the dissociation (which can, of course, also be observed with CO in some cases, depending on the strength of the metal-carbon interaction) is a relatively complicated process which is difficult to treat theoretically. One serious problem is to calculate the actual trajectory of an impinging molecule on its way to the chemisorption

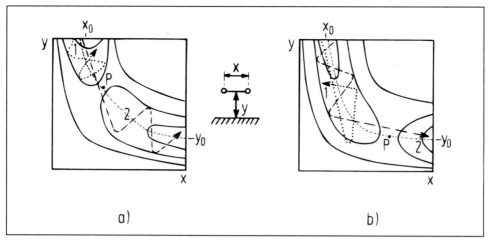

Fig. 3.18. Projection of the two-dimensional potential energy diagram pertinent to the dissociation reaction on the surface. The diatomic molecule (interatomic distance $= x$) approaches the surface (coordinate y). For large distances y the characteristic molecular bond length x_0 is observed (potential well at top part of a) and b)). As y becomes smaller, x is more and more stretched until, beyond the saddle point P, the dissociation is accomplished resulting in a deep potential energy well of the complex: adatom–surface atom with bond length y_0 (rigth-hand side of a) and b)). Two different situations are shown: in a) the activation barrier is located relatively far away from the surface (larger y- and smaller x-values), that is to say, in the entrance channel, and mainly *translational* energy is required to surmount the barrier; b) illustrates the situation, where the barrier is closer to the surface (small y-coordinate), and a *vibrational* excitation of the (stretched) molecule leads to dissociation in the exit channel. The reaction coordinate following the easiest reaction path is represented in each case by the thin dashed line (2). Dotted lines (1) describe an unsuccessful attempt to cross the barrier; the bold dashed lines a successful attempt whereby vibrational excitation is involved. The entire problem resembles very much the famous Polanyi rules whereafter exoergic reactions of type A + BC → AB + C with an *early* barrier (*late* barrier) request *translational* (*vibrational*) energy of the reactants [109].

potential minimum. Here, a two-dimensional representation is much more suited to illustrate the situation (Fig. 3.18). If we denote the internuclear distance in the molecule by x and the distance of the molecular axis to the surface by y, we have a small x at large y values, and as the molecule gets closer to the surface (decreasing y) x finally increases to such an extent that the molecule breaks apart. Here, we should remember the discourse about transition state theory of Sect. 2.6 – we may well regard the trapped molecule at the cross-over point P of the two potential energy curves as representing the transition state of the reaction

$$H_2 + 2\,Me \longrightarrow 2\,Me\text{-}H, \qquad\qquad 3.14$$

and can then more precisely formulate

$$H_{2(g)} \xrightarrow[\substack{\text{surface}\\\text{site(s)}}]{k_1} \{H_{2(ad)}\}^{\ddagger} \xrightarrow[\substack{\text{surface}\\\text{site(s)}}]{k_2} 2H_{(ad)}\,, \qquad\qquad 3.15$$

whereby certain surface *sites* are required for these reaction steps to occur. Particularly, the rate of the dissociation step (k_2) may be very structure- and site-dependent. Also, the *electronic* structure of the surface plays a dominant role, as we shall see below.

A theoretical description of the dissociation process has been attempted many times: here we refer to promising concepts offered for H_2 dissociation by Melius et al. [106], by Nørskov and Stoltze [107], or Harris and Andersson [108]. Following Nørskov and

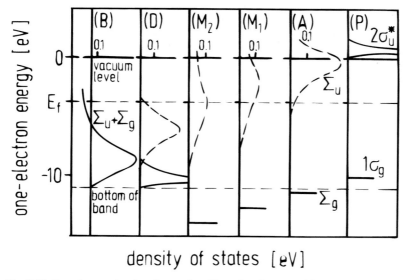

Fig. 3.19. One-electron density of states for a H_2 molecule approaching a Mg(0001) surface along the reaction coordinate. Capital letters in the top of the figure indicate extrema on the potential energy surface: from right to left, (P) physisorbed state (practically unperturbed H_2 molecule); (A) activation barrier for molecular adsorption; (M) molecularly adsorbed state (here, the density of states is shown both in the outer (M_1) and inner (M_2) part of the well); (D) activation barrier for dissociation; (B) two separated H atoms chemisorbed in the two-fold bridge site. Dotted lines denote the width of the metal's conduction band (E_F = Fermi level); positions of the molecular orbitals of the H_2 molecule are indicated by arrows and state symbols. After Nørskov and Stoltze [107].

Stoltze, for H_2 interaction with a surface, the various one-dimensional potential energy curves can be calculated in a one-electron density of states approximation. The situation is depicted in Fig. 3.19. Far from the surface, we have the typical sharp (empty) antibonding $2\sigma_u^*$ level and the filled bonding $1\sigma_g$ state separated by more than 10 eV of the unperturbed H_2 molecule. As this entity approaches the surface orbital interactions gradually gain importance, leading, in the first instance, to a broadening of the respective H_2 levels. Furthermore, a downward shift of both MO's occurs, the decisive step being the pulldown of the antibonding H_2 state. As it is shifted to below the metal's Fermi level E_f, it can be successively filled with electrons, thus weakening the H-H and strengthening the Me-H bond. The yet remaining H-H interaction must still be overcome in order to completely dissociate the molecule, which gives rise to a more or less pronounced activation barrier, E_{ad}^*. The dissociated state exhibits a single adsorbate-induced resonance well below E_f. The above description is by no means restricted to H_2, rather, all simple molecules have anti-bonding levels which must fill during dissociation. For transition metals the interaction with the d electrons generally facilitates this process and lowers the activation energy barriers thus resulting in a low-lying cross-over point P (non-activated adsorption).

Harris [110] has given a simple quantum chemical explanation of the beneficial role of the unfilled d shell of the transition metals. In simple s electron or noble metals the direct orbital repulsion (called Pauli repulsion) between the filled hydrogen $1\sigma_g$ orbital and the filled metallic s states (which requires these states to orthogonalize) leads to an appreciable rise of the total energy of the system as the H_2 molecule approaches the surface, whereas in case of transition metals the *d* holes provide an easy escape route for the *s* electrons which can be transferred to the empty *d* states at the common Fermi level thus circumventing the Pauli repulsion.

Whereas spontaneous dissociation of this kind occurs on most transition metal surfaces (experimental hints are a rapid uptake of adsorbed gas and a decrease of the sticking probability with increasing temperature), there are cases where the cross-over point P of the two potential energy curves of Fig. 3.17 is above the zero energy level, as illustrated in Fig.3.20. This means that an incoming gas molecule must possess a certain amount of kinetic energy in order to overcome the barrier of Fig. 3.20 of height E_{ad}^{*} (*activated adsorption*). For metals with filled and low-lying *d* bands (Cu, Ag, Au) or typical *sp* electron metals (Be, Mg, Al etc.) appreciable activation barriers are the rule, making these metals inert for hydrogen chemisorption. A frequently studied case is H_2 interaction with copper surfaces, where the existence of an activation barrier is known for a long time [111-113]. Experimentally, activated adsorption is indicated by slow rates of adsorption and an increase of gas uptake with temperature which provides the gas molecules with kinetic energy. It is only mentioned here that a question currently under discussion is whether or not, besides translational degrees of freedom, vibrational excitation of the impinging molecules is essential for rapid dissociation [113,114].

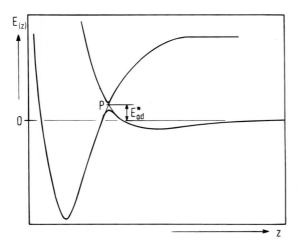

Fig. 3.20. One-dimensional potential energy diagram for dissociative adsorption involving an activation energy barrier of height E_{ad}^{*}. The cross-over point explained in the legend of Fig. 3.17 is now located above the energy zero line and slows down spontaneous dissociation.

3.2.2 The Multi-particle Interaction and the Formation of Ordered Adsorbate Phases

For the very first particle that adsorbs on an ideal surface (which we assume to be chemically and crystographically clean) all possible adsorption sites will provide *identical* binding conditions. In other words, all binding sites can be described by the same potential energy curve of Figs. 3.16 or 3.17 and will yield the same energy of adsorption. If we look at the energetic situation *parallel* to the surface (Fig. 3.21a) all the identical adsorption sites are separated from each other by comparatively small activation barriers, namely, barriers for surface diffusion which vary with the periodicity of the crystal surface to give rise to a sinusoidal potential in *x, y* -direction. A particle trapped on the surface has, depending on the temperature of the crystal, a certain amount of kinetic (mainly

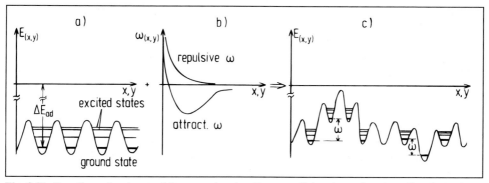

Fig. 3.21. One-dimensional potential energy situation $E_{(x,y)}$ parallel to the surface (x, y direction), showing the modulated adsorption potential: a) empty surface with a single particle bound with adsorption energy E_{ad}; b) pairwise interaction potential $\omega(x,y)$ between two adsorbate atoms indicating the repulsive situation (upper curve) and the attractive case (lower curve). c) superposition of $E_{x,y}$ and $\omega_{(x,y)}$ leading to adsorbate-induced energetic heterogeneity (weakening of adsorbate-surface bond in the case of repulsive, strengthening in the case of attractive interaction potentials ω).

translational) energy and can either be in the ground state or in the excited state of the potential shown in Fig. 3.21a. At low enough temperatures, most of the adsorbed particles will be in the ground state and are not able to "hop" or migrate to adjacent sites, because their thermal energy is too small to overcome the respective activation barrier for surface diffusion. This is the situation of *immobile* adsorption with residence times of particles in a certain site that are extremely long. Correspondingly, for higher thermal energies of the adsorbed particles and/or smaller activation barriers for diffusion, site exchange or "hopping" events can occur much more frequently, resulting in a *mobile* adsorbed layer with merely short or very short residence times of particles in certain sites.

The mean residence time in a certain site on the surface τ' depends exponentially on the activation energy of diffusion, via the equation

$$\tau'_{(T)} = \tau'_0 \exp\left(\frac{\Delta E^*_{diff}}{kT}\right) , \qquad\qquad 3.16$$

where τ'_0 represents a kinetic pre-exponential factor. τ'_0 contains the duration of a single vibrational period (10^{-13} s) and a probability factor that accounts for the number of "escape routes" from a given site. Hence, τ'_0 can range from 10^{-7} to 10^{-13} s. ΔE_{diff} on the other hand, is roughly $1/10$ of the heat of adsorption and is approximately 1–5 kJ/mol for physisorption and ~5–20 kJ/mol for chemisorption systems.

Obviously, the magnitude of ΔE^*_{diff} governs quite sensitively how long a particle will stay in a certain site at a given temperature. With typical τ'_0 and ΔE^*_{diff} parameters, τ' ranges at 300 K in the order of microseconds. In many cases, due to a relatively small ΔE^*_{diff} value, one finds a complete mobility of the adsorbate at room temperature, because the mean residence time in a certain site is too short, site exchange processes via hopping are so frequently occurring that a given particle is as often located in a site as it is between sites. This has strong consequences for the formation of adsorbate phases with long-range order, which reflect the mutual particle-particle interactions (these will be dealt with further below). Only at sufficiently low temperatures do the particles reside long enough in their periodic sites and give rise to noticeable amplitudes in a diffraction experiment (cf., Chapter 4).

The diffusion of adsorbed particles (which we shall address in a separate short section, cf., Sect. 3.3.4) depends very much on the local binding energy situation and thus on the overall surface concentration of the adsorbate, because the particles can and will interact with each other, whereby the physical origin lies in the modification of the charge distribution of the solid in the vicinity of an adsorbed atom or molecule. Usually, charge is withdrawn from the metal-metal bonds of the substrate environment to make the chemisorption bond, in other cases (alkali metal adsorption) charge flows to the metal, and image charge effects perturb the electronic structure of the metal surface region. The net result is always similar: the two-dimensional potential energy surface around an adsorbed particle becomes bent, and either a smaller or a larger adsorption energy results for the adjacent adsorption sites. In terms of mutual lateral interactions, this means that either repulsive or attractive forces are exerted by a given adsorbed atom to its neighbors. These modifications are shown in Fig. 3.21b,c. A significant consequence of these mutual interaction energies (denoted as ω) is the formation of long-range order within an adsorbed layer which leads to periodical two-dimensional arrays of particles of the type discussed before in the context of adsorbate structure. From a wealth of low-energy electron diffraction (LEED) experiments (cf., Chapter 4) it has become clear that the build-up of ordered adsorbed layers is the rule rather than the exception. The order simply reflects the periodicity of favorable (or unfavorable) bonding conditions on a surface. The physical origin of the particle-particle interactions is of quantum chemical nature. Two neighboring adsorbed atoms can either interact via the substrate (through-bond interaction) or via direct orbital repulsion. In the first case so-called indirect interactions are exerted that can operate over fairly large distances and are repulsive or attractive depending on the kind of charge modification of the metallic solid [115-117]. The second case has already been addressed in the context of hydrogen dissociation: if two orbitals are so close that they penetrate each other, they must orthogonalize, which drives up the total energy. To give an example, adsorbed CO molecules cannot be brought closer together than about 3 Å, which roughly corresponds to their van-der-Waals diameter [93].

A very interesting and frequently studied aspect of long-range order phenomena are the two-dimensional phase transitions which can only be touched here. Similar to bulk thermodynamics first-order or continuous phase transitions can also occur in two-dimensional layers at certain coverages and temperatures. The determination of phase diagrams, critical temperatures and exponents provides an elegant way to obtain, for example, interaction energies ω. As an example, we refer to a work by Park et al. [118], who studied the adsorption of oxygen on a Ni(111) surface by means of LEED and other methods. Figure 3.22 displays the phase diagram that the authors obtained for the O/Ni(111) system. They also evaluated the critical exponents and were able to classify the type of phase transition within the so-called universality classes. Here we would enter the field of statistical mechanics which is, however, not the topic of this book, instead, we list references for the interested reader [119, 120]. It is worthwhile, at this point, to refer again to thermodynamics, because phase transitions can very well be treated by this discipline, and there is an intimate connection between the microscopic and the statistical thermodynamic viewpoint. What we deal with here is essentially the temperature dependence of structure (surface structure in our context). At $T = 0$ K all particles are located in the minima of the potential energy curves and they only possess their zero-point energy. Any increase of temperature supports the entropy term $T\Delta S$ in the Gibbs-Helmholtz-equation (Chapter 2, Eq. 2.84) and necessarily introduces disorder and phase transformations. Quite generally, phase transitions can be classified with respect to the temperature dependence of the thermodynamic functions [121]. One distinguishes first-order phase

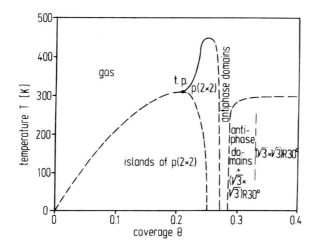

Fig. 3.22. Experimentally determined phase diagram for oxygen adsorbed on a Ni(111) surface. Solid lines indicate continuous phase transitions, while dashed lines indicate those of first order. There is a tricritical point, denoted as t.p. After Park et al. [118].

transitions and continuous transitions. From ordinary thermodynamics the reader is certainly familiar with the "normal" transitions which lead to melting, boiling, etc.. The Gibbs energy ΔG and hence the chemical potential $\mu = (\partial G/\partial n)_{P,T,n_i}$ is a continuous function of pressure and temperature. This is not necessarily so for the first derivatives, i.e., $(\partial \mu/\partial T)_P = -S$, or the expression $\mu - T(\partial \mu/\partial T)_P = H$. This means that for first-order transitions, both the entropy S and the enthalpy H are discontinuous right at the transition temperature T_c. Accordingly, the heat capacity $C_P = (\partial H/\partial T)_P$ reaches infinity at T_c. As opposed to this, continuous phase transitions are characterized by the fact that, e.g., the enthalpy H or the volume $V = (\partial \mu/\partial p)_T$ vary continuously at T_c. Only the heat capacity or the compressibility χ exhibit discontinuities at T_c because the second derivatives are discontinuous here. The consequences are the lack of latent heats and the occurrence of the so-called λ-transitions at T_c. Well-known three-dimensional examples are the transformation of β-brass or the break-down of the ferromagnetism in α iron. On surfaces, the distinction of first-order and continuous phase transitions is not so easy since latent heat or heat capacity measurements are usually difficult to perform. Nevertheless, there is a variety of examples for both types of phase transformations reported in the literature [122].

More important in view of practical physical surface chemistry is the particle-particle interaction in adsorbed layers with regard to the overall adsorption energy. If we remember the simple Langmuir model which led to the Langmuir isotherm, it was assumed therein that up to saturation, that is, occupancy of each adsorption site, the adsorption energy remained constant. In reality, however, this is never the case, rather the adsorption energy usually decreases at medium and high coverages due to the above-mentioned mutual repulsive interactions. An estimation of these interactions can be obtained from the dependence of the isosteric heat of adsorption, q_{st}, with coverage Θ. In Fig.3.23 we present three typical curves obtained for carbon monoxide adsorption on Pd(100) [75], for CO adsorption on Ni(111) [92], and for H_2 adsorption on Ni(110) [101]. In all these cases it turns out that the adsorption energy decreases strongly as saturation is approached. At small coverages, however, q_{st} is either constant, decreases, or increases with the particle concentration.

For coverage-dependent heat of adsorption, i.e., $dq_{st}/d\Theta \neq 0$, we mostly deal with the so-called induced heterogeneity of a surface which simply reflects the operation of the aforementioned particle-particle interactions. With H adsorbed on Ni(110) at and above

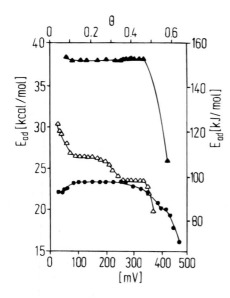

Fig. 3.23. Isosteric heats of adsorption E_{ad} (= q_{st}) as a function of adsorbate coverage Θ, for CO on Pd(100) (dark triangles, [70]), CO on Ni(111) (open triangles, [92]), and H on Ni(110) (circles, [101]). In the last case, the bottom scale applies, which represents the H-induced work-function change, a quantity that is proportional to the H coverage. A decrease of q_{st} with Θ indicates the operation of repulsive, an increase indicates attractive mutual interactions.

room temperature there is evidence for an initial increase in the heat of adsorption which is due to attractive lateral interactions. This means that adsorption into a given ensemble of already adsorbed particles is more favorable than adsorption onto the bare surface.

In a variety of cases, however, more than just one region of constant adsorption energy is observed in the experiment, whereby the two regions are well separated by a steplike decrease of q_{st}. Clearly, there occurs a successive population of adsorption sites with different binding conditions, in other words, we have two types of adsorption sites which differ with respect to the depth of the potential energy well. In thermal equilibrium there will be a Boltzmann distribution of the population of these sites which we may denote as a and b, and we have

$$N_a = N_b \exp(-\Delta E_{ad}/kT) \text{ , } (N = \text{particle numbers}) \qquad\qquad 3.17$$

if ΔE_{ad} stands for the energy difference of the two sites $q_{st,a} - q_{st,b}$, and partition function effects are neglected. This so-called *a-priori* heterogeneity is certainly important in practical catalytical reactions dealing with dispersed and heterogeneous surfaces. Effects of this kind also show up, however, in model experiments using single-crystal surfaces. Frequently, the initial heat of adsorption decreases by some 10% at still small coverages and then reaches a constant value. Such behavior was among others reported for a *stepped* palladium [104] and platinum surface [123] interacting with hydrogen, while the corresponding flat low index planes did not exhibit this initial decrease. Evidently, crystallographic defects must be made responsible for this behavior, and the *a-priori* energetic heterogeneity is manifested by the adsorbing atoms which function as sensitive probes of the sites with higher adsorption energy. Usually, these are the step and kink sites, because they can provide the adsorbed particle with a higher degree of coordination. Furthermore, it could be shown by the group of Comsa [124] that, for example, dissociation of dihydrogen effectively occurs on Pt(111) surfaces only at step and defect sites. Other groups reached the same conclusions [125, 126].

The isotope exchange reaction

$$H_2 + D_2 \leftrightharpoons 2HD \qquad\qquad 3.18$$

is greatly inhibited in the gas phase because of the large dissociation energy barrier for H_2 dissociation (432 kJ/mol). Even if the H_2 molecule is dissociated, there remains a small barrier of ~42 kJ/mol for the successive atom-molecule reaction

$$H + D_2 \leftrightharpoons HD + D. \qquad\qquad 3.19$$

With a stepped transition metal surface present the above reaction can take place rapid-lyeven below 10 0 K surface temperature as temperature-dependent isotope exchange measurements demonstrate. The limiting factor here appears to be surface diffusion of H (or D) atoms which sets in above 30 to 40 K surface temperature.

The particular catalytic activity of stepped surfaces was convincingly illustrated by angle-dependent molecular beam experiments in the group of Somorjai [127]. It was reported that the rate of the H_2/D_2 isotope exchange reaction on a stepped Pt surface depended sensitively on how the molecular beam was directed with respect to the direction of the steps. The situation is depicted in Fig. 3.24 taken from Salmeron et al. [127].

In the experiment a mixed molecular beam of H_2 and D_2 was incident on a Pt(332) surface. Figure 3.24 shows the production of HD as a function of the polar angle of incidence, for different (fixed) azimuthal angles Φ [$\Phi = 90°$ for curve a) and $\Phi = 0°$ for curve b)]. $\Phi = 90°$ means that the plane of molecular beam incidence is perpendicular to

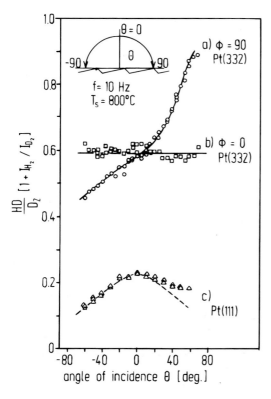

Fig. 3.24. HD (hydrogen deuteride) production as a function of angle of incidence Θ of a molecular beam consisting of H_2 and D_2 impinging on a Pt(332) surface. The production rate was normalized to the D_2 intensity I_{D_2}. The chopping frequency of the beam was $10\,s^{-1}$, the surface temperature $T_s = 800$ °C, the gas temperature $T_g = 25°C$. Curve a) was obtained for $\Phi = 90°$ (step edges perpendicular to the incident beam), curve b) for the projection of the beam on the surface being parallel to the step edges ($\Phi = 0°$), and curve c) refers to the reaction behavior on a non-stepped Pt(111) surface. After Salmeron et al. [127].

the step edges, as illustrated in the inset of Fig. 3.24. In curve b), the projection of the reactant beam to the surface is parallel to the direction of the step edges, i.e., $\Phi = 0°$. The experiments were performed with a beam modulation frequency (chopped beam) of $10\,\text{s}^{-1}$ with the surface temperature kept at 1100 K. The result of these experiments can be summarized as follows: for the beam incident perpendicular to the open step edge the activity (as measured by the HD production rate, normalized to the incident D_2 intensity) is a factor of two higher than if it approaches the "shielded" edge from above. Compared to a "flat" Pt(111) surface (curve c)), there is a generally higher activity of a factor 4. Apparently, the dependence of the H_2-D_2 exchange probability on the direction of approach is closely correlated with the structural anisotropy of the Pt surface, indicating a unique activity of so-called step sites of Pt for H-H bond breaking, whereby the site associated with the inner corner atom was found to be the most active. This certainly has to do with local coordination (overlap of wave functions of metal substrate and adsorbate particles) which may be one of the mysteries of the active centers in heterogeneous catalysis.

There is yet another consequence of the $q_{st}(\Theta)$ dependence in practice which concerns the drop of the heat of adsorption at high surface concentrations of adsorbate. Close to saturation q_{st} and hence the adsorbate binding energy can be reduced by as much as 50% to 80% of the initial value. These weakly bound species are often especially reactive – owing to their small binding energy to the substrate they can easily be transferred to neighboring functional groups of coadsorbates to give a desired reaction product. At high reactant pressures and not too elevated temperatures which are common in practice one can assume highly covered surfaces, and processes of this kind can play a significant role.

3.3 Surface Kinetics

We have, in Sect. 2.6, presented a short introduction to surface chemical kinetics which was based on the macroscopic picture, with emphasis on desorption phenomena. The adsorption was very briefly mentioned in the context of the kinetic derivation of the Langmuir adsorption isotherm (cf., Sect. 2.4), and we shall repeatedly refer to equations and definitions given there.

In the following, we are going to attempt a more microscopic (atomistic) understanding of various surface kinetic processes, namely, of trapping and sticking, of adsorption and desorption. Actually, this is a very demanding matter, for which there exists a wealth of current investigations. It is not at all possible, within the scope of this book, to treat the subject exhaustively and we must repeatedly refer to the relevant literature.

3.3.1 Trapping and Sticking

A gas phase atom or molecule that approaches a solid surface can "feel" a weak attractive van-der-Waals potential relatively far outside the actual surface. At room temperature and for normal incidence a gas atom has an appreciable amount of translational energy; with molecules also the rotational and (ground state) vibrational energy contents must be considered. The particle will finally collide with the surface and suffer momentum and eventual energy exchange with the solid. The phenomena of trapping and sticking are closely coupled to the shape of the effective surface potential felt by the particle and its residence time in this potential. This time can vary over many orders of magnitude. There may be the direct reflection at the repulsive branch of the potential which occurs in a time scale of less than 10^{-13} s whereby the particle "remembers" very well the direction of impact, there may be just exchange of momentum, but not of energy. Accordingly, a sharp specularly reflected beam will be obtained in a scattering experiment; this phenomenon is called "elastic scattering". A second situation can be envisaged if the incoming particle does feel the potential, carries out one or two vibrations in the potential, and is then reflected back into the gas phase with about the same energy that it had before. Here, the particle has almost forgotten its initial impact direction and there is exchange of momentum and of some (very small though) kinetic energy. This process is referred to as "direct inelastic scattering". It leads to a broadened angular distribution around the specularly reflected beam. The mean residence time of these particles is around 10^{-12} s^{-1} to 10^{-13} s^{-1}; they have actually "seen" the potential, and the aforementioned energy exchange comprises excitation of electron-hole pairs near the Fermi edge of the metal, the coupling to surface phonons, or transformations from translational to rotational energy states. The residence time in the potential is thereby too short for the particle to be considered as having really stuck on the surface, rather one refers to this process as "transient trapping".

Of much greater importance in our context is the situation where the impinging particle really succeeds to get into the ground state of the chemisorption potential and to stay at the surface for an appreciable time (which can range from microseconds to hours). These molecules have actually accommodated to the surface, their initial kinetic energy has been dissipated to the phonon bath of the solid surface, and they represent the typical adsorbed species that we have always dealt with before. In molecular beamscattering experiments these particles can be distinguished from all others because they come off the surface in a cosine distribution and no reflection angle is preferred. A typical example is given in Fig. 3.25, taken from a work by Engel and Ertl, which shows the spatial dis-

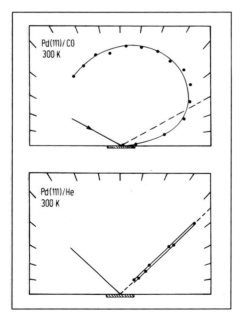

Fig. 3.25 Angular scattering distribution in the scattering plane for CO scattered from a Pd (111) surface held at 300 K (upper part) and for He under the same conditions (lower part). The angles of incidence were 60° and 45°, respectively. After Engel and Ertl [128]

tribution of carbon monoxide scattered off a Pd(111) surface [128]. The counter example is provided by the same authors, who also looked at helium scattering (He is known not to interact with Pd) and obtained a sharp specularly reflected beam, as expected, from elastically scattered particles. The sharpness of the specularly reflected He beam lobe can indeed be utilized to probe the crystallographic roughness of a surface. Comsa and coworkers compared a clean and perfectly smooth surface with a mirror that clouds upon adsorption of particles. In a He scattering experiment, disordered adsorption shows up by a decrease of the elastically scattered He intensity [129].

In macroscopic adsorption experiments, for example in thermal desorption studies, one usually exposes a given surface held at constant temperature T to a certain gas pressure P for a well-defined time t and compares the amount of gas taken up by the surface (σ_s = number of adsorbed particles per unit area) with the total number of gas particles that have actually struck the surface at the pressure chosen. The kinetic equations, as well as the definition of the "sticking probability", have already been presented in Chapter 2 (cf., Eq. 2.44). Here we give a supplementary definition of the term "exposure". The gas exposure simply is the product of P and t, with the dimension [$Nm^{-2}s$]. However, there is, for practical reasons, still the dimension Langmuir [L] being used in surface chemistry, whereby $1\,L = 10^{-6}\,Torr\cdot 1s = 1.33\,10^{-6}\,mbar\cdot 1s$. The reason is that if every gas molecule impinging on the surface sticks, an exposure of just 1 Langmuir would approximately lead to the adsorption of one complete monolayer.

As far as interaction dynamics are concerned, one of the most significant properties of an adsorption system is the initial sticking coefficient s_0, that is, the sticking probability at vanishing coverage. This is nothing other than the probability that an incident particle is finally chemisorbed after collision with a surface. Before we enter our (brief) discussion of s_0 and s we acknowledge the exhaustive review article on that subject by Morris et al. [130], who also present a long list of experimentally determined s_0 values for various adsorption systems. A small selection of such data is given here in Table 3.5.

Table 3.5. Some selected values of initial sticking probabilities for metal-gas interaction [130]

System	Initial sticking probability s_0	Reference
H/Ni(100)	0.06	[131, 132]
H/Ni(111)	≥0.01	[133]
H/Pt(111)	0.1	[134]
	≤0.0001	[124, 135]
H/Ni(110)	~1	[136]
	0.96	[137]
H/Rh(110)	~1	[64]
H/Ru(10$\bar{1}$0)	~1	[65]
H/Co(10$\bar{1}$0)	0.75 (±20%)	[105]
O/Cu(100)	0.03 (300K)	[138]
O/Ni(100)	1	[139]
O/Pt(111)	0.2	[140]
CO/Ni(111)	1	[92, 141]
CO/Pd(100)	0.6	[75]
CO/Pd(111)	0.96	[142]
CO/Ru(10$\bar{1}$0)	1.0	[96]
CO/Pt(111)	1.0	[143]

Apparently, the values range between unity and 10^{-6}, although for "normal" adsorption systems such as CO, H_2 or O_2 interacting with transition metal surfaces, $1 \geq s_0 \geq 0.1$. Careful single-crystal studies have shown that the crystallographic orientation of a surface governs the magnitude of s_0 sensitively. There is a clear trend that s_0 is substantially higher on atomically rough surfaces than on smooth samples, whereby this effect is more pronounced for hydrogen and nitrogen than for CO. This may be documented by the well-known fact that hydrogen-sticking probabilities on the open fcc surfaces with (110) orientation usually reach the value 1.0, whereas on the atomically flat (111) surfaces s_0 is often lower than 0.1 [100]. Other parameters, which are expected to influence the magnitude of s_0 are the collision angle with the surface, gas temperature T_g (i.e., kinetic energy of the incoming particle), distribution of the internal energy to the various degrees of freedom of a molecule, and surface temperature. The state of the energetic excitation of the molecule is essential in the case of activated adsorption (Sect. 3.2), because translationally or vibrationally excited particles can possibly surmount the respective activation barriers. This holds, e.g., for hydrogen adsorption on various low index copper surfaces [111, 113].

The message that is worth remembering for practical purposes here is that chemically reactive gases usually stick with relatively high probability on transition metal surfaces, whereby sticking is generally more effective on atomically rough surfaces, which can obviously provide good energy accommodation. Carbon monoxide and nitric oxide stick quite effectively ($0.2 < s_0 < 1$) on many transition metal surfaces regardless of their crystallographic orientation. However, on open surfaces, these gases tend to dissociate. This dissociation is almost the rule for metals such as Fe, Mn, Cr, or V. Noble metals as well as sp-electron metals have very little activity for adsorbing N_2, H_2, or CO, for reasons outlined in Sect. 3.2. Oxygen, however, is adsorbed on these metals with fairly high probability. It appears that the whole pattern pertaining to whether or not a molecule sticks effectively or even gets into the dissociated stage is, unfortunately, rather complicated and requires a careful investigation in each case. Definitely, however, it is the beneficial role of surface defects (steps, kinks, dislocations) which can significantly contribute to accumulate chemically reactive species at surfaces.

3.3.2 Coverage Dependence of Sticking, Precursor States

Certainly, there will be no single reaction running at vanishing coverages. A discussion of sticking would, therefore, be totally incomplete without treating the coverage dependence, i.e., multi-particle effects. We have in Chapter 2 briefly touched the coverage-dependences of the adsorption and desorption rates. Here, we are going to present the underlying microscopic picture. Figure 3.26 is useful for describing and delineating all microscopic kinetic processes that may occur at a surface. Let us represent each process by means of its probability p. We base our considerations on a surface which is approximately half-covered with adsorbate atoms. We can then distinguish completely bare and locally more or less densely covered areas. A particle that arrives from the gas phase may hit the surface either on an empty or an occupied adsorption site. Accordingly, its probability for chemisorption p_{ch} will be different in each case. In the simple Langmuir model, p_{ch} will be 1 in the first, and 0 in the second case. This means that s would drop immediately with the adsorption of the first particles, according to the expression $s_0(1-\Theta)$ for molecular adsorption, which was already discussed in Chapter 2. However, it is a well-known fact in adsorption kinetics that s often remains constantly high over an

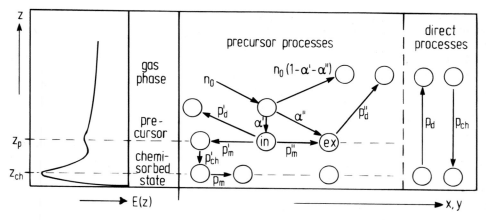

Fig. 3.26. Schematic representation of the possible kinetic processes occurring at a transition metal surface as a function of the distance z perpendicular to the surface. Direct processes are indicated in the right-hand part of the diagram by the probabilities p_{ch} for chemisorption and p_d for desorption. In the middle, precursor, reflection, migration, and adsorption phenomena are considered by their respective probabilities: $n_0 =$ number of impinging gas molecules, $\alpha' =$ fraction trapped in the intrinsic, $\alpha'' =$ fraction trapped in the extrinsic precursor state. p'_d desorption from the intrinsic, p''_d desorption from extrinsic precursor. p'_m and p''_m denote the probabilities for migration in the intrinsic and extrinsic precursor state, respectively, and p_m that of the particle's migration in the chemisorbed state via the probability p_{ch}. The potential energy situation pertinent to these processes is indicated in the left part of the figure.

appreciable coverage range until the sticking decreases fairly abruptly at medium coverages and reaches zero at saturation. Some typical curves of this kind are comprised in Fig. 3.27 for CO adsorption on various transition metal surfaces. A possible explanation for this unusual non-linear behavior is the existence of a so-called *precursor* state, which means a molecularly trapped (but not yet fully accommodated) particle in a weak potential well somewhat outside the surface. Kisliuk [144,144a] was one of the first to draw attention to precursor states in adsorption kinetics, and he derived a statistical model for

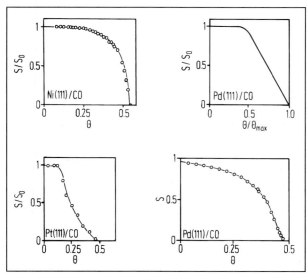

Fig. 3.27. Relative sticking probabilities for CO adsorption onto various (111) transition metal surfaces at room temperature. Upper panel left: Ni(111)/CO [92]; right: Pd(111)/CO [94]. Lower panel left: Pt(111)/CO [155]; right: Pd(111)/CO [142] (molecular beam study). In this latter case the *absolute* sticking probability $s(\Theta)$ was measured

it. While trapped in the precursor, the molecule is only weakly held and therefore fairly mobile. Apparently it can diffuse across the surface and search for an empty adsorption site. For usual precursor lifetimes, which are in the 10^{-6} s range, there is a good chance to find such site if the surface areas already covered are not too large. Because the molecular precursor potential will be locally affected wherever particles are preadsorbed, one must, for precise considerations, distinguish between *intrinsic* and *extrinsic* precursor states [130]. The first exists at empty surface sites and the latter at sites filled with a chemisorbed particle.

Returning to Fig. 3.26, we can introduce various probabilities, namely, for physisorption into the intrinsic and extrinsic precursor α' and α'', for migration of particles trapped in the precursor state (p'_m, p''_m for intrinsic and extrinsic precursor, respectively), p'_{ch} for chemisorption from the intrinsic precursor, and p'_d (p''_d) for desorption into the gas phase via the intrinsic (extrinsic) precursor. Using rate constants (k_i) and populations (A_i), we may formulate the schemes:

a) for molecular (non-dissociative) adsorption, at vanishing desorption
   (A = adsorbing species)

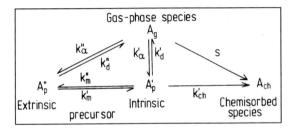

and b) for desorption (vanishing adsorption)

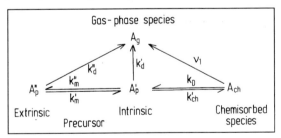

to give just one example. Quite similar schemes can be worked out for dissociative adsorption and desorption [130].

With known probabilities and rate constants, it is possible to arrive at expressions for macroscopic rate of adsorption or desorption which, of course, must be fully equivalent to the formulae given in the overall macroscopic treatment of Chapter 2 (Eqs. 2.44 and 2.45). The only difference that appears is a new $s(\Theta)$ function, which accounts for the precursor existence. Following the model considerations of Kisliuk [144,144a], one can combine the decisive probabilities to a new precursor constant K which reads

$$K = \frac{p''_d}{p'_{ch} + p'_d} , \qquad\qquad 3.20$$

and one obtains, for molecular adsorption, the function [144]

$$s(\Theta) = s_0 \frac{1}{1 + \frac{\Theta}{1-\Theta}K} \; . \qquad\qquad 3.21$$

$K = 1$ then corresponds to the (linear) Langmuir behavior, where $s_0(1-\Theta)$. $K > 1$ leads to concave curves (smaller s values at a given coverage) and $0 < K < 1$ results in typical convex precursor relations, whereas $K = 0$ yields $s(\Theta) = s_0$ independent of coverage. An example is presented in Fig. 3.28. Obviously, K is largely determined by the probability p_d'', which must become smaller than the sum of p_{ch}' and p_d' in order to reach the condition $K < 1$. It should be added that similar considerations and derivations can be put forth for dissociative adsorption. For the purposes of heterogeneous surface reactions, the existence of weakly bound precursor states may be of some significance. At higher reactant pressures, even weak (physisorbed) states may become considerably populated, so that the rate constants associated with adsorption into as well as transition and desorption from the precursor state become rate-limiting.

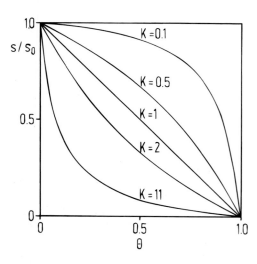

Fig. 3.28. Calculation of various sticking coefficient-coverage dependences as predicted by the Kisliuk model for molecular adsorption, according to Eq. 3.21 [144,144a].

There is yet another factor, which can and will modify the coverage dependence of the sticking probability, namely, adsorbate-induced changes of the surface structure, the adsorbate-induced reconstruction. As we learned in Sect. 3.1, it may very well be that a certain local or overall critical concentration of adsorbate makes the surface reconstruct, whereby the new configuration of substrate atoms can provide a more (or less) effective energy accommodation and hence sticking. We shall return to this point in the context of oscillating surface reactions in Chapter 5.

All in all, surface kinetical processes, such as adsorption, usually depend very much on the surface concentration of adsorbed particles. The example of the precursor model could illustrate that a microscopic analysis of the surface processes can help people understand the macroscopic behavior.

3.3.3 Pre-exponential Factor and Coverage Dependence of Desorption

Desorption is, of course, the reversal of the adsorption process. The relations between adsorption and desorption, that is entering and leaving the surface potential energy well, has led many investigators to spend some effort on developing kinetic and statistical models [130, 145-149]. The famous concept of *detailed balancing* is just mentioned here. In a somewhat naive interpretation, it states that those particles that stick very effectively must leave the surface with a correspondingly low efficiency.

When dealing with desorption, we confine ourselves to the process of *thermal* desorption (the respective experimental spectroscopy will be presented in Chapter 4). The adsorbed particles are assumed to reside in the minima of the surface potential and to be in equilibrium with the surface (full accommodation). One of the problems is to treat the coupling of the particles' vibrations to the phonons of the substrate. More details on the microscopic view of the single desorption event may be found in the works by Gortel et al. [150] and Kreuzer and Gortel [151].

We remember the principal kinetic equation for thermal desorption (associative desorption):

$$r_{des}(\Theta, T) = \nu_{des}^{(1)}(\Theta) \cdot f(\Theta) \cdot \exp\left(-\frac{\Delta E_{des}(\Theta)}{RT}\right) . \qquad 3.22$$

Let us first briefly comment on the microscopic meaning of the pre-exponential factor ν. We had defined ν phenomenologically as a kinetic coefficient (rate constant at infinite temperature), cf., Sect. 2.4, Eqs. 2.46 and 2.48, and could express it in terms of transition-state theory by the molecular partition functions q_i (Sect. 2.6). The relation to thermodynamics could be established by introducing the activation entropy for adsorption, $\Delta S^{\ddagger}$ (Chapter 2, Eq.2.88).

Microscopically, ν_1 can be regarded as representing the total frequency of attempts of the adsorbate particle to move in the direction of the (desorption) reaction, i.e., to escape the chemisorptive potential. The exponential term of Eq. 3.22 then stands for the number of *successful* attempts (having the necessary activation energy). In this very simple picture, ν_1 would then equal the frequency of vibration of the adsorbed atom f_0 multiplied with the respective partition functions if it is assumed that it desorbs like an atom (first-order desorption of adsorbed noble gases may be taken as an example).

A fairly straightforward explanation of ν_2 describing a recombinative desorption, can be offered, for example, if the adsorbed gas is completely mobile in two dimensions. Then, ν_2 simply equals the collision frequency in this two-dimensional gas:

$$\nu_2 = d_A \sqrt{\pi k T / m} \left[\frac{m^2}{s}\right] , \qquad 3.23$$

where d_A = collision diameter [m], m = mass of A [kg].

In many cases, however, this assumption is certainly invalid, because, at higher adsorbate concentrations, island and cluster formation due to lateral interactions inhibits the mobility of the adparticles. Furthermore, precursor states may influence the elementary kinetic processes in a complex manner, which obscures a simple interpretation of the prefactor. Before we enter the discussion of coverage effects, we present in Table 3.6 a selection of some experimentally determined frequency factors for first- and second-order desorption reactions. For an interpretation of these values, we refer to the discourse of transition-state theory (Sect. 2.6).

For practical purposes, it is often useful to consider the coverage dependence of the

Table 3.6. Selection of experimentally determined frequency factors for first- and second-order desorption reactions [130]

System	Frequency factor at small coverages		Reference
H/Ni(100)	8×10^{-2}	cm^2s^{-1}	[101]
	2.5×10^{-1}	cm^2s^{-1}	[131, 133]
	3×10^{-0}	cm^2s^{-1}	[132]
H/Ni(111)	2×10^{-1}	cm^2s^{-1}	[101]
	2.3×10^{-2}	cm^2s^{-1}	[70]
H/Pd(100)	1×10^{-2}	cm^2s^{-1}	[103]
H/Pd(111)	1.3×10^{-1}	cm^2s^{-1}	[152]
H/Rh(111)	1.2×10^{-3}	cm^2s^{-1}	[153]
CO/Ru(0001)	$\sim 5 \times 10^{16}$	s^{-1}	[95]
CO/Pd(100)	3×10^{16}	s^{-1}	[75]
CO/Ir(111)	2.4×10^{14}	s^{-1}	[154]
CO/Ni(111)	10^{17}	s^{-1}	[141]
CO/Pt(111)	4×10^{15}	s^{-1}	[155]
O/Ag(110)	4×10^{14}	s^{-1}	[156]
O/Ir(110)	3.5×10^{-3}	cm^2s^{-1}	[157]
N/Ru(10$\bar{1}$0)	5×10^{12}	s^{-1}	[158]

desorption phenomena, which can govern the progress of a surface reaction to a great extent. While the macroscopic kinetic equations have already been presented in Sect. 2.6 (Eqs. 2.69-2.88) and discussed in terms of transition-state theory, we shall now be concerned with a microscopic understanding of the coverage dependence of the desorption rate. Besides the explicit $f(\Theta)$ function, any such coverage dependence can come about for two different reasons: i) The pre-exponential factor v_{des} of Eq. 3.22 is coverage-dependent, and ii) the activation energy for desorption ΔE_{des}^* depends on coverage.

The microscopic model behind the function $f(\Theta)$ simply is that the number of desorption events is directly proportional to the number of particles in the adsorbed state. For molecular (associative) desorption, this means that $f(\Theta)$ becomes Θ (first order desorption), whereas for dissociative adsorption and molecular desorption two individual particles must encounter each other before they can recombine and leave the surface as a molecule. The probability for this recombination is proportional to the square of Θ, hence $f(\Theta)$ becomes Θ^2 (second order desorption). In a few cases, other coverage functions than first- or second-order are observed, particularly on surfaces with adsorbate islands or high concentrations of crystallographic imperfections.

Cassuto and King [159], Alnot and Cassuto [160], and Gorte and Schmidt [146] have pointed out that the existence of any precursor state in the *adsorption* path can also be relevant for the *desorption* reaction and modify the coverage function $f(\Theta)$. Likewise, expressions for the rate of desorption can be derived, which again contain sort of a precursor constant. This, in turn, is composed of the various probabilities, which we have discussed in the context of the adsorption kinetics. However, we abandon presenting the relatively complicated formulae, which can be looked up, for example, in the review article by Morris et al. [130].

The coverage dependence of the pre-exponential factor, $v(\Theta)$, and the coverage dependence of the activation energy for desorption $\Delta E_{des}^*(\Theta)$ shall be discussed together, because these quantities are intimately related with each other via the so-called *compensation* *effect*, which we shall deal with later. For associative desorption, the variation of v_2 with coverage can be explained in terms of a random-walk surface-diffusion process as the rate-limiting step [161,161a]. It is immediately seen that this kind of diffusion will be greatly affected if the adsorbed particles are subject to mutual interaction forces with energy ω. Island formation can take place if ω is attractive, and directed-walk may be superimposed on random-walk phenomena. Accordingly, the respective kinetic formulae must be modified to include the ω-related effects. Even more obvious seems to be the influence of coverage-dependent activation energy for desorption. As in adsorption, one must distinguish between a-priori and induced energetic heterogeneity. With "real" surfaces, certainly the *a-priori* heterogeneity determines the overall desorption phenomena. That is to say, if a linear temperature ramp is applied to the system, for example in a temperature-programmed desorption experiment, the various filled potential wells of different depths are successively emptied as thermal energy is supplied to the system, starting with the shallowest and ending with the deepest wells. Likewise, at constant temperature, particles desorb from shallow potentials at a much higher rate than from deep ones. On ideal single crystal surfaces, on the other hand, the induced heterogeneity should be dominating as provoked by the mutual particle-particle interactions at medium- and high-surface concentrations. The net effect, however, will be exactly the same in this case, namely a reverse proportionality between desorption rate and adsorbate bond strength.

A discussion of the abovementioned compensation effect is now in order. This interesting phenomenon consists of an apparent coupling between the pre-exponential factor v and the activation energy for desorption ΔE_{des}^*. In a sense high prefactors are associated with high ΔE_{des}^* values and vice versa. This has the surprising consequence that an increase in the activation energy for desorption does not lead to the expected decrease of the rate constant and hence the rate, because there occurs at the same time an increase of the pre-exponential factor, which more or less *compensates* the influence of the change in the exponent! A typical example is shown in Fig.3.29. It concerns the CO desorption from a Ru(0001) surface as measured by Pfnür et al. [95]. Obviously, v_{des} follows ΔE_{des}^* in every detail.

In practical catalysis, the phenomenon of a compensation effect has been known for a long time and has been discussed repeatedly. Constable [162] probably was the first to discover it in the context of his study of dehydrogenation of ethanol over copper catalysts. Cremer [163] treated the problem of the compensation effect in catalysis in some more general manner and explained it in terms of a relation between activation enthalpy and activation entropy (which is implicit in the pre-exponential factor). For the simple thermal desorption reaction from surfaces there was an early work by Armand and Lapujoulade [164] followed by an extensive consideration of Alnot and Cassuto [160]. Both groups were able to establish a logarithmic relation between the frequency factor and the activation energy for desorption, of the kind

$$v = v_0 \exp\left(\frac{\Delta E_{des}^*}{RT_i}\right) , \qquad\qquad 3.24$$

where T_i represents the so-called *isokinetic* temperature.

The meaning of T_i can be deduced from Fig. 3.30 in which the rate of desorption is plotted vs inverse temperature. For quite different systems it appears to be a constant. At T_i, any pairs of ΔE_{des}^* and v will lead to the same rate of desorption. The compensation effect shows up quite clearly: small slopes cause a low v; large slopes cause high prefactors. So far, a reasonable physical interpretation of the compensation effect has not been given. If we believe Alnot and Cassuto [160], it is, however, nothing but an artifact,

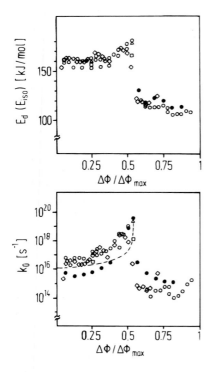

Fig. 3.29. Illustration of the operation of a compensation effect by a CO adsorption study on Ru(0001) [95], where (upper panel) adsorption energies obtained by various methods (different symbols) are contrasted to the pre-exponential factors k_0 determined for the same CO coverages (monitored by the relative work-function change $\Delta\Phi/\Delta\Phi_{max}$) The formation of an ordered ($\sqrt{3} \times \sqrt{3}$) R 30° LEED structure results in both cases in a sharp rise around $\Delta\Phi/\Delta\Phi_{max}$ = 0.5.

caused by improper data-evaluation procedures using the simple Arrhenius plot of ln(rate) vs $1/T$. They stress that precursor states in the desorption path must be taken into account, which then results in deviations from linearity of an ln(rate) vs $1/T$ plot.

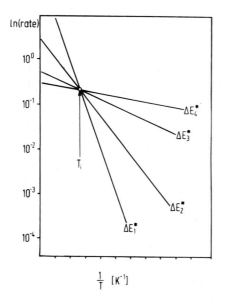

Fig. 3.30. Schematic Arrhenius diagram explaining the compensation effect. Apparently, small activation energies ΔE^* (small slopes) cause small values of ln(rate) and vice versa. Rates and activation energies coincide at the isokinetic temperature T_i.

A final word should be devoted to the role of thermal desorption processes in heterogeneous reactions. As pointed out in the introductory chapter, it is the thermal desorption reaction, which finally removes the reacted particles of the product from the surface, thus clearing the way or better the adsorption sites for the subsequently adsorbing reactants. If there were no desorption, the surface would be blocked shortly after the beginning reaction, leading to complete poisoning. A description of experimental thermal desorption techniques will be presented in Chapter 4.

3.3.4 Surface Diffusion

As repeatedly mentioned previously, the phenomenon of surface diffusion plays a central role in adsorption phenomena. It can primarily affect the substrate, but also the adsorbate structure, and principally, one must distinguish two kinds of diffusion processes (which can occur at the same time). First, the atoms of the substrate material can move, particularly at elevated temperatures, because the surface tends to lower its free energy content. This process can occur during thermal annealing or in the course of exothermic heterogeneous reactions, whereby in the latter case the extent of substrate atom diffusion is also governed by the turnover of the respective reaction. Diffusion, which involves substrate material, is usually quite an unwanted process in heterogeneous catalysis, because it leads to the sintering phenomena or to the healing of crystallographic defects, resulting in an overall loss of active surface area of porous materials. Furthermore, bulk impurity atoms can segregate at the surface and poison active sites. The diffusion of substrate atoms has been investigated frequently in the past, among others by Bonzel [165-167], Butz and Wagner [168,168a], Ehrlich [169], and Hölzl and coworkers [170,171]. Due to the rather high activation energies required for substrate atom displacements, temperatures of up to 1000 K have to be employed to obtain reasonable rates of diffusion. Frequently, surface-reconstruction phenomena are often based on a collective migration of surface atoms and hence involve substrate diffusion.

Within the framework of this chapter, however, we shall be more concerned with the second type of surface diffusion, which does not require as vigorous temperature conditions, namely, diffusion that occurs within the adsorbate phase. Pioneering investigations in this field were performed among others by Gomer and his group [172,173]. Similar to the first type, this diffusion is closely related to *mobility*. While mobility in the physisorbed state (via precursor intermediates) usually improves the rate of chemisorption, mobility in the chemisorbed state can help establish thermodynamic equilibrium, in which all particles have reached the minima of the potential energy curve and thus reside in periodic sites. It is self-evident that only in this situation the maximum degree of long-range order will exist, and in a diffraction experiment (cf. LEED, Chapter 4) intense "extra" spots will be obtained. Furthermore, surface diffusion helps to overcome lateral concentration gradients due to non-equilibrium clustering phenomena, which frequently occur at very low temperatures. There exists a vast literature on the topic surface diffusion, and we can give only a selection here [130, 169, 170]. The mechanism of diffusion can be thought of as individual "hops" of particles from their original site to empty adjacent (nearest neighbor) sites. The hopping frequency depends in the usual manner exponentially on the temperature, via

$$D_{(T)} = D_0 \exp\left(-\frac{\Delta E_{\text{diff}}^*}{RT}\right) .$$

3.25

This equation is closely connected to the residence time of a particle in a given site (Eq. 3.16). D is the diffusion coefficient at temperature T, ΔE^*_{diff} the activation energy for diffusion (the height of the barriers in x-y direction of Fig. 3.21) and D_0 the pre-exponential factor called diffusivity.

D_0 can be expressed in terms of transition state theory assuming the transmission coefficient $\varkappa$ to be unity as [130]

$$D_0 = \frac{\lambda^2}{2\alpha} \frac{kT}{h} \exp\left(\frac{\Delta S^*_{\text{diff}}}{R}\right) , \qquad 3.26$$

where λ stands for the jump distance, α for the symmetry number (= 2 for a *two-dimensional* problem), and ΔS^*_{diff} for the activation entropy of surface diffusion.

A decisive parameter of Eq. 3.25 is the activation energy for surface diffusion ΔE^*_{diff} which is approximately one-tenth the value of the adsorption energy. For normal chemisorption systems, such as CO on Pd, this means that $\Delta E^*_{\text{diff}} \approx 3 - 4$ kcal/mol = 15 – 20 kJ/mol. In the physisorption regime, this barrier height is is certainly lower by an order of magnitude.

The diffusion coefficient $D_{(T)}$ can be experimentally determined by field emission [175, 176] or by laser desorption experiments [177-179]. In the first case, there is a direct observation of the diffusing boundary in the field electron microscope possible [172], while the latter technique is well suited for single crystal surfaces covered with adsorbate. A short laser pulse is shot onto the surface, and due to local heating all the particles adsorbed within the zone of impact of the light beam ($\sim 1\,\text{mm}^2$) desorb, thus creating an empty spot on the surface. Subsequent laser pulses fired on the same spot after well-defined time intervals t allow conclusions to be made about the rate of surface diffusion into the bare zone. The diffusion coefficient $D_{(T)}$ and the mean-square displacement $\bar{x}$ are interrelated by

$$\bar{x} = (Dt)^{1/2} \qquad 3.27$$

This equation is a "random walk" relation and assumes no concentration dependence of D when diffusion occurs out of a boundary of adsorbate. In these cases, therefore, D is sort of an average diffusion coefficient over the range of coverage of the boundary. In the same way, ΔE^*_{diff} is taken to be coverage-independent. As long as the particles do not form a phase with long-range order, these assumptions may represent valid approximations. However, as pointed out previously, mutual particle-particle interaction forces are the rule in adsorbate layers, and it is expected that the respective interaction energies ω can greatly modify the distribution of particles across the surface and hence the diffusion via vacancies. This can be taken care of by introducing a coverage-dependent diffusion coefficient, and the diffusion problem must be treated using the one-dimensional form of Fick's second law

$$\left(\frac{\partial N}{\partial t}\right)_x = \frac{\partial}{\partial x}\left(D_{(N)}\frac{dN}{dx}\right)_t , \qquad 3.28$$

where N is the number of adsorbed particles per unit area (coverage), t the diffusion time, x the distance moved at particle number N, and $D_{(N)}$ the coverage-dependent diffusion coefficient.

A general solution of the partial differential equation (Eq. 3.28) has been worked out [180] so that diffusion coefficients could be determined. We should add here that D's coverage dependence comes about by a coverage dependence of ΔE^*_{diff} and Eq. 3.25 must actually be rewritten:

74

$$D_{(T,\Theta)} = D_0 \exp\left(-\frac{\Delta E_{\text{diff}}^*(\Theta)}{RT}\right),$$

3.29

whereby $\Delta E_{\text{diff}}^*(\Theta)$ can be expressed within the quasi-chemical approximation, as evaluated by Fowler and Guggenheim [181]:

$$\Delta E_{\text{diff}}^*(\Theta) = \Delta E_{\text{diff},0}^* + \frac{Z\omega}{2}\left\{1 - \frac{(1 - 2\Theta)}{\sqrt{1 - 4\Theta(1 - \Theta) \cdot B}}\right\},$$

3.30

where $B = 1 - \exp(\omega/RT)$ represents the short-range order parameter, Z the number of nearest neighbor sites, and ω the lateral interaction energy.

This expression must be inserted into Fick's law (Eq. 3.28). The partial differential equation can then be solved numerically with some mathematical effort [182], which is, however, beyond the scope of our brief overview. In Table 3.7, we instead present some experimentally determined diffusion coefficients taken from the review article by Bowker et al. [130].

Table 3.7. Some experimentally determined diffusion coefficients of adsorbates [130]

System	D_0 [cm^{2}s^{-1}]	Reference
Cs on W(110)	0.23	[183]
K on W (tip)	10^{-4}... 10^{-6}	[184]
N on W(110)	0.014	[185]
O on W(110)	0.04 ... 0.25	[182]
H on Ni(100)	2.5×10^{-3}	[178]
D on Ni(100)	8.5×10^{-3}	[178]
H on W (tip)	1.8×10^{-5}	[186]
D on Pt(111)	8×10^{-2}	[179]
CO on Pt(111)	10^{-2}... 10^{-3}	[179]

A final word should be devoted to the role of diffusion in a heterogeneous reaction in general. As we have pointed out in the introductory chapter (1.2) there may be various physical processes rate-limiting for the overall velocity of a surface reaction, namely, i) trapping and sticking, ii) adsorption, iii) surface reaction, and iv) desorption of product molecules. So far we have mainly considered surface diffusion as a vital elementary process in steps ii – iv. In view of the practical conditions in which heterogeneous reactions are carried out, namely, at pressures in the kPa and MPa range, two additional reaction steps can come into play. Strictly speaking, they do not have so much to do with *surface* diffusion, but with diffusion or mass transport in the vicinity of the surface. Frequently, these steps determine the overall macroscopic reaction rate. The respective diffusion phenomena comprise the transport of reactants to the catalyst surface and the transport of products away from it, particularly under viscous flow conditions. When either of these processes is slower than the chemical reaction rate, the total rate will be governed by the rate of arrival of reactants and of the removal of products, respectively. A reaction of this kind is called diffusion-limited. Here, the microscopic diffusion parameters are not really relevant anymore; rather the more macroscopic parameters such as flow rate, viscosity, geometrical dimensions of the reactor, and porosity of the catalyst material become operative and determine the scenery. Diffusion limitation is indicated by one of the three characteristics. i) The rate is no longer directly proportional to the weight (surface) of the catalyst material, a power law with exponent $0 < n < 1$ is obtained instead. ii) The rate can be greatly affected by changes in the flow rate conditions. iii) The temperature coefficient of the reaction is low, pointing to very small apparent activation energies. Diffusion limitation can also be a problem with porous catalyst material. However, in this case, changes in the flow rates do not affect the rate of diffusion inside the pores. Here we touch again on practical questions as to the optimum macroscopic morphology of catalyst particles for a given reaction, which are not to be addressed here.

References

1. Kittel C (1966) Introduction to Solid State Physics, 3rd edn. Wiley, New York
2. Kleber W (1959) Einführung in die Kristallographie, 3. Aufl. VEB Verlag Technik, Berlin
3. Ziman JM (1972) Principles of the Theory of Solids, 2nd edn. Cambridge University Press, Cambridge
4. Woolfson MM (1970) An Introduction to X-ray Crystallography, Cambridge University Press, Cambridge
5. Wilson AJC (1962) X-ray Optics. Methnen's Monographs on Physical Subjects. Wiley, New York
6. Wood EA (1964) Crystals and Light. Van Nostrand, Princeton, NJ
7. Raaz F (1975) Röntgenkristallographie. De Gruyter, Berlin
8. Atkins PW (1982) Physical Chemistry, 2nd edn. Oxford University Press, Oxford
9. Moore WJ (1963) Physical Chemistry, 4th edn. Longman, London
10. Alonso M, Finn EJ (1978) Fundamental University Physics, ninth printing, vol 3, Quantum and Statistical Physics. Addison-Wesley, Reading, MA
11. Binnig G, Rohrer H (1984) Scanning Tunneling Microscopy. Physica 127 B:37–45
11a. Behm RJ, Hösler W (1986) In: Vanselow R, Howe R (eds) Chemistry and Physics of Solid Surfaces VI. Springer, Berlin, p 361
12. Wintterlin J, Brune H, Höfer H, Behm RJ (1988) Atomic Scale Characterization of Oxygen Adsorbates on Al(111) by Scanning Tunneling Microscopy. Appl Phys A47:99–102
13. Müller K (1986) Relaxation and Reconstruction of Solid Surfaces. Ber Bunsenges Phys Chem 90:184–190
14. Adams DL, Nielsen BH, Andersen NJ (1983) Quantitative Analysis of LEED Measurements. Physica Scripta T 4:22–28
15. Davis LH, Noonan JR (1983) Multilayer Relaxation in Metallic Surfaces as Demonstrated by LEED Analysis. Surface Sci 126:245–252
16. Andersen NJ, Nielsen HB, Petersen L, Adams DL (1984) Oscillatory Relaxation of the Al(110) Surface. J Phys C: Solid State Phys 17:173–192
17. Noonan JR, Davis HL (1984) Truncation-induced Multilayer Relaxation of the Al(110) Surface. Phys Rev B29:4349–4355
18. Sokolov J, Shi HD, Bardi U, Jona F, Marcus PM (1984) Multilayer Relaxation of Body-centered-cubic Fe(211) J Phys C: Solid State Phys 17:371–383
19. Sokolov J, Jona F, Marcus PM (1984) Trends in Metal Surface Relaxation. Solid State Commun 49:307–312
20. Gauthier Y, Baudoing R, Joly Y, Gaubert C, Rundgren J (1984) Multilayer Relaxation of Ni(110) Analysed by LEED and Metric Distances. J Phys C: Solid State Phys 17:4547–4558,
21. Adams DL, Moore WT, Mitchell KAR (1985) Multilayer Relaxation of the Ni(311) Surface: A New LEED Analysis. Surface Sci 149:407–422
22. Noonan JR, Davis HL (1981) LEED Analysis of Ag(110) – Second Interlayer Spacing Expansion. Bull Am Phys Soc 26:224

23. Davis HL, Zehner DM (1980) Structure of the Clean Re(10$\bar{1}$0) Surface. J Vac Sci Technol 17:190–193
24. Christmann K (1987) Relaxation and Reconstruction: How Metal Surfaces Respond to the Chemisorption of Gases. Z Phys Chem NF 154:145–178
25. King DA (1983) Clean and Adsorbate-induced Surface Phase Transitions on W(100) Physica Scripta T4:34–43
26. Chan CM, van Hove MA (1986) Confirmation of the Missing-row Model with Three-layer Relaxations for the Reconstructed Ir(110)-(1x2) Surface. Surface Sci 171:226–238
27. Christmann K, Heinz K (eds) (to appear in 1991) Reconstruction of Solid Surfaces. Springer, Berlin
28. Alan G (1987) Atomic Surface Relaxation. Progr Surf Sci 25:43–56
29. Inglesfield EJ (1985) Reconstructions and Relaxations on Metal Surfaces. Progr Surf Sci 20:105164
30. Foiles SM (1987) Reconstruction of fcc (110) Surfaces. Surface Sci 191:L779–L786
31. Somorjai GA (1981) Chemistry in Two Dimensions: Surfaces. Cornell University Press, Ithaca, NY
32. Lang B, Joyner RW, Somorjai GA (1972) Low Energy Electron Diffraction Studies of High Index Crystal Surfaces of Platinum. Surface Sci 30:440–453
33. van Hove MA, Somorjai GA (1980) A New Microfacet Notation for High-Miller-Index Surfaces of Cubic Materials with Terrace, Step and Kink Structures. Surface Sci 92:489–518
34. Benard J (ed) (1983) Adsorption on Metal Surfaces. Studies in Surface Science and Catalysis 13, Elsevier, Amsterdam, p 11
35. Muetterties EL, Rhodin TN, Band E, Brucker CF, Pretzer WR (1979) Clusters and Surfaces. Chem Rev 79:91–137
36. Burch R (1988) In: Páal Z, Menon PG (eds) Hydrogen Effects in Catalysis. Dekker, New York
37. Bond GC (1987) Heterogeneous Catalysis, Principles and Applications, 2nd edn. Oxford Science Publications, Clarendon Press, Oxford, p 72
38. Sachtler WMH, Dorgelo GJH, Jongepier R (1965) Phase Composition and Work Function of Vacuum-deposited Copper-nickel Alloy Films. J Catal 4:100–102;
38a. Sachtler WMH, Dorgelo GJH, Jongepier R (1965) The Surface of Copper-Nickel Alloy Films. II Phase Equilibrium and Distribution and their Implications for Work Function, Chemisorption, and Catalysis. J Catal 4:665–671
39. Sachtler WMH, van Santen RA (1977) Surface Composition and Selectivity of Alloy Catalysts. Adv Catal Rel Subj 26:69–119
40. Ponec V (1975) Selectivity in Catalysis by Alloys. Catal Rev Sci Eng 11:41–70
41. Ponec V (1983) Catalysis by Alloys in Hydrocarbon Reaction. Adv Catal Rel Subj 32:149–214
42. Sinfelt JH (1983) Bimetallic Catalysts. Wiley, New York
43. Dalla Betta RA, Cusamano JA, Sinfelt JH (1970) Cyclopropane-hydrogen Reaction over the Group VIII Noble Metals. J Catal 19:343–349
44. Sinfelt JH, Carter JL, Yates DCJ (1972) Catalytic Hydrogenolysis and Dehydrogenation over Copper-nickel Alloys. J Catal 24:283–296
45. Sinfelt JH (1973) Supported "Bimetallic Cluster" Catalysts. J Catal 29:308–315
46. Sinfelt JH, Lam YL, Cusamano JA, Barnett AE (1976) Nature of Ruthenium-copper Catalysts. J Catal 42:227–237
47. Prestridge EB, Via GH, Sinfelt JH (1977) Electron Microscopy Studies of Metal Clusters: Ru, Os, Ru-Cu, and Os-Cu. J Catal 50:115–123
48. Sinfelt JH, Via GH, Lytle FW (1980) Structure of Bimetallic Clusters. Extended X-ray Absorption Fine Structure (EXAFS) Studies of Ru-Cu Clusters. J Chem Phys 72:4832–4844;
48a. Sinfelt JH, Via GH, Lytle FW (1982) Structure of Bimetallic Clusters. Extended X-ray Absorption Fine Structure (EXAFS) Studies of Pt-Ir Clusters. J Chem Phys 76:2779–2789
49. Christmann K, Ertl G, Shimizu H (1980) Model Studies on Bimetallic Cu/Ru Catalysts. I. Cu on Ru(0001). J Catal 61:397–411
50. Shimizu H, Christmann K, Ertl G (1980) Model Studies on Bimetallic Cu/Ru Catalysts. II. Adsorption of Hydrogen. J Catal 61:412–429

51. Yates JT, Peden CFH, Goodman DW (1985) Copper Site Blocking of Hydrogen Chemisorption on Ruthenium. Catal 94:56–580
52. Goodman DW, Peden CFH (1985) Hydrogen Spillover from Ruthenium to Copper in Cu/Ru Catalysts: A Potential Source of Error in Active Metal Titration. J Catal 95:321–324
53. Wood EA (1964) Vocabulary of Surface Crystallography. Appl Phys 35:1306–1312
54. Park RL, Madden HH (1968) Annealing Changes on the (100) Surface of Palladium and their Effect on CO Adsorption. Surface Sci 11:188–202
55. Nicolle J (1954) Die Symmetrie und ihre Anwendungen, Deutscher Verlag der Wissenschaften, Berlin
56. Cotton FA (1971) Chemical Applications of Group Theory, 2nd edn. Wiley, New York
57. Sachtler WMH (1973) Surface Composition of Alloys in Equilibrium. Le Vide 164:67–71
58. van der Planck P, Sachtler WMH (1967) Surface Composition of Equilibrated Copper-nickel Alloy Films. J Catal 7:300–303;
58a. van der Planck P, Sachtler WMH (1968) Interaction of Benzene with Protium and Deuterium on Copper-nickel Films with Known Surface Composition. J Catal 12:35–44
59. Balandin AA (1958) The Nature of Active Centers and the Kinetics of Catalytic Dehydrogenation. Adv Catal Rel Subj 10:96–129;
59a. Balandin AA (1969) Modern State of the Multiplet Theory of Heterogeneous Catalysis. Adv Catal Rel Subj 19:1–210
60. Mate CM, Somorjai GA (1985) Carbon Monoxide induced Ordering of Benzene on Pt(111) and Rh(111) Crystal Surfaces. Surface Sci 160:542–560
61. Morrison J, Lander JJ (1966) Ordered Physisorption of Xenon on Graphite. Surface Sci 5:163–165
62. Lee J, Cowan JP, Wharton L (1983) He Diffraction from Clean Pt(111) and (1×1)H/Pt(111) Surface. Surface Sci 130:1–28
63. Lindroos M, Pfnür H, Feulner P, Menzel D (1987) A Study of the Adsorption Sites of Hydrogen on Ru(001) at Saturation Coverage by Electron Reflection. Surface Sci 180:237–251
64. Ehsasi M, Christmann K (1988) The Interaction of Hydrogen with a Rhodium(110) Surface. Surface Sci 194:172–198
65. Lauth G, Schwarz E, Christmann K (1989) The Adsorption of Hydrogen on a Ruthenium (10$\bar{1}$0) Surface. J Chem Phys 91:3729–3743
66. Oed W, Puchta W, Bickel N, Heinz K, Nichtl W, Müller K (1988) Full-Coverage Adsorption Structure of H/Rh(110). J Phys C 21:237–234
67. van Hove MA (1979) Surface Crystallography and Bonding. In: Rhodin TN, Ertl G (eds) The Nature of the Surface Chemical Bond. North Holland, Amsterdam, pp 277–311
68. van Hove MA, Tong SY (1979) Surface Crystallography by LEED. Springer, Berlin
69. Marcus PM, Jona F (eds) (1984) Determination of Surface Structure by LEED. Plenum Press, New York
70. Christmann K, Behm RJ, Ertl G, van Hove MA, Weinberg WH (1979) Chemisorption Geometry of Hydrogen on Ni(111): Order and Disorder. J Chem Phys 70:4168–4184
71. Lindroos M, Pfnür H, Menzel D (1987) Investigation of a Disordered Adsorption System by Electron Reflection: H/Ru(001) at Intermediate Coverages. Surface Sci 192:421–437
72. Heinz K (1989) In: Proc ESF Workshop on "Reconstructive or Asymmetric Adsorption on fcc (100) Metal Surfaces", Erlangen, p 63
73. Wuttig M, Franchy R, Ibach H (1989) Oxygen on Cu(100) – A case of an Adsorbate induced Reconstruction. Surface Sci 213:103–136
74. Imbihl R, Behm RJ, Ertl G, Moritz W (1982) The Structure of Atomic Nitrogen Adsorbed on Fe(100). Surface Sci 123:129–140
75. Behm RJ, Christmann K, Ertl G, van Hove MA (1980) Adsorption of CO on Pd(100). J Chem Phys 73:2984–2995
76. Lehwald S, Rocca M, Ibach H, Rahman TS (1985) Surface Phonon Dispersion of c(2×2)S on Ni(100). Phys Rev B31:3477–3485

77. van Hove MA, Tong SY (1975) Chemisorption Bond Lengths of Chalcogen Overlayers at a Low Coverage by Convergent Perturbation Methods. J Vac Sci Technol 12:230–233
78. Salwén A, Rundgren J (1975) The Structure of the p(2×2) Tellurium Overlayer on the (001) Surface of Copper investigated by LEED. Surface Sci 53:523–537
79. Christmann K, Penka V, Chehab F, Ertl G, Behm RJ (1984) Dual Path Surface Reconstruction in the H/Ni(110) System. Solid State Commun 51:487–490
80. Cattania MG, Penka V, Behm RJ, Christmann K, Ertl G (1983) Interaction of Hydrogen with a Palladium (110) Surface. Surface Sci 126:382–391
81. Barker RA, Estrup PJ (1978) Hydrogen on Tungsten (100): Adsorbate-induced Surface Reconstruction. Phys Rev Lett 41: 1307–1310
82. Debe MK, King DA (1979) The Clean Thermally Induced W{OO1}(1×1) (2×2)R45° Surface Structure Transition and its Crystallography. Surface Sci 81:193–237
83. Heinz K, Müller K (1982) LEED Intensities – Experimental Progress and New Possibilities of Surface Structure Determination. In: Höhler G (ed) Structural Studies of Surfaces. Springer, Berlin p 39 ff
84. Willis RF (1986) Critical Fluctuations on Solid Surfaces. Ber Bunsenges Phys Chem 90:190–197
85. Frenken JWM, Krans RL, van der Veen JF, Holub-Krappe E, Horn K (1987) Missing-row Surface Reconstruction of Ag(110) Induced by Potassium Adsorption. Phys Rev Lett 59:2307–2310
86. Behm RJ, Thiel PA, Norton PR, Ertl G (1983) The Interaction of CO and Pt(100) I. Mechanism of Adsorption and Pt Phase Transition. J Chem Phys 78:7437–7447
86a Behm RJ, Thiel PA, Norton PR, Ertl G (1983) The Interaction of CO and Pt(100) II. Energetic and Kinetic Parameters. J Chem Phys 78:7448–7458
87. Christmann K, Ertl G (1973) Interaction of CO and O_2 with Ir(110) Surfaces. Z Naturf 28a:1144–1148
88. Lennard-Jones JE (1932) Processes of Adsorption and Diffusion on Solid Surfaces. Trans Faraday Soc 28:333
89. Dunken HH Lygin VI (eds) (1978) Quantenchemie der Adsorption an Festkörperoberflächen, Verlag Chemie, Weinheim
90. Hoffmann R (1988) Solids and Surfaces – A Chemists View of Bonding in Extended Structures. Verlag Chemie, Weinheim
91. Slater JC (1974), The Self-consistent Field for Molecules and Solids. McGraw-Hill, New York
92. Christmann K, Ertl G, Schober O (1974) Adsorption of CO on a Ni(111) Surface. J Chem Phys 60:4719–4724
93. Tracy JC, Palmberg PW (1969) Structural Influences on Adsorbate Binding Energy. I. Carbon Monoxide on (100) Palladium. J Chem Phys 51:4852–4862
94. Ertl G, Koch J (1970) Adsorption von CO auf einer Palladium(111) Oberfläche. Z Naturf 25a:1906–1911
95. Pfnür H, Feulner P, Engelhardt HA, Menzel D (1978) An Example of "Fast" Desorption: Anomalously High Pre-exponentials for CO Desorption from Ru(001). Chem Phys Lett 59:481–486
96. Lauth G, Solomun T, Hirschwald W, Christmann K (1989) The Interaction of Carbon Monoxide with a Ruthenium(10$\bar{1}$0) Surface. Surface Sci 210:201–224
97. Blyholder G (1964) Molecular Orbital View of Chemisorbed Carbon Monoxide. J Phys Chem 68:2772–2778
98. Tracy JC (1972) Structural Influences on Adsorption Energy. III. CO on Cu(100). J Chem Phys 56:2748–2754
99. Tracy JC (1972) Structural Influences on Adsorption Energy. II. CO on Ni(100). J Chem Phys 56:2736–2747
100. Christmann K (1988) Interaction of Hydrogen with Solid Surfaces. Surface Sci Repts 9:1–163
101. Christmann K, Schober O, Ertl G, Neumann M (1974) Adsorption of Hydrogen on Nickel Single Crystal Surfaces. J Chem Phys 60:4528–4540
102. Rinne H (1974) Absolutmessungen der Adsorption von Wasserstoff an der Nickel(111)-Fläche und an ultradünnen Nickelfilmen, PhD thesis, Universität Hannover

103. Behm RJ, Christmann K, Ertl G (1980) Adsorption of Hydrogen on Pd(100). Surface Sci 99:320–340
104. Conrad H, Ertl G, Latta EE (1974) Adsorption of Hydrogen on Palladium Single Crystal Surfaces. Surface Sci 41:435–446
105. Ernst KH (1990) Die geometrischen und elektronischen Strukturen der Adsorbatphasen von Wasserstoff auf der Kobalt(1010)-Oberfläche. PhD thesis, Freie Universität Berlin
105a Ernst KH, Christmann K (1991) to be published
106. Melius CF, Moskowitz JW, Mortola AP, Baillie MB, Ratner MA (1976) A Molecular Complex Model for the Chemisorption of Hydrogen on a Nickel Surface. Surface Sci 59:279–292
107. Nørskov JK, Stoltze P (1987) Theoretical Aspects of Surface Reactions. Surface Sci 189/190:91–105
108. Harris J, Andersson S(1985) H_2 Dissociation at Metal Surfaces. Phys Rev Lett 55:1583–1586
109. Polyani JC (1987) Some Concepts in Reaction Dynamics. Science 236:680–690
109a. Manz J (1989) Mode Selective Bimolecular Reactions. In: Molecules in Physics, Chemistry and Biology, vol III. Kluwer, New York, pp 165–404
110. Harris J (1988) On the Adsorption and Desorption of H_2 at Metal Surfaces. Appl Phys A47:63–71
111. Balooch M, Cardillo MJ, Miller DR, Stickney RE (1974) Molecular Beam Study of the Apparent Activation Barrier Associated with Adsorption and Desorption of Hydrogen on Copper. Surface Sci 46:358–392
112. Kubiak GD, Sitz GO, Zare RN (1985) Recombinative Desorption Dynamics: Molecular Hydrogen from Cu(110) J Chem Phys 83:2538–2551
113. Anger P, Winkler A, Rendulic KD (1989) Adsorption and Desorption Kinetics in the System H_2/Cu(111) H_2/Cu(110) and H_2/Cu(100). Surface Sci 220:1–17
114. Berger HF, Leisch M, Winkler A, Rendulic KD (1990) A Search for Vibrational Contributions to the Activated Adsorption of H_2 on Copper. Chem Phys Lett 175:425–428
115. Grimley TB (1967) The Indirect Interaction between Atoms or Molecules Adsorbed on Metals. Proc Phys Soc 90:751
116. Grimley TB, Torrini M (1973) The Interaction between Two Hydrogen Atoms Adsorbed on (100) Tungsten. J Phys C6:868–872
117. Einstein TL, Schrieffer JR (1973) Indirect Interaction between Adatoms on a Tight Binding Solid. Phys Rev B7:3629–3648
118. Park RL, Einstein TL, Kortan AR, Roelofs LD (1980) Multi-critical Phase Diagram of a Chemisorbed Lattice Gas System-O/Ni(111). In: Sinha SK (ed) Ordering in Two Dimensions. North Holland, Amsterdam, p 17–24
119. Hill TL (1962) Introduction to Statistical Thermodynamics, 2nd printing. Addison-Wesley, Reading, MA
120. Gebhardt W, Krey U (1980) Phasenübergange und kritische Phänomene. Vieweg, Braunschweig
121. Kortüm G (1972) Einführung in die Chemische Thermodynamik, 6th edn. Vandenhoeck und Ruprecht, Göttingen, p 335 ff
122. Bauer E (1987) Phase Transitions on Single-crystal Surfaces and in Chemisorbed Layers. In: Schommers W, von Blanckenhagen P (eds). Structure and Dynamics of Surfaces II. Springer, Berlin, pp 115–179
123. Christmann K, Ertl G (1976) Interaction of Hydrogen with Pt(111): The Role of Atomic Steps. Surface Sci 60:365–384
124. Poelsema B, Verheij LK, Comsa G (1985) Temperature Dependency of the Initial Sticking Probability of H_2 and CO on Pt(111). Surface Sci 152/153:496–504
125. Rendulic KD (1988) The Influence of Surface Defects on Adsorption and Desorption. Appl Phys A47:55–62
126. Russel, Jr. JN, Chorkendorff I, Lanzilotto AM, Alvey MD, Yates, JT Jr. (1986) Angular Distributions of H_2 Thermal Desorption: Coverage Dependence on Ni(111). J Chem Phys 85:6186–6191

127. Salmeron M, Gale RJ, Somorjai GA (1977) Molecular Beam Study of the H_2-D_2 Exchange Reaction on Stepped Platinum Crystal Surfaces: Dependence on Reactant Angle of Incidence. J Chem Phys 67:5324–5334

128. Engel T, Ertl G (1978) A Molecular Beam Investigation of the Catalytic Oxidation of CO on Pd, J Chem Phys 69: 1267–1281

129. Engel T (1979) Molekularstrahluntersuchungen der Adsorption und Reaktion von H_2, O_2 und CO an einer Pd(111) Oberfläche. Habilitationsschrift, Universität München

130. Morris MA, Bowker M, King DA (1984) Kinetics of Adsorption, Desorption and Diffusion at Metal Surfaces, In: Bamford CH, Tipper CFH, Compton RG (eds) Simple Processes at the Gas-solid Interface, vol 19, Elsevier Amsterdam, p 1–179

131. Lapujoulade J, Neil KS (1973) Hydrogen Adsorption on Ni(100) Surface Sci 35:288–301

132. Christmann K (1979) Adsorption of Hydrogen on a Nickel (100) Surface. Z Naturf A34:22–29

133. Lapujoulade J, Neil KS (1972) Chemisorption of Hydrogen on the (111) Plane of Nickel. Chem Phys 57:3535–3545

134. Christmann K, Ertl G, Pignet T (1976) Adsorption of Hydrogen on a Pt(111) Surface. Surface Sci 54:365–392

135. Poelsema B, Lenz K, Brown LS, Verheij LK, Comsa G (1986) Wasserstoff auf Pt(111): Haftwahrscheinlichkeit und Stufenkonzentration. Verh DPG 5:1378

136. Penka V, Christmann K, Ertl G (1984) Ordered Low-temperature Phases in the H/Ni(110) System. Surface Sci 136:307–318

137. Robota HJ, Vielhaber W, Lin MC, Segner J, Ertl G (1985) Dynamics of Interaction of H_2 and D_2 with Ni(110) and Ni(111) Surfaces. Surface Sci 155:101–120

138. Hofmann P, Unwin R, Wyrobisch W, Bradshaw AM (1978) The Adsorption and Incorporation of Oxygen on Cu(100) at $T \geq 300$ K. Surface Sci 72:635–644

139. Holloway PJ, Hudson JB (1974) Kinetics of the Reaction of Oxygen with Clean Nickel Single Crystal Surfaces. Surface Sci 43:123–140

140. Netzer FP, Wille RA (1978) Adsorption Studies on a Stepped Pt(111) Surface: O_2, CO, C_2H_4, C_2N_2. Surface Sci 74:547–567

141. Ibach H, Erley W, Wagner H (1980) The Pre-exponential Factor in Desorption – CO on Ni(111) Surface Sci 92:29–42

142. Engel T (1978) A Molecular Beam Investigation of He, CO, and O_2 Scattering from Pd(111) J Chem. Phys 69:373–385

143. Ertl G, Neumann M, Streit KM (1977) Chemisorption of CO on the Pt(111) Surface. Surface Sci 64:393–410

144. Kisliuk PJ (1957) The Sticking Probabilities of Gases Chemisorbed on the Surfaces of Solids. J Phys Chem Solids 3:95–101

144a. Kisliuk PJ (1958) The Sticking Probabilities of Gases Chemisorbed on the Surfaces of Solids. J Phys Chem Solids 5:78–84

145. King DA (1975) Thermal Desorption from Metal Surfaces: A Review. Surface Sci 47:384-402

145a. King DA (1977) The Influence of Weakly Bound Intermediate States on Thermal Desorption Kinetics. Surface Sci 64:43–51

146. Gorte R, Schmidt LD (1978) Desorption Kinetics with Precursor Intermediates. Surface Sci 76:559–573

147. Schönhammer K (1979) On the Kisliuk Model for Adsorption and Desorption Kinetics. Surface Sci 83:L633–L636

148. Péterman LA (1972) Thermal Desorption Kinetics of Chemisorbed Gases, Progress in Surface Science, vol 3. Pergamon Press, Oxford, pp 2–61

149. Menzel D (1975) Desorption Phenomena. In: Gomer R (ed) Interactions on Metal Surfaces. Springer, Berlin, pp 101–142

150. Gortel ZW, Kreuzer HJ, Spaner D (1980) Quantum Statistical Theory of Flash Desorption. J Chem Phys 72:234–246

151. Kreuzer HJ, Gortel ZW (1986) Physisorption Kinetics.Springer, Berlin
152. Conrad H (1976) Wechselwirkung von Gasen mit einer Pd(111)-Oberfläche. PhD thesis, Universität München
153. Yates, JT Jr, Thiel PA, Weinberg WH (1979) The Chemisorption of Hydrogen on Rh(111) Surface Sci 84:427–439
154. Zitdan PA, Bereskov GK, Baronin Al (1976) An XPS and UPS Investigation of the Chemisorption of CO on Ir(111) Chem Phys Lett 44 :528–532
155. Campbell CT, Ertl G, Kuipers H, Segner J (1981) A Molecular Beam Investigation of the Interactions of CO with a Pt(111) Surface. Surface Sci 107:207–236
156. Bowker M (1980) The Effect of Lateral Interactions on the Desorption of Oxygen from Ag(110) Surface Sci 100 :L472–L474
157. Taylor JL, Ibbotson DF, Weinberg WH (1979) The Chemisorption of Oxygen on the (110) Surface of Iridium. Surface Sci 79:349–384
158. Klein R, Shih A (1977) Chemisorption and Decomposition of Nitric Oxide on Ruthenium. Surface Sci 69:403–427
159. Cassuto A, King DA (1981) Rate Expressions for Adsorption and Desorption Kinetics with Precursor States and Lateral Interactions. Surface Sci 102:388–404
160. Alnot M, Cassuto A (1981) Analysis of Computed TPD Curves Involving a Precursor State: Influence of the Method on the Parameters of the Adsorbate. Surface Sci 112:325–342
161. Tamm PW, Schmidt LD (1969) Interaction of H_2 with (100) W. I. Binding States. J Chem Phys 51:5352–5363
161a. Tamm PW, Schmidt LD (1970) Interaction of H_2 with (100) W. II. Condensation. J Chem Phys 52:1150–1160
162. Constable FH (1952) The Mechanism of Catalytic Decomposition, Proc Roy Soc (London) A108:355
163. Cremer E (1955) The Compensation Effect in Heterogeneous Catalysis. Adv Catal Rel Subj 7:75–91
164. Armand G, Lapujoulade J (1967) Le facteur preexponentiel en cinetique de desorption gaz-solide. Surface Sci 6:345–361
165. Bonzel HP (1971) Auger Electron Spectroscopy Study of a Sulfur-Oxygen Surface Reaction on a Cu(110) Crystal. Surface Sci 27:387–410
166. Bonzel HP, Ku R (1973) Adsorbate Interactions on a Pt(110) Surface. II: Effect of Sulfur on the Catalytic CO Oxidation. J Chem Phys 59:1641–1651
167. Bonzel HP, Latta EE (1978) Surface Self-diffusion on Ni(110): Temperature Dependence and Directional Anisotropy. Surface Sci 76:275–295
168. Butz R, Wagner H (1979) Surface Diffusion of Pd and Au on W Single Crystal Planes I. Spreading Behavior of Pd and Au Layers. Surface Sci 87:6984
168a Butz R, Wagner H (1979) Surface Diffusion of Pd and Au on W Single Crystal Planes II. Anisotropy of Palladium Surface Diffusion due to the Influence of Substrate Structure. Surface Sci 87:85–100
169. Ehrlich G (1974) In: Jayadenaah TS, Vanselow R (eds) Surface Science: Recent Progress and Perspectives. CRC Press, Cleveland, Ohio
170. Hölzl J, Porsch G (1975) Contact Potential Difference Measurements on Polycrystalline Ni during and after Deposition with Evaporated Ni in the Temperature Range $230° \leq T \leq 450°C$, Thin Solid Films 28:93–106
171. Schrammen P, Hölzl J (1983) Investigation of Surface Self-diffusion of Ni Atoms on the Ni(100) Plane by means of Work Function Measurements and Monte Carlo Calculations. Surface Sci 130:203–228
172. Gomer R, Hum JK (1957) Adsorption and Diffusion of Oxygen on Tungsten. Chem Phys 27:1363–1376
173. Gomer R, Wortman R, Lundy R (1957) Mobility and Adsorption of Hydrogen on Tungsten. J Chem Phys 26:1147–1164
174. Drechsler M (1982) In: Binh TV (ed) Surface Mobilities. Plenum Press, New York

175. Chen JR, Gomer R (1979) Mobility of Oxygen on the (110) Plane of Tungsten. Surface Sci 79:413–444
176. Wang SC, Gomer R (1985) Diffusion of Hydrogen, Deuterium, and Tritium on the (110) Plane of Tungsten. J Chem Phys 83:4193–4209
177. Ertl G, Neumann M (1972) Laser-induzierte schnelle thermische Desorption von Festkörper-Oberflächen. Z Naturf 27a:1607–1610
178. Mullins DR, Roop B, Costello SA, White JM (1987) Isotope Effects in Surface Diffusion. Hydrogen and Deuterium on Ni(100) Surface Sci 186:67–74
179. Seebauer EG, Kong ACF, Schmidt LD (1988) Surface Diffusion of Hydrogen and CO on Rh(111): Laser-induced Thermal Desorption Studies. J Chem Phys 88:6597–6604
180. Boltzmann L (1934) Wien Ann 53:53
181. Fowler RH, Guggenheim EA (1939) Statistical Thermodynamics, Cambridge University Press, Cambridge
182. Bowker M, King DA (1980) Oxygen Diffusion on Tungsten Single Crystal Surfaces: Secondary Electron Emission Studies. Surface Sci 94:564–580
183. Love HM, Wiederick HD (1969) Cesium Diffusion at a Tungsten Surface, Can J Phys 47:657–663
184. Schmidt L, Gomer R (1965) Adsorption of Potassium on Tungsten. J Chem Phys 42:3573–3598
185. Polak A, Ehrlich G 11977) Surface Diffusion of Gas Chemisorbed on a Single Crystal Plane: N on W(110). J Vac Sci Technol 14:407
186. Lewis R, Gomer R (1969) Adsorption of Hydrogen on Platinum. Surface Sci 17:333–345

4 Some of the Surface Scientist's Tools

In the preceding chapter we provided the reader with a whole variety of detailed physical and chemical information about the state of surfaces and adsorbates. Certainly, on several occasions, the question has arisen about how this information was obtained for each case, and in the following, we are going to answer at least some of those questions as we present a selection of appropriate experimental methods.

There is one common feature of the respective experimental techniques, namely, their *surface sensitivity*. Since, for a given piece of bulk material, the number of bulk atoms always exceeds the number of surface atoms by many orders of magnitude, surface sensitivity really is the key property of any such analytical method. Surface sensitivity can be attained in various ways, whereby the penetration depth or escape depth of impact or off-scattered particles is one of the most prominent physical properties. As we shall see later, most surface analytical tools use *electrons* as probing particles, and the escape depth of electrons with variable kinetic energy emitted from solid, particularly metallic, material is essentially the basis of any surface sensitivity here. The electron-escape depth is relatively independent of the kind of metal and can be determined in various ways, for example, by probing the characteristic electron emission of a substrate metal which is then uniformly coated with a different kind of material. Simultaneously, the decay of the original emission is monitored as a function of overlayer thickness. In Fig. 4.1, we present the so-called universal mean free path (λ) curve for electrons in solids as a function of their kinetic energy [1]. Each point in that curve refers to an individual λ measurement. It is quite obvious that the curve exhibits a pronounced minimum of 1–3 monolayers (3–10 Å) around 10–100 eV electron energy, which is the reason why this particular energy range is usually chosen in surface electron spectroscopy experiments.

Another quite important criterion for the applicability of an analytical method is that it is *non*-destructive. In other words, the surface properties should be retained after the analysis has been made. Unfortunately, there are various methods that actually do damage

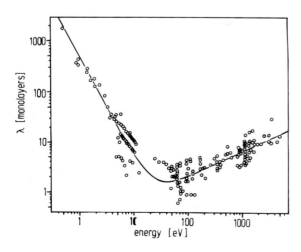

Fig. 4.1. Universal inelastic mean free path (imfp) curve for electrons in solids as a function of their kinetic energy (eV). After Seah and Dench [1].

the surfaces, at least partially, and one must therefore distinguish *destructive* and *non-destructive* surface analytical techniques.

Before we enter a detailed description of selected methods, we should, once again, emphasize the importance of the *pressure gap* pertinent to what we call surface analysis (cf., Chapter 1). Accordingly, the surface analytical tools must be subdivided into two classes, viz., those operating at pressures below $\sim 10^{-4}$ mbar, and those functioning at higher and much higher pressures. Unfortunately, by far the most standard surface analytical methods are based on unperturbed particle impact and detection and thus require Knudsen (molecular flow) conditions, where the mean free path of the gas molecules or electrons is larger than the dimensions of the reactor. Moreover, since it is demanded to analyze (or characterize) *well-defined* systems, it is mandatory to establish ultra-high vacuum conditions. It follows immediately from kinetic theory that the number of gas particles, $\dot{N}_s$, striking a surface area of 1 cm^2 per second is given by

$$\dot{N}_s = {}^1 N \sqrt{\frac{RT}{2\pi M}} = 2.634 \times 10^{22} \frac{P}{\sqrt{MT}} \,, \qquad 4.1$$

where ${}^1 N$ equals the number of gas molecules per cm^3, and P the gas pressure in [mbar].

Assuming an average molecular weight of $M = 28$ [g Mol^{-1}] (which could refer to molecular nitrogen or carbon monoxide), one can, from Eq. 4.1, construct a plot of the number of impact particles/s and cm^2 against gas pressure, as is reproduced for three kinds of gases with different molar mass in Fig. 4.2. It is perhaps surprising that around 10^{-6} mbar ambient pressure there are as many particles colliding with the surface in 1 s as are necessary to build up a complete monolayer (we recall that this fact led us to define the exposure in "Langmuir", cf., Sect. 3.3.1). Even at 10^{-10} mbar pressure there are still about 10^{10} to 10^{11} particles hitting a 1 cm^2 surface in 1 s, and evidently, contamination problems can arise with long-term surface analyses. Unfortunately, there is not enough space to comment on the *production* of ultra-high vacuum conditions which is achieved by all

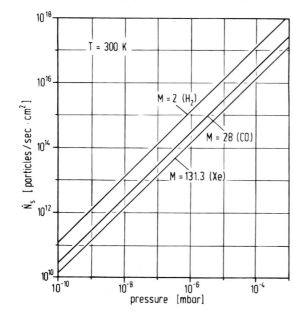

Fig. 4.2. Collision frequency $\dot{N}_s$ (particles per second and unit area) with a surface as a function of gas pressure for three different kinds of particles.

sorts of simple, and more sophisticated and powerful vacuum pumps, along with the use of stainless-steel reaction chambers, system bake-out procedures, etc. Rather, we refer to the relevant textbooks or monographs which deal with vacuum technology [2-6]. Preference here is given to the book by Dushman [3], which is a compilation of the whole body of vacuum physics, including details of pumping speed and pressure measurements, of sample handling and cleaning, as well as of all kinds of material aspects as far as vacuum compatibility is concerned.

In surface analysis, there is principally only a limited number of analytical tools available, in that matter can interact only with particles and/or radiation, with thermal energy, or with electrical and magnetic fields.

Some years ago, Benninghoven [7] gave (along with an overview of the surface analytical methods) a matrix that compiles most of the techniques that can be and are still applied to surface and interface analysis. This scheme is reproduced in Fig. 4.3. Horizontally, the excitation sources are displayed and vertically, the emitted particles are listed, whereby a distinction is made between photons, electrons, ions, and neutral particles. Although this matrix includes many methods it is, nevertheless, incomplete; for example, work-function change measurements or scanning tunneling microscopy are not considered. Furthermore, the "resolution" of the matrix is not sufficient, i.e., the fields 12, 21, and 22 could easily be subdivided into elastic and inelastic processes, and into low- and high-energy regimes, etc. Nevertheless, one immediately realizes that just in this matrix there are no less than 22 methods listed, and since most all of them have relatively complicated names, it has become quite common in surface analysis to use abbreviations. In Table 4.1 we give a short selection of commonly used abbreviations in surface science.

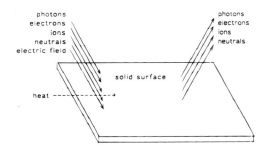

excitation $\longrightarrow$						
	photons hv	electrons e⁻	ions i	neutrals N	heat kT	electric field F
emission ↓ hv	reflection fluorescence	electron induced photo-emission	ion induced photo-emission	neutral induced photo-emission	thermal radiation	
e⁻	Photo- and Auger electrons EXAFS	secondary electron emission (includ. inelastic electrons)	ion induced electron emission	neutral induced electron emission	thermal electron emission	field electron emission
i	photo-desorption of ions	electron induced desorption of ions	secondary ion emission	neutral induced ion emission	thermal desorption of ions	field ion emission
N	photo-desorption of neutrals	electron induced desorption of neutrals	cathodic sputtering	cathodic sputtering	"thermal desorption" (of neutrals)	

Fig. 4.3. Schematic overview of the principally possible excitation and emission spectroscopies with a solid surface. After Benninghoven [7].

86

Table 4.1. Some abbreviations and acronyms frequently used in surface chemistry and surface analysis

AES	Auger electron spectroscopy
APS	appearance potential spectroscopy
ARUPS	angle resolved ultraviolet photoelectron spectroscopy
BIS	bremsstrahlung isochromat spectroscopy
BZ	Brillouin zone
DIET	desorption induced by electronic transition
DOS	density of states
EDC	electron distribution curve
EELS	electron energy loss spectroscopy
ELS	electron loss spectroscopy
ESCA	electron spectroscopy for chemical analysis
ESD	electron stimulated desorption
ESDIAD	electron stimulated desorption angular distribution
EXAFS	extended x-ray absorption fine structure
EXELFS	extended electron loss fine structure
FEM	field electron microscopy
FIM	field ion microscopy
HREELS	high resolution electron energy loss spectroscopy
ILEED	inelastic low energy electron diffraction
INS	ion neutralization spectroscopy
IPE	inverse photoemission
IRAS	infrared reflection-absorption spectroscopy
ISS	ion scattering spectroscopy
LEED	low energy electron diffraction
LEEM	low energy electron microscopy
LEIS	low energy ion scattering
NEXAFS	near edge x-ray absorption fine structure
PSD	photon stimulated desorption
RHEED	reflection high energy electron diffraction
SBZ	surface Brillouin zone
SDOS	surface density of states
SERS	surface enhanced Raman scattering
SEXAFS	surface extended x-ray absorption fine structure
SIMS	secondary ion mass spectrometry
STM	scanning tunneling microscopy
TDS	thermal desorption spectroscopy
TDMS	thermal desorption mass spectroscopy
TEM	transmission electron microscopy
TON	turn-over number
TPD	temperature programmed desorption
UHV	ultra-high vacuum
UPS	ultraviolet photoelectron spectroscopy
XANES	x-ray absorption near edge structure
XPS	x-ray photoelectron spectroscopy

Of course, we are also aware of the wealth of literature covering the field of surface and interface analysis. We refer the reader who is interested in other methods, or in certain apparative or theoretical details, to the relevant literature; here, we do not attempt a complete and exhaustive description of all the currently existing methods. A synopsis of modern surface analytical techniques is given, among others, in the book by Woodruff and Delchar [8] or in the Handbook of Surface Analysis [9]. Also in the monographs of Ertl and Küppers [10] or of Prutton [11] there are various such methods presented. Every year there are, worldwide, several status seminars on the methodical aspects, and it is extremely difficult for any surface scientist to really keep up with all the technical developments. Certainly, there is a clear trend to computerize and to automatize the surface analytical methods whenever possible, therefore it becomes increasingly difficult to track the physical principles behind such methods. In industry, surface analyses have been performed routinely for a number of years, preferentially based on AES, ESCA or SIMS techniques. Now, instead of discussing the methods in the sequence as given by Benninghoven's matrix it seems more appropriate to use a different organization principle and to confine ourselves to a few important and frequently exploited analytical tools.

We shall list, in this sequence, selected and representative methods for analyzing *geometrical surface structure*, *electronic surface structure*, *surface chemical composition*, and *surface thermodynamical* (energetic) and *kinetic* properties in order to make clear how the microscopic information provided in Chapter 3 can be obtained. In all cases, it is tacitly assumed that a clean surface as well as adsorbate phase properties are to be analyzed.

4.1 Determination of Geometrical Surface Structure

The task here is to get information about bond lengths and bond angles of the atoms or molecules present in the surface region. These may comprise substrate atoms as well as adsorbate particles. Furthermore, the determination of surface periodicity (long-range order) is one of the goals of surface structure analysis. Quite generally, we may distinguish "real space" and "diffraction" methods; historically, certainly the diffraction methods have attracted the main interest, LEED in particular, while in recent years direct imaging techniques have also been developed (scanning tunneling microscopy (STM), high-voltage transmission electron microscopy (TEM), LEED microscopy (LEEM), ion scattering, etc.). We again recall the difficulty to restrict analysis to the topmost atomic layer, due to the electron mean free path-energy correlation of Fig. 4.1. Actually, there are very few techniques which really probe the structure of the outermost atomic layer, for instance, STM, field ion microscopy (FIM) or Penning ionization spectroscopy [10].

For practical purposes it is relevant as to whether one deals with polycrystalline or single crystalline samples. Monocrystalline structure should extend over areas greater than $\sim 1\,\mathrm{mm}^2$ (the typical width of probing electron beams) in order to produce a well-defined diffraction pattern. Polycrystalline matter usually consists of grains or agglomerates with much smaller diameter, and within the width of the probing beam, there are no longer any well-defined phase relations between the scattered waves, thus leading to a loss of constructive interference maxima. It must be borne in mind that all *diffraction* methods require large *single* crystalline surface areas. Unfortunately, these are not common in practical chemical applications (except perhaps in semiconductor technology where relatively large single crystals of silicon are used for wafer production).

Rather, polycrystalline powders, pellets or foils dominate, e.g., in practical catalysis, as pointed out in Chapter 2. Here, the ordinary diffraction methods are no longer very helpful, since most of them are not compatible with high-pressure conditions. However, in recent years other powerful analytical techniques have been developed which can probe surface structure on a more microscopic scale; these include x-ray absorption techniques or tunneling microscopy which will be dealt with further on. Despite these developments, it is not at all easy to obtain clear structural information from "practical" materials, and this is the reason why, in this area, structure-sensitive methods are not as popular as analyses that probe the surface chemical composition.

Our model approach, however, relies very much on the use of single crystals, and most of the structure-sensitive tools here are based on an understanding of the microscopic or macroscopic diffraction physics. This is why we present diffraction methods in the first place and focus on the well-established diffraction of low-energy electrons (LEED), including the most recent development of LEED microscopy LEEM.

4.1.1 Electron Diffraction

Low-energy electron diffraction (LEED): The physical principles and the technical realization of the LEED method have been repeatedly described in the literature [10-18]. It is not the intention here to present all important details of the method, but rather the reader shall be exposed to the basic physical principles, as well as the standard experimental set-up of LEED.

The method itself dates back to experiments carried out by Davisson and Germer in 1927 [19,20]. These authors studied the reflection of electrons from nickel targets in vacuo. These targets were recrystallized by heating, giving rise to anomalies in the angular distribution of the back-scattered electrons. Using a larger monocrystal, they observed maxima and minima in the angular distribution which were interpreted as being caused by constructive and destructive interference of the reflected electron waves. Their results nicely supported the earlier work of de Broglie on the wave theory of electrons. However, owing to the complexity of the experimental set-up at that time, it took another 30–40 years until low-energy electron diffraction patterns could be obtained routinely, thanks to the development of bakeable all-metal vacuum systems, reliable ceramic-to-metal seals, universal mechanical motion feedthroughs and improved low-energy electron-gun designs, as well as the use of powerful vacuum pumps and the accumulation of knowledge concerning the preparation and cleaning of single crystal surfaces.

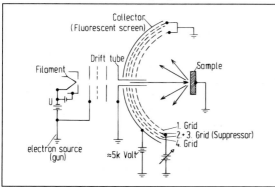

Fig. 4.4. Schematic set-up of a LEED experiment. Low-energy electrons are produced by a W cathode and focused onto the sample. The back-scattered electrons pass a grid system, which cuts off the inelastic fraction (suppressor) before the elastic (diffracted) electrons are post-accelerated onto a phosphorous screen (collector).

89

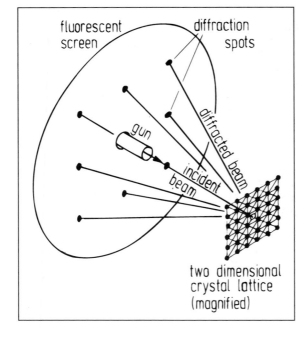

Before we enter a (brief) presentation of some basic diffraction physics, it is deemed useful to show the typical experimental arrangement. According to Fig. 4.4, a single crystal piece (area ~1 cm²) mounted on a mechanical manipulator (allowing motion in *x, y, z* coordinates and rotation around a perpendicular axis) inside a UHV chamber is required; an electron gun provides a well-collimated monoenergetic beam of slow electrons; a so-called LEED optics consists of at least three (or, better, four) highly transparent concentric hemispherical grids and a solid phosphorous screen behind. Usually, the surface normal passes the center of the electron gun and that of the spherical cap made up by the grids and the screen, whereby the crystal is positioned so that it is right in the center of the grid curvature. As is illustrated in Fig. 4.5, the electron wave originating from the gun hits the surface and is diffracted at its atomic grating. The elastically back-reflected electron waves interfere with each other, thus leading to diffraction maxima and minima. The maxima can be made visible on a fluorescent screen, by post-accelerating the slow electrons by a high electric field ($U \leq 7\,\text{keV}$). The second and third grids are kept on a potential slightly lower than the primary electron beam energy in order to cut-off the inelastically scattered electrons which do not carry relevant structural information and simply contribute to background intensity. (The first and fourth grids are held on ground potential and are only added to reduce field inhomogeneities.) Some typical LEED patterns are reproduced in Fig. 4.6. There are several new and sophisticated developments in LEED, among others a reverse-view system coupled with a TV camera and computer-controlled data acquisition (Video-LEED) [21–23], or an instrument with a high resolving power (SPA {spot profile analysis} LEED [24, 25]). The particularly interesting LEED microscopy will be dealt with further below, after a short discourse in theory.

The formation of a diffraction pattern can be rationalized as follows: The incoming electron beam of energy U [Volts] is described by a plane electron wave of length

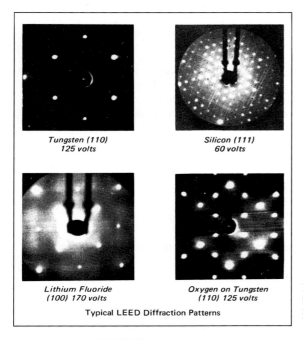

Tungsten (110)
125 volts

Silicon (111)
60 volts

Lithium Fluoride
(100) 170 volts

Oxygen on Tungsten
(110) 125 volts

Typical LEED Diffraction Patterns

Fig. 4.6. Examples of typical LEED patterns from surfaces with different symmetry (courtesy Intevac, formerly Varian).

$$\lambda = \frac{h}{mv} = \sqrt{\frac{150}{U[\text{Volt}]}} \ [\text{Å}] \qquad\qquad 4.2$$

and gives rise to interference phenomena on gratings of atomic scale, in close analogy to x-ray diffraction. Note that an electron beam of 150 eV energy has a wave length of 1 Å. The most important difference compared to x-rays, however, consists in a much smaller penetration depth of the low-energy electrons (10 eV–500 eV energy, corresponding to 0.55 Å < λ < 3.8 Å) which ensures that the backscattered electrons originate only from the surface.

The one-dimensional situation is illustrated in Fig. 4.7, which shows a chain of atomic scatterers of mutual distance a_1 which is hit by an electron wave at angle φ_0.

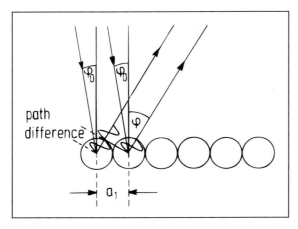

Fig. 4.7. Scattering of a plane electron wave at a one-dimensional periodic chain of atoms. The path difference for reinforcement of adjacent scattered beams (angle φ) is indicated.

The condition for constructive interference (path difference between neighboring waves = integral number of wavelengths) states (n = diffraction order):

$$a_1(\sin \varphi - \sin \varphi_0) = n \cdot \lambda \qquad\qquad 4.3$$

The two-dimensional surface can now be regarded as an array of parallel rows of atoms along [h, k] direction with distance $d_{h,k}$, and the corresponding diffracted beam will appear in the same plane as the incoming beam, perpendicular to the direction [h,k]. Again, in close analogy to Eq. 4.3, there will be interference maxima at angles φ given by

$$d_{h,k}(\sin \varphi - \sin \varphi_0) = n \lambda \qquad\qquad 4.4$$

Most often, the LEED experiment is carried out under normal incidence conditions ($\varphi_0 = 0°$), and Eq. 4.4 simplifies to

$$\sin \varphi = \frac{n \lambda}{d_{n,k}} = \frac{n}{d_{n,k}\sqrt{\frac{150}{U[\text{Volt}]}}} \cdot \qquad\qquad 4.5$$

Accordingly, a certain beam will appear for the first time at an electron energy of

$$U_0 = \frac{150}{d_{h,k}^2} \qquad\qquad 4.6$$

It is now quite important that the formation of a surface layer with a new or altered periodicity will give rise to additional or altered interference maxima, since the sensitivity of LEED is not restricted to the outermost layer; also, the second and third layers are imaged, as illustrated in Fig. 4.8, and the respective LEED pattern contains both the substrate maxima *and* additional spots caused by the overlayer. If the "grating" distance here is denoted by $d'_{h,k}$, one obtains

$$\sin \varphi' = \frac{n \lambda}{d'_{h,k}} , \qquad\qquad 4.7$$

a condition which is fulfilled simultaneously to Eq. 4.5, and additional diffraction spots appear on the LEED screen. These "extra" spots can be easily associated with ordered

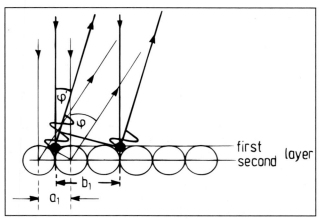

Fig. 4.8. Scattering of a plane electron wave (normal incidence φ_0) at a one-dimensional grating consisting of substrate atoms (open circles, distance of adjacent scatterers = a_1) and an adsorbate layer on top (black circles, distance between adjacent scatterers = b_1). The reinforced scattered electron beams leave the surface at different angles φ and φ', respectively, thus leading to the formation of 'extra' LEED spots.

adsorbate layers or (in some cases), also with reconstructed surfaces; they help to identify adsorbate periodicities and coverages.

A more elegant description of the diffraction physics can be obtained using the Laue formalism in two dimensions, based on the concept of *reciprocal space*. If s and s_0 denote the unit vectors for the directions of the scattered and incident beam, respectively, the interference conditions on a two-dimensional lattice (distance a_1, of scatterers in h-direction, distance a_2 in k-direction, electron wave length λ) read:

$$a_1(s - s_0) = h\,\lambda \qquad\qquad\qquad 4.8a$$

$$a_2(s - s_0) = k\,\lambda \qquad\qquad\qquad 4.8b$$

The two equations 4.8 must be solved simultaneously for all possible s at given s_0; the solution is found as

$$\frac{s - s_0}{\lambda} = ha_1^* + ka_2^* = g\ , \qquad\qquad\qquad 4.9$$

where the $a_1{}^*$ and $a_2{}^*$ are the unit mesh vectors, and g is a translation vector of the so-called reciprocal lattice which is related with the real-space lattice via the conditions

$$a_1 \cdot a_1^* = 1 \qquad\qquad\qquad 4.10a$$

$$a_1 \cdot a_2^* = 0 \qquad\qquad\qquad 4.10b$$

$$a_2 \cdot a_2^* = 1 \qquad\qquad\qquad 4.10c$$

$$a_2 \cdot a_1^* = 0 \qquad\qquad\qquad 4.10d$$

This means that $a_1{}^*$ is always perpendicular to a_2 and has the length

$$|a_1^*| = \frac{1}{|a_1| \sin \alpha}\ ; \qquad\qquad\qquad 4.11a$$

correspondingly, $a_2{}^* \perp a_1$, and

$$|a_2^*| = \frac{1}{|a_2| \sin \alpha}\ , \qquad\qquad\qquad 4.11b$$

whereby α denotes the angle between the real space lattice vectors a_1 and a_2.

These relations can be used to construct the reciprocal lattice from the real space lattice and vice versa. This is illustrated, for two differently chosen unit cells, by means of Fig. 4.9.

One can immediately realize, using the definition in Eq. 4.10, that Eq. 4.9 represents the only possible solution of the interference condition (Eq. 4.8).

In a LEED experiment, one determines the direction of the diffracted beams (vector s) as points of intersection with the hemispheres of the LEED optics which are made visible as bright spots on the fluorescent screen. The well-known Ewald construction can serve to illustrate the interference condition [26]. Here it suffices to say that the LEED pattern obtained in a diffraction experiment is a direct representation of the reciprocal lattice. It must be added that, so far, only single diffraction events were taken into account. In reality, however, multiple scattering phenomena are the rule, those being predominantly double scattering, meaning that a scattered electron wave can be scattered a second time before leaving the surface. Therefore, the diffraction pattern that originates from a substrate lattice (vectors a_1 and a_2) plus an overlayer lattice (vectors b_1 and b_2) leads to an

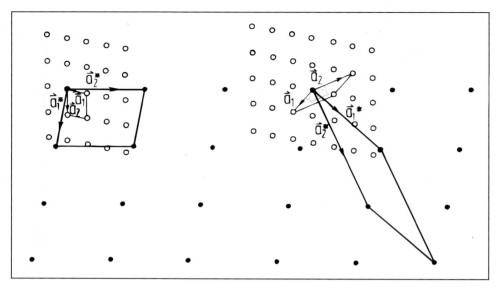

Fig. 4.9. Example of transformation of real space to reciprocal space for two differently chosen unit meshes a_1, a_2. Open circles represent the reciprocal, full circles the real lattice points. After Woolfson [29].

effective reciprocal lattice and, hence, a diffraction pattern which can be described by the expression:

$$g = h_1 a_1^* + k_1 a_2^* + h_2 b_1^* + k_2 b_2^* . \qquad 4.12$$

Apparently, the diffraction pattern does not only contain the periodicities of the individual reciprocal lattices of the two layers, but also their linear combinations. This can cause interference maxima which are not produced by either of the layers alone. It is possible to show that these spots only appear with complex LEED structures (coincidence lattices and incoherent structures), but do not play a role with the normal simple overstructures.

Furthermore, in many cases there are various domains or islands of a certain structure coexisting on a macroscopic surface. If the diameter of these domains is smaller than the coherence length Δx of the electrons (this will be explained further on), then there occurs a superposition of the respective scattering *amplitudes* of the individual domains with certain phase shifts that result in reflex splitting, reflex broadening or streak formation. However, if the mean island diameter exceeds Δx the individual scattering *intensities* are superimposed and the characteristic spot patterns are observed simultaneously.

A word must be added to comment on the coherence length of the LEED electrons, Δx. This quantity is, among others, limited by the energetic width of the primary electron beam, ΔE, and the beam divergence, i.e., the deviation from parallelism, $\beta \Delta x$ can be approximately expressed by

$$\Delta x \approx \frac{\lambda}{2\beta \sqrt{1 + (\Delta E/E)^2}} , \qquad 4.13$$

and is, for normal LEED instruments, of the order of 100–200 Å. The already mentioned SPA LEED system [24] provides a much better coherence length of several thousand Ångstroms and can be used to analyze beam profiles with respect to domain sizes, surface crystallographic defects, etc. with high resolution.

We turn now to the very purpose of LEED, namely, the actual determination of surface structure. So far, it has only become clear that the geometry of a LEED pattern carries the information of the substrate (or overlayer) symmetry, but one must keep in mind that any displacements of the adsorbate lattice parallel to the surface lattice result in *identical diffraction patterns*. Furthermore, primitive and non-primitive unit meshes cannot be distinguished except in some special cases. As in x-ray crystallography, it is essential to consider the LEED *intensities* and their dependence on the electron wave length, that is, on their kinetic energy. The simplest approach is to apply kinematic diffraction theory, in analogy to x-ray diffraction, and to calculate the structure amplitude $F_{h,k}$. This implies that an incident wave suffers only a *single* scattering event. Actually, however, slow electrons interact strongly with matter, thus making *multiple* scattering events most likely. As a consequence, kinematic theory is seldom appropriate to correctly describe the intensity-energy dependence, and there is a need for dynamical LEED theories [15, 18, 27, 28] that treat the complex problem of intralayer and interlayer scattering of spherical or plane electron waves in the surface region of the crystal. For the sake of simplicity and easy physical understanding it is sufficient to briefly touch on kinematic LEED theory; a full account of the dynamical treatment is given, among others, in Pendry's book [15].

We define the wave vectors for the incident and scattered wave, respectively, as

$$k_0 = \frac{2\pi}{\lambda} s_0 ,$$ 4.14a

and

$$k = \frac{2\pi}{\lambda} s ,$$ 4.14b

whereby

$$k - k_0 = \Delta k = \frac{2\pi}{\lambda} (s - s_0) = 2\pi g .$$ 4.15

The intensity for diffraction on a periodic two-dimensional array (unit mesh vectors a_1 and a_2) is given by

$$I \propto |F_{hk}|^2 \frac{\sin^2(\frac{1}{2} a_1 N_1 \Delta k)}{\sin^2(\frac{1}{2} a_1 \Delta k)} \cdot \frac{\sin^2(\frac{1}{2} a_2 N_2 \Delta k)}{\sin^2(\frac{1}{2} a_2 \Delta k)} ,$$ 4.16

in which the sinus square terms denote the interference function (lattice factor) due to the presence of the lattice, i.e., the periodic repetition of $N_1 \cdot N_2$ unit cells. $F_{h,k}$ stands for the kinematic structure factor. It is obvious that reinforcement of the diffracted electron waves occurs only in those directions k that satisfy both Laue equations (cf., Eq. 4.8) which can be rewritten as (h, k = Miller indices)

$$\frac{1}{2} a_1 \Delta k = h\pi ,$$ 4.17a

and

$$\frac{1}{2} a_2 \Delta k = k\pi .$$ 4.17b

For a primitive unit mesh containing a single atom j at location r_j (cf., Chapter 3) the structure factor $F_{h,k}$ reads

$$F_{hk} = f_j \exp(i\Delta k r_j) \qquad 4.18$$

(f_j = atomic scattering factor for electrons), whereby

$$r_j = u_j a_1 + v_j a_2 + w_j \qquad 4.19$$

(the coordinates u_j, v_j extending in the surface plane and measured in fractions of a_1, and a_2, respectively, the coordinate w_j taking care of a displacement of the atom perpendicular to the surface spanned by a_1, a_2 given in absolute units of length).

Considering a non-primitive unit mesh containing n atoms, we have for *normal* beam incidence

$$F_{hk} = \sum_{j=1}^{n} f_j \exp\left(2\pi i \left\{ hu_j + kv_j + (1 + \cos\varphi)\frac{w_j}{\lambda} \right\}\right) \qquad 4.20$$

(φ = diffraction angle to surface normal as defined before).

The dependence of the intensity of a given LEED spot on the electron wave length (which we remember is defined by the voltage V applied to the electron beam of the LEED gun) actually contains all the structural information, as mentioned before. Such a typical LEED "I,V" curve is shown in Fig. 4.10 for a Pt(111) curve [30]. It can be measured by a moveable Faraday cup inside the LEED optics or by a TV camera that displays the pattern on a monitor, whereby the intensity of selected beams can be electronically followed and stored as a function of energy ("Video-LEED" [21–23]). Any changes produced by adsorbed gas (hydrogen in our case) can likewise be monitored.

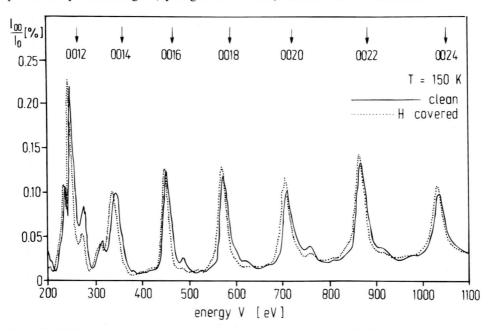

Fig. 4.10. LEED intensity-energy (voltage) curve of the specularly reflected (0,0) beam, normalized to the incident intensity I_0 for the clean (full line) and H-covered (dotted line) Pt (111) surface. The primary Bragg maxima and their diffraction order are indicated by arrows. After [30].

When looking at a typical I,V curve it is immediately realized that there is no constant intensity, but rather there are relatively sharp maxima ("Bragg maxima") separated by low intensity regions. The reason is that we do not have an ideal two-dimensional lattice consisting of a single layer, but the electron waves do at least partially enter the crystal (cf., Fig. 4.10), with the consequence that, in addition to interference between electron waves scattered off adjacent atoms *parallel* to the surface, interference also occurs with scattered waves originating from second, third etc. layers *perpendicular* to the surface. This happens whenever the third Laue condition for the respective layer distance is satisfied (l = Miller index):

$$a_3(s - s_0) = l \cdot \lambda \qquad\qquad 4.21$$

where a_3 is the vector of the (bulk) lattice unit cell (spanned by a_1, a_2, a_3).

The diffraction can also be regarded to occur at the individual layers of the solid with mutual distance d_{hkl}, as described by the well-known Bragg equation

$$2d_{hkl} \sin \Theta = n \cdot \lambda , \qquad\qquad 4.22$$

Θ being the angle between the electron wave impact direction and the surface.

Therefore, the aforementioned maxima of the I,V curve are called (primary) "Bragg" maxima. In reality, not only these primary Bragg maxima appear in the I,V curves, but also additional smaller peaks known as "secondary" Bragg maxima. They are due to multiple scattering (dynamical) effects.

To conclude the explanation of the LEED method we briefly describe how a structural analysis is usually carried out. First, a series of experimental I,V curves is measured for a variety of beams. Thereafter an electron scattering calculation is carried out for various plausible adsorption geometries, in which multiple scattering is explicitly considered. The thus obtained theoretical I,V curves are compared with the experimental ones and the geometrical parameters of the best fit curve are believed to reflect the "true" surface structure. The agreement between theoretical and experimental $I,$ V curves is judged by the so-called reliability (r) factor. r is a mathematical function which compares peak maxima positions, curve shapes, intensities etc. and can range between zero and 1, whereby $r = 0$ means perfect agreement. r values greater than ~0.5 indicate disagreement between theory and experiment.

The present state of the art is r factors as small as 0.05 for clean, unreconstructed metal surfaces, and $0.2 < r < 0.4$ for adsorbate layers.

Electron microscopy: Closely related to LEED is the *microscopy* of electrons. Depending on their kinetic energy and the particular scattering geometry, we may distinguish "low energy electron microscopy (LEEM)" and "high energy electron microscopy", which is normally known as TEM (transmission electron microscopy). As LEED, LEEM is carried out in the reflection mode, while high energy electrons ($E \geq 100$ keV) can penetrate thicker layers and form a *transmission* image. We recall that there are principally two kinds of optical images of an object possible (Fig. 4.11): A primary image, namely, the diffraction pattern, and a secondary image known as the real image of the object.

If a *LEED* microscopy is to be carried out, there is need for other focussing elements which collect the diffracted beams to a real image of the respective structure. Bauer has been working on a LEED microscope for many years, and in 1985, he came up with a successful solution that is schematically illustrated in Fig. 4.12 [31, 32]. The heart of the electron optics is the cathode lens which, along with a collimator lens, an intermediate lens and a projective lens allows imaging of a surface by means of its reflected electrons. A relatively high electric potential (~25 keV) has to be applied to the cathode lens which

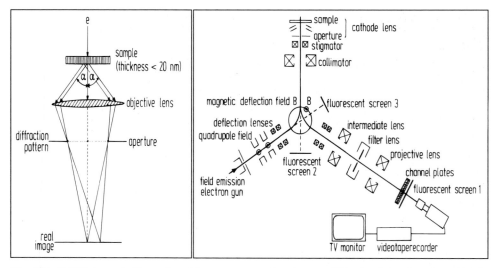

Fig. 4.11. (left) Schematic illustration of the occurrence of an electron-diffraction image in a transmission-electron microscope.

Fig. 4.12. (right) Experimental set-up required for low energy electron microscopy (LEEM). After Teliepis and Bauer [31. 32].

bears some technical problems and makes the whole microscope somewhat bulky. In order to achieve a good optical image it is important to use a field emission electron gun, its brilliance is up to 10^6 times greater than that of a conventional tungsten-tip cathode.

The currently attainable magnification allows to image structure elements of the order of 40 Å to 60 Å diameter, which is possible simply by changing the focal point of the microscope to also obtain ordinary diffraction, that is, LEED patterns. The LEED microscope has been proven particularly useful to monitor silicon surfaces and to follow the growth kinetics of reconstructive phase transformations, for example, the formation of the well-known Si(111)-7×7 structure from the unreconstructed (1×1) surface. Other examples are molybdenum and tungsten surfaces with adsorbed oxygen layers. In all these cases the formation and growth of the nuclei can be nicely followed by means of the microscope. (For more details the reader is referred to the thesis work of Teliepis, who helped to develop the microscope [33]).

In high-energy electron microscopy one can, as mentioned before, obtain transmission patterns, but with a lack of surface sensitivity. However this can be regained if a very small angle of incidence ("grazing" incidence) is chosen. Here, we refer to the so-called RHEED technique (reflection high-energy electron diffraction) in which medium energy electrons ($1 < E_p < 5$ keV, E_p = primary beam energy) are diffracted at the surface region of a crystal and made visible on a fluorescent screen as more or less streaky reflexes [34,35]. While RHEED cannot really compete with the well-established LEED technique, high-resolution electron microscopy has again moved to the center of interest because electron microscopists have succeeded in imaging surface structures with atomic resolution [36,37]. Semiconductor, oxide, and clean metal surfaces can be seen directly with transmission electron microscopy (TEM). Figure 4.13, which is taken from the article by Smith [38], gives two examples which are particularly revealing for chemists in that it is possible to directly follow oxidation processes with TEM (Fig. 4.13 a), as well

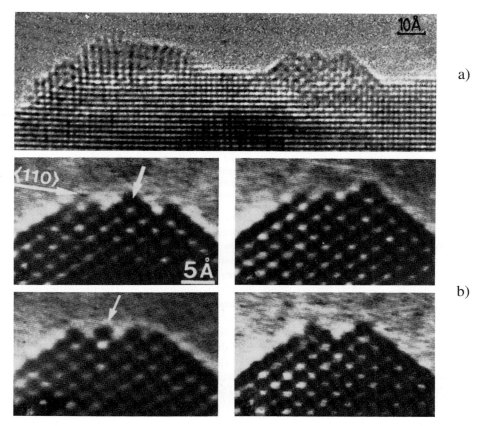

Fig. 4.13. Two examples for the power of high-resolution transmission electron microscopy: **a)** surface of a Pd crystal showing the development of oxide growth during observation. **b)** Single frame images from a videotape showing the edge of a small Au crystal with atom hopping. The time proceeds from left to right. The arrow points to an individual atom. After Smith [38]. Reproduced by permission.

as reconstruction phenomena (e.g., with clean Au(110)-1×2 surfaces). Furthermore, even migration and hopping processes of individual atoms can be monitored. In Fig. 4.13b a hopping event of a gold atom at the edge of a small Au crystal as recorded by videotape technique is illustrated [38]. Problems still encountered in this type of microscopy are achieving and sustaining real UHV conditions (to avoid sample contamination) and minimizing electron beam damage on the samples which can easily occur with the high energy beams (~100–500 keV) at larger current densities.

It should be added here that in surface structure analysis diffraction of light ions and neutral particles (hydrogen, helium) also plays an important role (see [39–44]).

4.1.2 Field-Electron and Field-Ion Microscopy (FEM and FIM)

Both techniques for imaging surface structure with atomic resolution date back to Müller who invented the field techniques for surface microscopy. We do not attempt to present an exhaustive description of the methods, which can be found in many textbooks and

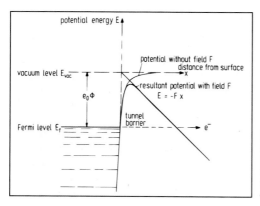

Fig. 4.14. Potential energy situation at a surface of an *sp* metal using the free-electron model with and without application of a high external electric field *F*. The possibility for metal electrons to tunnel through the barrier (field emission) is indicated.

review articles [45–50]; instead, only some of the latest applications of FEM and FIM for surface structure analysis will be briefly reviewed.

The physical basis of field emission microscopy techniques is the occurrence of the tunneling effect if the potential well at a surface is bent and narrowed by a high electric field, as illustrated in Fig. 4.14. Electrons from the Fermi level of the metal can escape by tunneling through the potential barrier rather than by surmounting it. The tunneling probability in part depends exponentially on the thickness of the barrier, and to make this thin, very high electric fields are required, i.e., of the order of several 10^7 Volts/cm. Experimentally, these high field strengths can be obtained by using metal tips with a small radius of curvature and by applying a high accelerating voltage of about 20–30 keV, as schematically sketched in Fig. 4.15. In the field *electron* microscope electrons are extracted from the tip and electrostatically accelerated towards the screen, where they produce bright patterns on a dark background. These patterns correspond directly to the

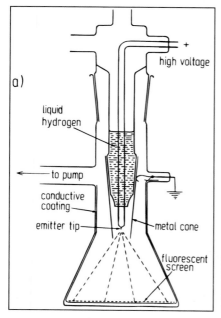

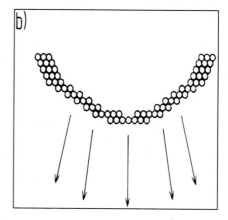

Fig. 4.15a) Experimental realization of a simple field-electron microscope, as proposed by Erwin Müller [45]; **b)** perpendicular cut through the emitting tip with atomic resolution, whereby the arrows indicate the direction of the emitted electrons.

tip structure or, more precisely, to the lateral distribution of the surface work function. It is clear that areas with small work function will emit electrons more effectively than those of higher work function. In the field *ion* microscope the tip is cooled to liquid hydrogen temperature (10–20 K) and helium gas at low pressure is admitted to the microscope tube. When He atoms hit the surface they become ionized under the high field conditions. The He$^+$ ions then travel outward in a straight line to the screen from the tip-surface atom. The magnification factor is enormous, it equals the ratio of the area of the tip to the area of the screen, and atomic resolution is easily obtained.

There are many practical applications of the FEM and particularly of the FIM method in surface physics and chemistry. It is the FIM technique that first enabled a direct observation of surface diffusion phenomena, and we refer to the original investigations of Gomer [48] and Ehrlich [51–54a]. By means of sophisticated imaging techniques it was possible, for example, to make the movements of single atoms, of doublets, triplets, etc., visible at W or Ir tips [54,54a]. Recently, FIM was combined with a video imaging technique which rendered the direct observation of the surface reconstruction of Ir(100) and Ir(110) possible [55, 56]. This is illustrated by means of Fig. 4.16, taken from the work by Witt and Müller [55]. For completeness, it is recalled that the presence of high electric fields can drastically influence the chemistry of the surfaces. Block has devoted much of his work to this issue [57]. However, it is felt that FEM and FIM have at present lost general interest, since the development of scanning tunneling microscopy (STM) offers another extremely suitable means to image the structure of all kinds of surfaces with atomic resolution having the advantage of practically field-free conditions. This is why we shall give STM more priority in this book, as will be shown in the following section.

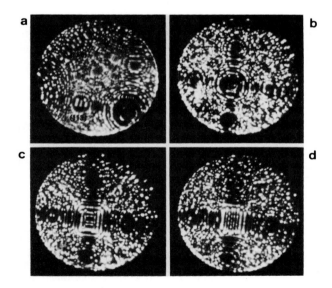

Fig. 4.16a) Surface of an Ir field-emission tip after heating and slight field evaporation, with the (110) and (113) surfaces showing 1×2 superstructures due to reconstruction; **b)-d)** formation of the (100) superstructure by heating to 900 (**b**), 1100 (**c**), and 1200 K (**d**). After Witt and Müller [55, 56].

4.1.3 Scanning Tunneling Microscopy (STM)

The invention and experimental realization of the scanning tunneling microscope by Binnig and Rohrer in 1983 [58, 59] (for which they were awarded the Nobel prize in physics in 1986) is certainly one of the most spectacular technical innovations of recent years, and has almost revolutioned the area of surface structure analysis. The impact on this field can be measured by the exponential increase of the number of related investigations since 1982, whereby the band width of applications includes biology, electrochemistry, surface structure analysis and lithography. The possible technical potential of STM, especially with regard to semiconductor technology, cannot be surveyed at present. For an introduction to the area of STM and the related recent developments the reader is referred to several review articles [60–66].

This section is organized as follows: First, some basic physics of the STM will be presented along with a description of the method; thereafter selected examples will be given of the performance and capacity of this ingeniously simple instrument.

The apparative principle consists of a very sharp tungsten tip which is, as one electrode of the tunnel junction, brought so close to a metal surface that electrons can tunnel from one metal to the other if a low potential (0.5–10 Volts) is applied to the system. As soon as the tunneling current starts to flow (which depends strongly on the distance d: tip – surface; see the following) there is a feedback circuit that regulates a servo drive mechanism to such a value of d that a preselected tunneling current is reached and held constant. At the same time, a lateral motion of the tip across the surface (x, y direction) is accomplished with small increments Δx or Δy of ~1–5 Å (lateral scans). The tip itself is mounted on a piezo tripod which allows, by respective high voltage adjustment of the piezos, well-defined small motion in x, y and z directions. The arrangement is schematically shown in Fig. 4.17. In the original version by Binnig and Rohrer [58], the coarse positioning of the sample was achieved by a so-called "louse" L (Fig. 4.18). Its body consists of a piezo plate (PP) with a sample holder on top that rests on three metal feet (MF), separated from the the metal ground plates (GP) by high dielectric-constant insulators (I). The feet are electrostatically clamped to GP by applying a voltage V_F. Elongating and contracting the "louse" with the appropriate clamping sequence of the feet (voltage V_F!)

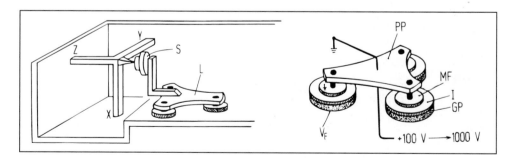

Fig. 4.17 (left). Principal experimental realization of scanning tunneling microscopy (STM) by means of a tunnel tip T mounted on a piezo tripod with legs x, y, z and the sample S fixed to the so-called louse L which can be mechanically driven and coarse-adjusted to the tip by piezo elements. After Binnig and Rohrer [58].

Fig. 4.18 (right). Detailed sketch of the louse used by Binnig and Rohrer [58], showing the electrical connections. PP = piezo plate, MF = metal feet, I = insulators with high dielectric constant, GP = ground plates, V_F = feet voltage.

allows motion of the device in any direction in steps between $100\,\text{Å}$ to $1\,\mu\text{m}$, with a frequency of 30 steps per second. Then the sample is brought to within about $100\,\text{Å}$ to the tip, and in this way the tunnel current servo-mechanism is activated, which finally results in tip-surface distances (tunnel gaps) as small as $0.2\,\text{Å}$.

The lateral resolution of the STM crucially depends on the properties of the metal tip, especially its curvature. In Binnig and Rohrer's original publication W or Mo wires of ~1 mm diameter were used; they were ground at one end at roughly $90°$. This yielded tips of overall radii of $10^{-6}\,\text{m}$, whereby the rough grinding process produced many sharper minitips. Owing to the strong sensitivity of the tunnel current on distance d, that minitip is automatically activated in the experiment to give the highest current, that is closest to the surface. Also, an in situ sharpening of the tip by gently touching the surface can bring the lateral resolution well within the $10\,\text{Å}$ range, additional application of high electric fields ($\sim 10^8\,\text{V}\,\text{cm}^{-1}$) for a certain time interval enabled resolutions even smaller than that value.

One of the main technical problems of STM is the vibrational damping and decoupling from the surroundings. The original STM was mounted in a relatively bulky device that used mechanical springs, etc., but more recent developments such as Viton rings, for example, allow a much smaller device to be constructed, the so-called "pocket-size" STM [67] which can, meanwhile, be purchased from vacuum companies.

Turning to the physics of STM we can only repeat that its principle is really straightforward. Consider, for example, Fig. 4.19, which shows, on an atomic scale, the tip and the facing surface. The circles represent the respective atoms; the broken lines indicate the exponentially decreasing equipotential surfaces between the two tunnel electrodes. Note that their distance is of the order of only some atomic diameters. Under these geometrical conditions, the electron wave functions of the two electrodes overlap so that when a (slight) potential difference V_T is applied the electrons can flow from one electrode to the other. The resulting tunnel current density I_t depends exponentially on the gap width d. For the one-dimensional case, one can formulate, according to Bethe and Sommerfeld [68],

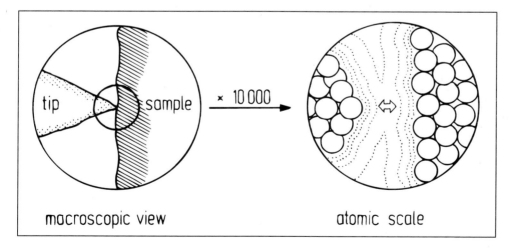

Fig. 4.19. Illustration of the tunnel-tip surface junction on the macroscopic (left-hand side) and atomic (right-hand side) scale. After Wintterlin [67].

103

$$I_t = \frac{3}{8} \frac{e_0^2}{\hbar} \frac{\varkappa}{\pi^2 d} V_t \exp(-2\varkappa d) , \qquad\qquad 4.23$$

where

e_0 = unit charge 1.602×10^{-19} As,

$\hbar$ = Planck's constant 1.054×10^{-34} Js,

$\varkappa$ = $\sqrt{2m_0\Phi/\hbar^2} \approx \frac{1}{2}\sqrt{\Phi}$ = characteristic decay curve of the wave function
in the potential barrier,

Φ = average barrier height (surface potential) in [eV], and

d = gap distance in [Å].

In these units, $2\varkappa$ becomes approximately $1.025\sqrt{\Phi_{\text{eff}}}$. This means that with Φ_{eff} on the order of some [eV], the tunnel current changes by a factor of $\sim$10 for every Ångstrom of d, which leads to a very high sensitivity or space resolution perpendicular to the surface. In the scanning mode, the electronic control unit adjusts a voltage V_z to the p_z piezo drive to keep the tunnel current I_t constant while the tip is scanned (via voltages V_x and V_y applied to the piezos p_x and p_y, respectively) across the surface. The electronic situation in the tunnel junction can be visualized from the potential diagram depicted in Fig. 4.20. It is evident from Eq. 4.23 that there is (besides d) another decisive quantity, namely the "effective" work function Φ_{eff}, which can be set equal to the arithmetic average of the effective work functions of the tip and the investigated surface:

$$\Phi_{\text{eff}} = \frac{1}{2}\left(\Phi_{\text{tip}}^{\text{eff}} + \Phi_{\text{sample}}^{\text{eff}}\right) . \qquad\qquad 4.24$$

The effective work functions Φ_{eff} are not identical to the work functions in a field-emission or photoelectric experiment. In STM, the work functions are reduced somewhat due to interactions between the tunnel electrons and the electrons inside the two solids. These interactions can be described classically as image potential effects, and lead to a

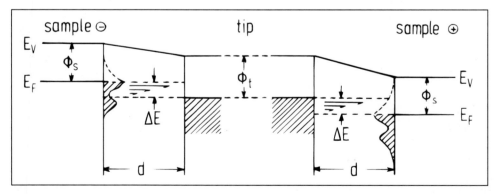

Fig. 4.20. Electrostatic potential situation in the tunnel junction for negatively (left-hand side) and positively (right-hand side) polarized sample. In the first case, the electrons tunnel from occupied states of the sample to unoccupied states of the tip (negative tunnel current); in the second case, tip electrons tunnel to empty states of the sample (positive tunnel current). The occupied electron states are indicated by solid lines, unoccupied states by broken lines. After Gritsch [80].

(distance-dependent) reduction of the barrier height. Note that the work function is defined as the work required to bring an electron from the Fermi level of the solid to a distance 10^{-4} cm outside the surface. In an STM experiment, however, there are much smaller distances involved, and the shape of the potential barriers can play a significant role. Nevertheless, it is the work function (or part of it) that is largely responsible for the tunneling current, and a lateral scan of the surface basically yields its work function structure. As pointed out by Binnig and Rohrer [58], one can delineate work function structures and "true" surface topography by modulating the tunnel distance d by Δd while scanning at a frequency higher than the cut-off frequency of the servo control unit. The modulation signal $J_d = \Delta \ln I_t/\Delta s \approx \sqrt{\Phi_{\text{eff}}}$ then directly monitors the system's effective work function.

We have to add here that there exists a convention with respect to the sign of the tunnel current. It is chosen so that it always designates the potential of the *sample*. Negative tunnel voltages (V_{t_0}) then mean electron emission from the sample, whereas with V_{t_0} electrons stem from the tip and tunnel into the sample.

As far as a theoretical description of the phenomena leading to the tunnel current is concerned, there are, at present, some promising concepts, although in general any such theory must be very complicated. Not only must it take into account the electronic structure of both the tip and the sample near the Fermi level (it is clear that the density of the overlapping wavefunction states in the region of the tunnel junction plays a decisive role), but also the topographical structure of the sample with its crystallographical imperfections (adatoms, steps, and kinks), which contribute to the tunnel current. Instead of presenting more detailed theoretical descriptions of the STM process, we refer to the literature [69–74].

Turning to the practical application of the STM technique in the area of surface-structure analysis, its power can perhaps best be demonstrated by showing scans from reconstructed metal and semiconductor surfaces. It should be mentioned here that precisely this STM allowed a final decision to be made on the long-discussed problem of the (7×7) reconstruction of the silicon (111) surface [75,76]. Other less spectacular examples are the (1×2) reconstructed Pt(110) [77] and Au(110) [78] surfaces. From LEED studies, it has been argued [79] that this latter reconstruction was of the missing-row type, and this could be confirmed nicely by STM. In Fig. 4.21, we present an example taken from the thesis work by Gritsch [80]. Clearly, the close-packed Au rows in [1$\bar{1}$0] direction can be resolved to show the strong corrugation of this reconstructed surface. Also, lattice defects, such as facets and kinks, can easily be monitored. STM's high resolution renders it possible to distinguish monoatomic steps, for example, from larger steps. On Al(111), steps of various height could be resolved by Wintterlin [81]. This is illustrated in Fig. 4.22.

To image adsorbed particles with STM is more difficult for various reasons. One reason is the high mobility of most of the adsorbed gases at room temperature. (Here we should add that so far temperature-dependent measurements cannot be performed with STM due to serious thermal drift problems. That is to say, the sample contracts or expands as a function of temperature so that it is impossible to follow these (comparatively large) motions with STM and find a given surface area again, which would be necessary in order to follow, e.g., temperature-dependent structure changes.) There are two ways to circumvent these problems. One method is to conduct the STM experiment at such low temperatures (< 10 K) that any adsorbate is actually immobile. Here one of the most recent developments is the 4 K STM designed by Eigler and Schweizer [82], where individual xenon atoms adsorbed on Pt(111) could be made visible with large mag-

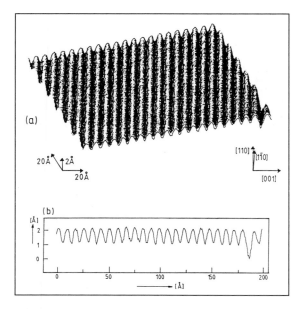

(a)

20Å↖↑2Å
20Å

[110]↑‖[1̄10]
[001]

(b)
[Å]
2
1
0

0   50   100   150   200
→ [Å]

Fig. 4.21. Perspective view of the STM image of a reconstructed Au (110)-1×2 surface. While the densely packed rows of the reconstructed surface can be resolved (distance of adjacent rows 8.16 Å), it is not possible to resolve the individual gold atoms parallel to the rows in [1̄10)] direction. Tunnel voltage V_t was −130 mV; tunnel current $I_t = 40$ nA. The bottom part of the figure visualizes the surface corrugation in [001] direction perpendicular to the rows. After Gritsch [80].

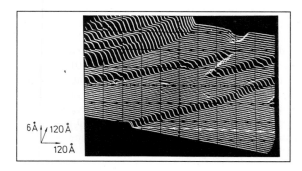

6Å↓↙120Å
120Å

Fig. 4.22. STM image of an Al(111) surface showing steps of monatomic, diatomic, and triatomic height. This surface produced a very sharp and bright (1×1) LEED pattern. After Wintterlin et al. [81].

nification. The other method is to pack the adsorbate layer so densely that the molecules no longer possess any translational degrees of freedom. In this way, Gritsch et al. [83] succeeded in imaging adsorbed CO molecules on a Pt(110) surface. At saturation, these molecules form a non-primitive $(2×1)p1g1$ structure, for which a zig-zag row configuration along [1̄10] direction was concluded from LEED and UPS experiments [84]. As Fig. 4.23 illustrates, the zig-zag topography of the CO layer could indeed be monitored by STM investigation.

Despite these – admittedly – spectacular observations (we are reminded of the fascinating STM photographs obtained by Wintterlin [81] showing aluminum (111) surfaces with extremely high resolution, cf., Fig. 3.4), it is thought there is an even higher potential of the STM method, namely, to monitor directly the mechanism(s) of surface processes, such as adsorption, restructuring, and chemical etching, in situ.

Very recent investigations have disclosed, for example, the CO-induced lifting of the inherent Pt(110) 1×2 reconstruction as a local process [83]. Pt atoms in the direct vicinity of the adsorbed CO molecule are displaced in a manner depicted in Fig. 4.24. Even more

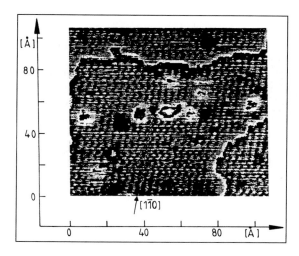

Fig. 4.23. STM image of a Pt(110) surface covered with a monolayer of CO molecules (which form a (2×1)p2mg superstructure with mutually inclined molecules). Clearly, the well known zigzag chains of adsorbed CO can be distinguished in the [1$\bar{1}$0] direction. Irregular monatomic steps and other defects are responsible for the perturbations of the image. Tunnel current I_t was 110 nA. After Gritsch [80].

exciting are STM investigations of metal-vapor growth processes (for instance, observing a layer-by-layer growth of copper deposited on ruthenium (0001) surface [85]) and the outstanding in situ measurements of electrochemical metal-deposition and surface-etching processes in the condensed (liquid) phase [86]. It is one of the great advantages of the tunnel microscope that it obviously can be employed not only in the UHV environment, but also at higher gas pressures and even at the liquid-solid interface. This potential opens up a whole variety of practical chemical applications about which one, at present, can only speculate. To repeat, the interpretation of an STM image has the advantage of arguing in real space, unlike the diffraction methods, which always yield the Fourier transform. Nevertheless, one must bear in mind that even the STM image is not a full replica of the "real" world. Rather, charge densities and tunnel probabilities are involved, which, however, correspond in many cases directly to our view of the atomic structure of matter.

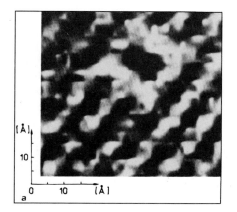 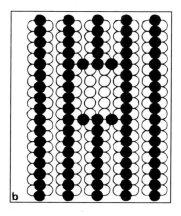

Fig. 4.24. Mechanism of the CO-induced removal of the Pt(110) – 1×2 reconstruction as imaged by STM. An adsorbed CO molecule displaces four atoms of a row sidewards and forms a local (1×1) domain as a hole of ca. 15 Å × 15 Å area. **a)** direct STM image, **b)** schematic structure model. Dark circles represent Pt atoms of the top rows. After Gritsch et al. [83].

4.1.4 (Surface) Extended X-Ray Absorption Fine Structure ((S)EXAFS)

In this section it will be shown that even phenomena known for a fairly long time can be profitably used for surface analysis if data are carefully evaluated.

It was in the 1930s when the complex structure of a solid's x-ray absorption intensity as a function of energy (extending up to 1000 electron volts above the absorption threshold) was realized. Due to the shortcomings of the experimental equipment at that time (low intensity x-ray tubes, unsatisfactory detection and counting circuitry) the interpretation of these phenomena remained dormant until in 1970 Lytle and collaborators [87, 88] repeated some measurements and realized that the analysis of the fine structure of x-ray absorption edges (EXAFS) could provide valuable bulk and surface-structure information for solids. It was underlined that backscattering and diffraction of ejected photoelectrons from the atoms in the direct environment of the photoionized atom gave rise to the fine structure observed. Similarly, the idea was borne that a careful analysis of the diffraction maxima's position on the energy scale could in turn be used to deduce geometrical parameters (atomic distances).

The use of intense synchrotron radiation sources instead of unfiltered x-ray (Bremsstrahlung) tubes beginning in 1972 was a real breakthrough in the field of EXAFS [89]. Since then, this method has entered the field of structure analysis and has become competitive to LEED, because it can also probe the structure of polycrystalline disordered material. It became possible for the first time to obtain access to reliable distance parameters even of practical catalyst particles and promoter distributions in highly diluted substances having overall concentrations of as low as 10^{13} atoms/cm^2. In the following discussion we shall briefly present the basic physics of EXAFS and supply the reader with some selected practical examples. From the wealth of literature, we provide only a selection here [90–93]. The following description is based on a short review presented some years ago by Eisenberger and Kincaid [90].

The primary process in x-ray absorption is the photoionization of a given atom, whereby, depending on photon energy $\hbar\omega$, excitation of outer or inner shell electrons is achieved. In case of EXAFS, mostly inner-shell (K-shell) electrons are removed. There is a minimum photon energy required for this process to occur ($\hbar\omega = E_K$ where $E_K = K$ shell-binding energy). For photons with greater energy, the balance for the kinetic energy of the ejected photoelectron E_{kin} reads

$$E_{kin} = \hbar\omega - E_K \ . \qquad\qquad 4.25$$

Photons with lower energy cannot ionize the respective shell, so there is a real threshold of excitation when $\hbar\omega$ reaches E_K for the first time. In an x-ray transmission spectrum, we observe at this point the so-called K absorption edge. If we describe the absorptive power of material by an absorption coefficient μ, the initial x-ray intensity I_0 is damped according to

$$I = I_0 \exp(-\mu d) \ , \qquad\qquad 4.26$$

where d stands for the thickness of the material.

The photon energy range for K-shell excitation varies, according to the different K-shell electron-binding energies, largely with the atomic number of the elements Z. For low Z elements, say from sulfur to copper ($16 < Z < 29$), this energy lies between ~3 keV and 16 keV. For iodine excitation ($Z = 53$), for example, photons of approximately 35 keV energy are needed. A synchrotron storage ring with its almost continuous radiation and a crystal x-ray monochromator is required to provide an intense monochromatic beam of x-rays; monochromator crystal rotation then provides tunable radiation.

108

In order to understand the principle of EXAFS, we must distinguish now between matter in the monoatomic dilute state (e.g., noble gases) and in the polyatomic state (diatomic gases, such as chlorine or bromine on the side of diluted material, and metal crystals, e.g., $(Cu)_n$ with $n \geq N_L$ on the side of condensed matter).

If we follow the x-ray absorption behavior of a Kr atmosphere, for example, there is a sharp rise of μ right at the absorption edge followed by an almost monotonic decay above this edge. Quite revealing now is a comparison with Kr's neighbor, bromine Br_2. Its absorption curve has the same overall shape, but exhibits additional wiggles on the decay side of the x-ray transmission spectrum. This is illustrated in Fig. 4.25. It is precisely

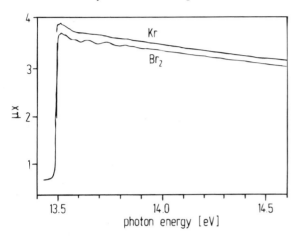

Fig. 4.25. Energy-dependent x-ray absorption of Kr gas (upper curve, no EXAFS) compared with the spectrum of diatomic Br_2 (lower curve, EXAFS wiggles superimposed). The thickness of the samples x is plotted on the ordinate. After Eisenberger and Kincaid [90].

these wiggles that EXAFS is all about, and they are called EXAFS oscillations. Their physical origin is an interference of the outgoing photoelectrons ejected from neighboring atoms. Consider, for example, part of a cubic copper crystal (Fig. 4.26) and assume the center atom to be hit by the x-ray photon. The emitted photoelectron wave is a spheri-

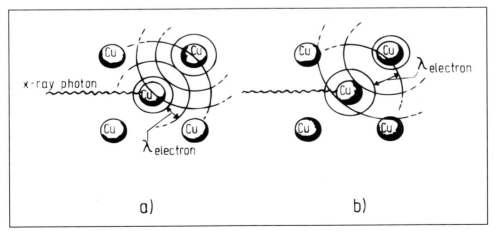

Fig. 4.26. The sinusoidal EXAFS pattern can be understood as the changing interference pattern between the outgoing wave and the wave scattered at the absorbing site (Cu atom) with the varying photoelectron wavelength (energy): **a)** indicates constructive interference, **b)** destructive interference between outgoing and scattered wave. After Eisenberger and Kincaid [90].

cal wave with length $\lambda_e = 2\pi/k$ (k = wave vector), whereby λ and hence k depends on the kinetic energy

$$k = \sqrt{\frac{2m_e(\hbar\omega - E_0)}{\hbar^2}} \,, \qquad\qquad 4.27$$

where m_e = electron mass, and E_0 = threshold energy, where electrons from the respective shell are emitted for the first time.

This wave now is back-scattered from the four neighboring Cu atoms, and evidently there is an interference between the original outgoing and the reflected waves, which can be constructive or destructive depending on the path difference. The phase shift is solely determined by the distance R_j of the excited atom to its neighbors and λ_e, as well as by the propagation of the electron between the absorbing and the scattering atoms. A complicating factor, therefore, is that, according to Eq. 4.25, photoelectrons of various wavelengths can interfere, each leading to a different phase shift and hence to a different amount of interference. Furthermore, there are not just first neighbors, but also second, third, and more neighbors, which can cause (although with decreasing efficiency) back-scattering. If we define the function $\chi(k)$ as the fractional modulation of the absorption coefficient caused by EXAFS interference and the difference $\Delta\mu = \mu(k) - \mu_0(k)$, we obtain

$$\chi(k) = \frac{\Delta\mu}{\mu_0} =$$

$$= \sum_j \frac{-N_j|f_i(k,\pi)|}{k \cdot R_j^2} \cdot e^{-2\sigma_j^2 k^2} \cdot e^{-2\lambda^{-1}k_j} \cdot \sin[2kR_j + \delta_j(k)] \qquad 4.28a$$

$$= \sum_j A_j(k) \cdot \sin[2kR_j + \delta(k)] \,, \qquad\qquad 4.28b$$

where

$\mu(k)$	= oscillatory part of the x-ray absorption coefficient,
$\mu(k_0)$	= absorption coefficient for an isolated atom,
N_j	= number of scattering atoms at distance R_j (shell atom),
$f_i(k,\pi)$	= electron scattering amplitude in the backward direction of the atom j,
$e^{-2\sigma_j^2 k^2}$	= the Debye-Waller factor accounting for thermal vibrations or static disorder with root-mean square fluctuations σ_j,
$e^{-2\lambda^{-1}R_j}$	= damping due to inelastic photoelectron scattering,
λ	= electron mean free path,
$\sin[2kR_j + \delta_j(k)]$	= the sinusoidal interference function in which $\delta_j(k)$ represents the phase shift.

As can be seen from Eq. 4.28, each shell of neighboring atoms contributes a sine function multiplied by a slowly varying amplitude function $A_j(k)$.

The complicated equation 4.28 simply tells us that EXAFS is a relatively complex process, and the task is clearly to extract the structural information, i.e., determine the values of R_j in the first instance, $A_j(k)$, N_j, and others.

110

The first step in EXAFS analysis is always to obtain the experimental energy dependence of the absorption coefficient, i.e., the curve $\mu(E)$. Usually, the smooth part of this function is taken as μ_0; the remaining wiggling part gives access to $\Delta\mu$, hence $\chi(k)$ can be deduced. There is, however, ambiguity concerning the *exact* value of the threshold energy and thus E_K, which is circumvented by making E_0 a variable parameter in later stages of the analysis. Then, by some additional procedures, for instance, background subtraction, the EXAFS oscillations can be extracted, usually being plotted in the form of a $k^3 \cdot \chi(k)$ function vs the wave vector k [Å^{-1}] in order to give smaller oscillations at larger k more weight. This is documented in Fig. 4.27a. The next step then is a Fourier transformation, which isolates the contributions of the different shells of neighbors (Fig. 4.27b). The largest peak of the Fourier transform always corresponds to the first shell scattering

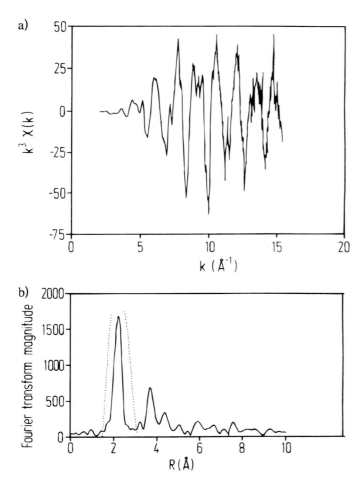

Fig. 4.27a) Processed absorption spectrum of powdered crystalline Ge. A value for E_0 has been chosen and a smooth background has been subtracted from the raw spectrum. The result, after division by the smooth background, has been multiplied by k^3 to generate the function $k^3 \chi(k)$ as a function of the wave vector k, defined by Eq. 4.27. **b)** Fourier transform spectrum of the data shown in a) (full line) together with the filter function (dotted line) used to isolate the first-shell scattering contribution. Other outer shells are also visible by the maxima in the curve. After Eisenberger and Kincaid [90].

contribution; several other smaller peaks are due to more distant shells of neighbors. The advantage of Fourier transform data handling is that a filter function can be employed to isolate just the first shell contribution. All higher sine-wave frequency parts are filtered out, and the single peak in the Fourier transform spectrum corresponds to a well-behaving sine function with unique amplitude and phase. This function can then be obtained by backtransformation from the Fourier transform. It carries two different pieces of information. The *phase function* contains the terms $2\,k\,R_j$ and $\Delta_j\,(k)$. If the latter were known, R_j could be obtained easily with high accuracy by subtraction. Here it helps to consider $\Delta(k)$ as an empirical function independent of the chemical environment and characteristic of a given pair of atoms. The idea of chemical transferability of phase shifts $\Delta(k)$ has been quite successfully applied to many different materials, and distances as accurate as $R_j \pm 0.01$ Å have been evaluated. The *amplitude function* $A_j\,(k)$, yielded from the Fourier transform-filtering analysis, provides information about N_j, number of nearest atoms, chemical nature of the atoms involved, electron mean free path, and amount of order or disorder.

Before we turn to some applications, a few words about *surface* EXAFS are worthwhile. As pointed out in the introductory part of this chapter, predominant surface sensitivity is achieved if the energy of the probing or detected electrons is approximately between 10 and 100 eV. In a typical EXAFS experiment, electrons having much higher energies are involved, and the method is not per se particularly surface sensitive. Here, one can, however, use electrons, which are emitted in a secondary process, to monitor the x-ray absorption edge. In particular, Auger electrons (cf., Sect. 4.3.1) associated with the decay of the core hole after photoionization can be used rather than the direct measurement of the x-ray absorption coefficient [94]. Thus, real surface sensitivity (SEXAFS) is attained. In most cases, however, the ordinary EXAFS technique still allows surface-structure analysis if particles with a large surface-to-volume ratio are investigated. Fortunately, most of the practically important catalyst materials belong to this category.

The EXAFS method can be successfully exploited in various fields ranging from biology to surface chemistry. Within the framework of this book, application in heterogeneous catalysis and surface analysis are to be emphasized. Sinfelt and coworkers [95–100] have consequently applied EXAFS to structural characterization of bimetallic-supported catalyst materials, for example Cu on silica (SiO_2) and Cu + Ru on SiO_2 surfaces, the great advantage being that the method can be used under the catalyst's working conditions. Hence, structural changes occurring in the course of a reaction can be monitored directly. An example for the usefulness of EXAFS in heterogeneous catalysis is provided by Sinfelt's work on silica-supported copper-ruthenium catalysts [100]. For the efficiency of these materials, the local structure, that is to say, the kind and number of Cu and Ru neighbors, is of great importance. The question arises as to whether there are Cu (Ru) atoms as first-shell ligands or an effective metal atom ensemble exclusively consists of Ru (Cu) atoms. Figure 4.28 summarizes some of Sinfelt's results [100]. It presents a comparison of Cu on SiO_2 (upper part) and Cu/Ru on SiO_2 (lower part). In the left part a) the normalized absorption coefficient $k\chi(k)$ near the Cu K-absorption edge (EXAFS oscillations) are shown (here, the $\chi(k)$ function was only multiplied by k, not by k^3 as described before, a procedure also possible). Then a Fourier transformation and filtering over the range 0.170×10^{-9} m $\leq 0.310 \times 10^{-9}$ m was carried out. Figure 4.28b presents the back transform function $k\,\chi(k)$ vs k in the respective wave-vector range for the Cu and Cu/Ru catalyst. It comes out quite clearly that both functions differ in amplitude and phase. These differences can be directly attributed to the fact that in the Cu/Ru catalyst particles the Cu atoms possess nearest neighbors consisting of Cu and Ru, while in the

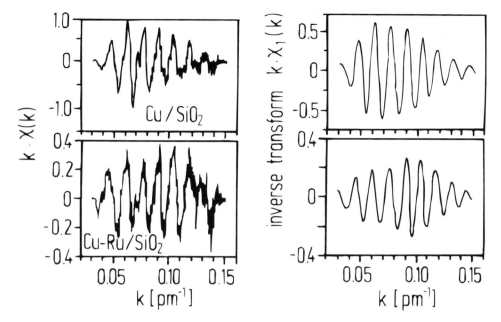

Fig. 4.28. EXAFS data of the Cu K-absorption edge from Cu/silica and Cu + Ru/silica catalysts: **a)** displays the function $k\chi(k)$ vs k as obtained by data processing; **b)** shows the corresponding inverse transform spectra over the window $170 \leq r \leq 310 \times 10^{-12}$ m. After Sinfelt et al. [100].

Cu/SiO$_2$ material there are only Cu atoms present. These findings are evidence that an atomistic view is justified. Then small Ru particles are covered with a kind of chemisorbed Cu atoms and lend these noble metal atoms some of their peculiar catalytic activity (cf., Sect. 3.1.1). In other words, a direct vicinity of Cu and Ru atoms is responsible for the interesting catalytic reactivity of these bimetallic materials. It should be noted that it is not necessary in these investigations for the materials to be in their single-crystalline state! With regard to further chemical-catalytical applications of EXAFS it is only mentioned here that even complicated molecular structure can be resolved. An example is provided by Wilkinson's catalyst [101], chlorotris(triphenylphosphine)rhodium, $((C_6H_5)_3P)_3RhCl$, which is active in dehydrogenation if it is used in a supporting medium consisting of polystyrene cross-linked with divinylbenzene. Its structure could be analyzed particularly with respect to its polymer-bound network, as described in detail in the review article by Eisenberger and Kincaid [90].

For reasons mentioned above, SEXAFS is the appropriate method when one aspires to obtain high-precision surface-structure parameters, for example, of adsorbed atoms on single-crystal surfaces. Here, we briefly present some of the results obtained by Citrin et al. [94] with the adsorption system iodine on silver, which were obtained by determining the absorption edge via the Auger electron yield, i.e., by measuring the Auger intensity as a function of photon beam energy incident on Ag(111) covered by 0.1 monolayers of iodine. In addition to the basic information about the average first-shell Ag-I distance (which offers a somewhat higher degree of precision than in the previous LEED work [102]), the absolute configuration of the iodine atom (I$_2$ adsorbs dissociatively on Ag) was accessible by measuring the spectrum with two different polarizations of the photon beam set normal and parallel to the surface. Since that study, many EXAFS and SEXAFS investigations have been carried out in the field of surface-structure analysis. Naturally, not all of them can be listed here. Instead, we just want to mention that even the shape of the x-ray absorption intensity very close to the absorption edge can be analyzed and exploited to determine the local structure of an adsorption complex (NEXAFS or XANES technique). For more details, we refer to the literature [103, 104].

4.1.5 High-Resolution Electron-Energy-Loss Spectroscopy (HREELS)

We conclude our brief excursion into structure sensitive tools with an extremely versatile method, high-resolution electron-energy-loss spectroscopy (HREELS), whose description would deserve a whole chapter. In the first instance, the HREELS method probes vibrations of adsorbed molecules and surface atoms in general. It can be used favorably to identify the chemical nature of surface species by means of an analysis of their vibrational modes. Therefore, it has much in common with conventional infrared spectroscopy. However, due to the excitation mechanisms of the surface vibrations by means of the used low-energy electrons, which will be explained further below, another quite important facet is HREELS' comparative sensitivity to the *orientation* of an adsorbed molecule on the surface, which we shall emphasize here. The real breakthrough of HREELS in surface analysis was achieved by the pioneering investigations of Ibach and his collaborators in the mid-1970s [105–108a]. They designed an appropriate high-resolution electron-energy-loss spectrometer, which provided a monochromatic electron beam (about 5–10 meV half-width) and contained an energy analyzer capable of detecting scattered low-energy electrons with a similar resolution.

Practically all theoretical and instrumental aspects of HREELS are elucidated in great detail in the monograph of Ibach and Mills [109]. Some other important review articles can also be recommended for further reading [110, 111]. Our presentation of the HREELS technique is organized so that first some basic physical principles of inelastic low-energy electron-scattering are presented, along with a comment on important selection rules (which make HREELS a structure-sensitive tool!). This is followed by a short summary of the instrumentation required to obtain high-quality EEL spectra. We conclude by presenting some practical examples taken from Ibach's and our own work.

Low- energy electrons ($1\,\text{eV} < E_0 < 10\,\text{eV}$) are emitted from a cathode and monochromatized electrostatically (resolution $\Delta E/E_0$ ca. $10^{-2} \ldots 10^{-3}$) and directed to the surface. Most of them are backscattered elastically, forming a strong elastic peak (0,0 LEED beam), but some electrons can excite surface vibrations. The required energies ($E_{\text{vib}} = h\nu$, where ν = the respective vibrations' ground state frequency) are relatively small, between say 50 meV and 500 meV. These electrons loose the respective energy and come off the surface with a correspondingly reduced kinetic energy:

$$E_{\text{kin}} = E_0 - E_{\text{vib}} \,. \qquad\qquad 4.29$$

If the energy distribution of all backscattered electrons is measured with sufficient resolution, one detects the so-called vibrational loss peaks at the respective positions on the energy axis. Then these can readily be identified and used for structural and chemical analysis of the surface species in a manner described in the following. Although we do not enter the quantum physics of vibrational excitation, it is nevertheless quite important to consider a vibrating adsorbed particle (see Fig. 4.29) as an oscillating dipole, which can and will interact with the wave field of the incident electrons. Strictly speaking, it is the *perpendicular* component of the dynamic-dipole moment, which is responsible for the interaction. This can easily be rationalized by means of Fig. 4.29 which shows, in a some-

114

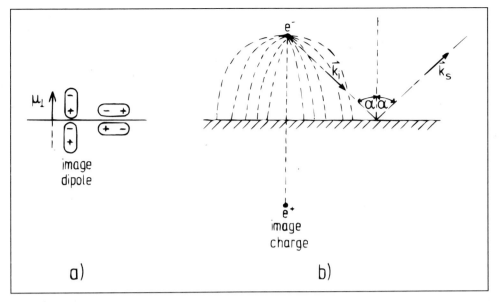

Fig. 4.29. Illustration of the dipole excitation mechanism in HREELS. **a)** symbolizes two cases of dipole orientation on a conducting surface. Only a dipole, whose dipole moment μ has a component perpendicular to the surface, is reinforced by image charge effects, while a dipole oriented parallel to the surface is totally compensated. **b)** indicates the electric field exerted by an electron e$^-$ on top of a conducting surface. The electron can be described by a plane wave (wave vector k_i), which is specularly reflected at the surface (angles α and α', respectively, wave vector k_s of the scattered electron wave).

what simplified and naive view, an electron idealized as a negative point charge at a distance R above the (metal) surface. This point charge exerts an image force on the metal electrons, which redistribute so as to form a positive image charge at distance $R' = R$ inside the metal. The associated dipole field is characterized by lines of force which are always directed perpendicularly to the surface, as illustrated in Fig. 4.29. A vibrational excitation of an adsorbed particle is hence only possible if its dipole moment has a component $\mu_\perp$ perpendicular to the surface (which likewise causes non-vanishing image charge effects, unlike any parallel component of μ, which leads to cancellation with its own image dipole).

This kind of interaction (which is of the long-range type) is called dipole scattering, and its cross-section is strongly peaked in the direction of specular reflection. That is to say, vibrational excitation occurs predominantly in forward scattering. The other important consequence is the so-called dipole selection rule, which follows from the fact that only vibrations with perpendicular components can be dipole-excited. The dipole selection rule offers an elegant means of determining the orientation of the respective bond on the surface. The HREELS dipole scattering is blind to vibrations that do not have this component of the dynamic dipole moment $\mu_\perp$. An outline of dipole scattering theory is provided, for example, by Newns [112].

The question arises how many vibrational modes of an adsorbed molecule can be excited by low-energy electrons. Although this seems a reasonable and simple question, its answer must be very complicated, because it involves complete information about the shape of the molecule, its normal vibrations, its adsorption site geometry, and its vibrational coupling to the substrate atoms. Within the framework of this section, we must

115

avoid trying to provide an exhaustive answer. Instead, we recommend Ibach's book for further reading [109]. Nevertheless, a few simple remarks can perhaps provide an introduction to this matter. Great simplification is achieved if one can neglect vibrational coupling between the substrate and adsorbate atoms. In other words, we consider the atoms of the solid as frozen, which represents a fairly good approximation if we deal with very light adsorbate atoms, for example, hydrogen on metals.

Let us first consider a single atom. Upon adsorption it loses its three degrees of translational freedom (provided we deal with localized adsorption) which are converted to vibrational modes – one mode describing the motion perpendicular to the surface, the other two the parallel motion. These two parallel modes are degenerate for a totally symmetric adsorption site (for example, a fourfold hollow site of an fcc (100) surface), but are nondegenerate for a site with lower symmetry (e.g., a bridge site on an fcc (110) surface). With dipole scattering only the perpendicular mode can be excited. In other words, only a single loss peak is observable in a HREELS experiment. This holds, by the way, for all similarly symmetric adsorption sites, for example those illustrated in Figs. 3.11 and 3.12a,b,d. On the other hand, one may also regard asymmetric binding sites (cf., Fig. 3.12c). In this case, all those vibrational modes, which possess a component perpendicular to the surface, can be dipole excited. Within the framework of the model outlined above, we can take a further step (Fig. 4.30). Following Ibach and Mills [109], we can calculate the vibrational frequencies $\omega_\perp$ and $\omega_\parallel$ within the approximation of a so-called central force model. It is assumed therein that the equilibrium position of the adsorbed particle is fixed by the condition $d\varphi(x)/dx = 0$, if $\varphi(x)$ denotes the central potential. The curvature of the potential is then given by the second derivative, i.e., $d^2\varphi(x)/dx^2$. Allowing d_0 to be the distance between the adsorbed atom and each of the neighboring substrate atoms, and m_s and m_{ad} the mass of the substrate atom and the adsorbate atom, respectively, and considering the case that m_s is much larger than m_{ad}, one may write for the two frequencies $\omega_\perp$ and $\omega_\parallel$, describing the motions perpendicular and parallel to the surface, respectively:

$$\omega_\perp = 2\cos\alpha\sqrt{\frac{d^2\varphi(d_0)}{dx^2}\Big/ m_{ad}} \qquad\qquad 4.30$$

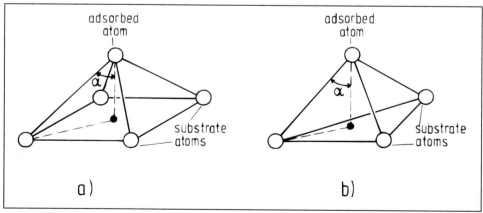

adsorbed
atom

adsorbed
atom

substrate
atoms

substrate
atoms

a)

b)

Fig. 4.30. Central force model applied to hollow adsorption sites of fourfold a) and threefold b) symmetry. α represents the bond angle. For these sites, there are three normal modes of adsorbed atom's vibration (one motion normal to the surface, two (degenerate) parallel motions). After Ibach and Mills [109].

and

$$\omega_{\parallel} = 2 \sin \alpha \sqrt{\frac{d^2 \varphi(d_0)}{dx^2} / m_{\text{ad}}} ,$$

4.31

where α represents the bond angle and $\varphi(d_0)$ stands for the central potential. Likewise, expressions can be written for finite substrate masses and different adsorption geometries [109, 113].

Turning to adsorbed molecules, all $3N$-6 normal vibrations must be taken into account as well as the frustrated translations and rotations due to the coupling to the substrate. Because usually rich loss spectra are observed for polyatomic adsorbed molecules, a comparison with the gas-phase infrared spectra as well as isotope substitution can help significantly in vibrational loss assignments. Furthermore, the dipole-selection rule may, for certain positions of the adsorbed molecule, greatly limit the number of observable vibrations.

Besides dipole scattering there is another excitation mechanism, the so-called impact-scattering, which does not show the pronounced directional dependence of dipole scattering. Impact scattering is due to a short-range type of interaction, because the scattering occurs only at very small distances upon an actual impact event between the incoming electron and the adsorbed particle. The extent to which this impact scattering occurs depends entirely on the shape of the electron-molecule interaction potential curve; during the scattering the electron may become transiently trapped in an unoccupied orbital of the hit particle. It can be shown that, in contrast to dipole scattering, there is no strong angular dependence of the impact cross-section. Accordingly, basically all vibrational modes are impact-active and can be detected even in off-specular directions, including the dipole-forbidden ones. A distinction between dipole and impact losses can, therefore, be made by plotting the scattering intensity vs the angle of detection. Only those peaks that are strongly peaked in specular direction are due to dipole active vibrations; the others are impact losses. There are, however, also selection rules, which govern the impact excitations, and these selection rules can be favorably used to implement structure sensitivity. Consider Fig. 4.31, which depicts typical scattering geometries of an electron energy-loss experiment with a face-centered cubic (110) single crystal surface. We must define two directions, namely i) the crystal azimuth (given in the x,y plane by the angle φ) and ii) the direction of the scattering plane of the electrons spanned by the emitting cathode, the reflection point on the surface and the electron detector position defined by the angle 2α. Arbitrarily, we can place the scattering plane in the xz position and introduce the electron wave vector $\boldsymbol{k}_i$ (which forms an angle α_i with the z axis) and $\boldsymbol{k}_s$ (angle α_s with z axis). One selection rule of impact scattering then states that if the scattering plane is parallel to a mirror plane of the respective vibration (here, the surface complex formed by the adsorbed particle and all substrate atoms participating in the bonding must be considered) all those modes are invisible that are antisymmetric with respect to this mirror plane [109, 114, 115]. This holds for all detection angles. The other rule considers the situation in which the scattering plane is perpendicular to a mirror plane. In this case, the cross-section for excitation of antisymmetric vibrations with respect to the mirror plane only reaches zero if the wave vector of the incident electron $\boldsymbol{k}_i$ is equal to that of the outgoing electron, $\boldsymbol{k}_s$. Due to the fact that $\boldsymbol{k}_i \neq \boldsymbol{k}_s$ principally for an inelastic process, this selection rule is not strictly obeyed; only for specular scattering geometry and large primary energies almost vanishing intensity is observed compared with the loss energies.

If a loss peak has been identified as impact active and there exists, due to the particular

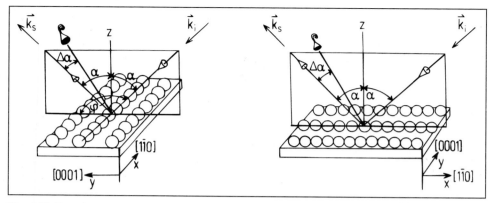

Fig. 4.31. Two simple cases of possible HREELS scattering geometries on a surface with twofold symmetry (here, a ruthenium (10$\bar{1}$0) surface is chosen). Left-hand side: electron-scattering plane *perpendicular* to the direction of the rows (angle $\varphi = 90°$), parallel to the [0001] direction. The incoming electron wave (wave vector $\mathbf{k}_i$) includes the angle α with the surface normal; the outgoing wave ($\mathbf{k}_s$) can be probed in the specular angle (α) or off-specular ($\alpha \pm \Delta\alpha$). Right-hand side: electron-scattering plane *parallel* to the rows ([1$\bar{1}$0] direction), angle $\varphi = 0°$.

adsorption geometry (C_s or C_{2v} symmetry), a mirror plane through the adsorption complex, a 90°-rotation of the scattering plane with respect to the crystal azimuth (defining the direction of the mirror plane) can make a particular loss disappear. There are several cases in the literature where, based on these impact selection rules, the local symmetry of the adsorption site could be determined. One nice example is provided by the H-on-Ni(110) system by Voigtländer et al., who could distinguish two different sites with C_s symmetry for the H saturation structure [115].

This application can also be illustrated by the 1×1-2H structure found in our own laboratory with a Ru(10$\bar{1}$0) surface [116]. As the structure model of Fig. 4.32 shows, all the H atoms are bound in sites with quasi-threefold symmetry, whereby, however, two kinds of

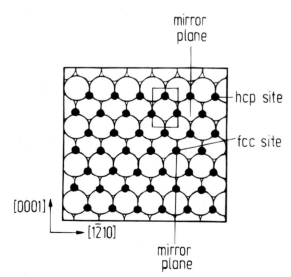

Fig. 4.32. Structure model of H atoms adsorbed on Ru(10$\bar{1}$0) forming a (1×1)-2H phase [116], as derived on the basis of HREELS impact selection rules. The direction of the mirror planes and the two types of sites are indicated in the figure.

118

sites must be distinguished: The so-called fcc sites (consisting of two atoms located in the troughs and one atom in a row) and the hcp (hcp = hexagonal close-packed) sites (two row atoms, one trough atom). Both sites have C_s symmetry, and for both of them there is a mirror plane in [0001] direction, that is, perpendicular to the rows. For a H atom in any such adsorption site, there should be two dipole-active losses (one vibration perpendicular to the surface and one parallel mode with a dipole-active component). The second parallel mode does not have this perpendicular component and is therefore dipole forbidden, it is, however, impact-active, as are all the other modes, too.

The loss spectra found in the two orthogonal azimuths indeed reveal evidence of two different adsorption sites, each of C_s symmetry. The loss spectra were recorded under off-specular conditions, in order to suppress the dipole-active losses, and for two orthogonal azimuthal directions, i.e., parallel and perpendicular to the troughs. Evidently, the spectrum taken in [$1\bar{2}10$] direction (scattering plane ⊥ mirror plane) is characterized by a wealth of losses (corresponding to all impact active modes of the two adsorption sites), whereas, in [0001] direction (scattering plane ∥ mirror plane), merely two strong losses were observed. All asymmetric modes have disappeared according to the first-impact selection rule explained above. This enabled us to develop the structure model of Fig. 4.32, in which both the hcp and the fcc sites of "quasi" threefold symmetry are occupied, thereby forming a very homogeneously packed hydrogen layer, which contains nominally two H atoms per Ru surface atom.

Among the various applications of HREELS in regard to structure sensitivity, we must not forget CO-adsorption chemistry. It is well known that carbon monoxide adsorbs readily on transition metal surfaces and can form terminal, bridge, threefold, and fourfold coordinated complexes with the underlying metal atoms. The CO-binding chemistry can well be described by the so-called "Blyholder" mechanism [117–119], whereby the binding to the surface occurs via the carbon end of the molecule so that there is overlap between the 5 σ CO-molecular orbital and the metal's wave functions, resulting in an electron donation from this MO to the metal and thus providing a stable Me-CO bond. On the other hand, there is a back-donation of metallic charge into the anti-bonding π^* MO's of the CO, which more or less weakens the C-O bond. It is assumed that this back-bonding effect becomes more pronounced as the coordination of the CO complex on the metal surface increases. The frequency of the CO-stretching vibration v_{CO} is an excellent monitor of the degree to which this bond weakens [126]: terminal configurations of the Me-CO complex shift v_{CO} from 2143 cm^{-1} (gas-phase value) only to the region between 2130 cm^{-1} and ~2000 cm^{-1}. Typical examples for terminally bound CO are Ru and Rh surfaces [120–123]. Bridge-bonded CO, which is found with some palladium surfaces [124, 125], is characterized by v_{CO} lying in the range 1880 cm^{-1} < v_{CO} < 2000 cm^{-1}, and higher CO-metal coordinations reveal v_{CO} frequencies smaller or much smaller than 1880 cm^{-1} [126, 127]. These extremely weakened stretching vibrations often precede a dissociation of CO, which can readily occur on the active surfaces of iron, chromium, tungsten, and related metals.

It is thus shown that the position of a vibrational frequency detected by means of HREELS can indeed provide conclusions not only about the surface and adsorbate structure, but also about surface chemistry (dissociation processes and surface chemical reactions).

The HREELS experiment requires, as mentioned above, an electron spectrometer, which provides sufficient resolution to observe vibrational losses. In Fig. 4.33, we present a schematic drawing of a typical spectrometer design. The instrument can be divided into four essential parts. First, there is the electron source composed of a tungsten hairpin

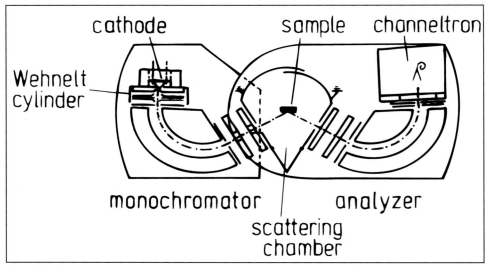

Fig. 4.33. Single-pass electron-energy-loss spectrometer consisting of a (fixed) electron source plus monochromator and a (rotatable) electron-energy analyzer plus detector (channeltron). A typical electron trajectory is indicated by the broken line. Both monochromator and analyzer represent 127° spherical condensers and provide, together with electrostatic focusing- and correction-lens elements, an overall energy resolution of better than 10 meV.

cathode, Wehnelt cylinder, and focusing lens elements. Adjacent to these lenses, there is an electron monochromator consisting of a 127° spherical condenser which, along with imaging lenses, focuses the electron trajectories at a point right in the center of the scattering chamber. In an actual loss experiment, the sample surface must be moved right into this center. The third part, the analyzer section, is basically a repetition of the monochromator – the back-scattered electrons are collected in another 127° spherical condensor and focused on the fourth part, the electron detector. Commonly, a channeltron is used to detect and amplify the electron current (which, for weak losses, can be lower than 10^{-16} A). Instead of 127° analyzers, hemispherical condensers are often used. Also, two-stage cylindrical capacitors in the monochromator and analyzer section are common. These two-stage instruments usually provide a superior resolution. It is possible, by carefully tuning the spectrometer and by using extremely well-stabilized electric-supply voltages, to bring the resolution well into the 2 meV regime. A particularly impressive example for the present state of the art is taken from Ibach [128], who investigated CO adsorption on a Ni(110) surface. A loss spectrum is reproduced in Fig. 4.34. Very recently, this resolution could even be improved to yield 0.93 meV, for CO adsorption on an Ir(100) surface. This was only possible by a careful computer-assisted calculation and optimization of the electron-optical elements of the loss spectrometer [129]. Although HREELS does not yet allow the high resolution that can easily be obtained in an infrared vibrational experiment, it has the advantage of a comparatively large sensitivity to even small amounts of adsorbate; surface concentrations as small as 10^{13} CO molecules/cm^2 can, for example, be detected well.

Among recent developments in the HREELS field we just want to draw attention to time-dependent HREELS, where dynamic effects of adsorption and surface reaction can be followed by monitoring spectra successively at a high rate. Several articles by Ho and coworkers should be acknowlegded in this context [130], as well as the work of Froitz-

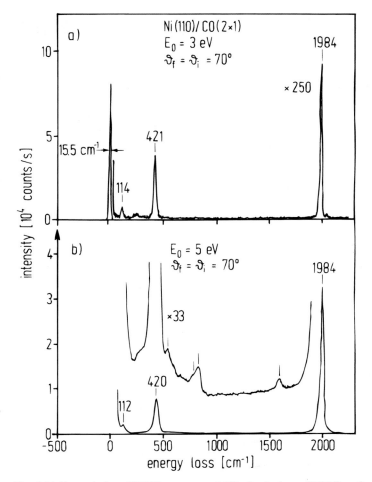

Fig. 4.34. Example for a HREEL spectrum of CO adsorbed on a Ni(110) surface, where the resolution has been tuned up to 15,5 cm^{-1} (≈ 2 meV). Only with this superior resolution can fine structure pertinent to the CO vibrational loss spectrum become visible. The upper part of the figure presents a spectrum with a magnification factor of 250; the lower spectra are further magnified and reveal additional spectral features. After Voigtländer et al. [128].

heim [131], who was able, by a careful experimental design, to perform state-selective HREELS. This provided valuable information about the adsorption and desorption kinetics of CO adsorbed in linear and bridge position on a platinum (111) surface. Figure 4.35 presents an example of monitoring the growth of these two CO species under exposure.

Although this chapter is devoted to surface structure, the very potential of HREELS lies in detecting vibrational modes. It is primarily a method for delivering chemical-analytical information, namely, allowing the identification of surface species by means of vibrational assignment. It is, for example, very useful in detecting all sorts of surface intermediates in the course of a chemical reaction, and we shall briefly return to this issue in the excursus about analysis of surface-chemical composition (cf., Sect. 4.3).

121

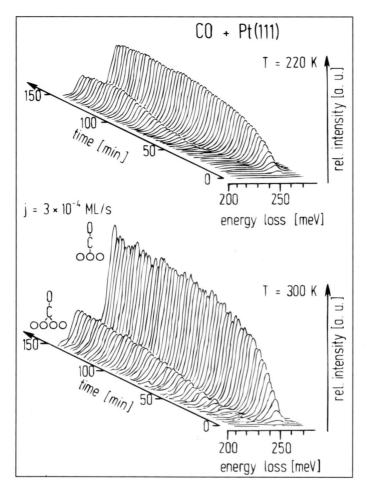

Fig. 4.35. Time-resolved HREEL spectra obtained from CO adsorbed on a Pt(111) surface, at $T = 220$ K (upper part) and at $T = 300$ K (lower part). Shown is a sequence of spectra after different time intervals, whereby only the loss-energy region between 200 and 280 meV is displayed, in which the stretching vibration v_{CO} of the two observed CO coordinations (on-top site around 250 meV, bridge site around 220 meV) appear. This experiment allows the simultaneous observation of the occupation of different adsorption sites and is, hence, suitable for investigating site-specific adsorption kinetics as a function of temperature and coverage. After Froitzheim and Schulze [131]

4.2 Determination of Surface Electronic Structure

Closely related to surface geometrical structure is, as mentioned before, the electric-charge distribution in the surface region. Actually, the combination and overlap of the individual orbitals of the atoms in a solid determine the macroscopic shape of the crystal and hence of the surface. In surface analysis, there are various experimental tools that probe specifically the electronic structure of the surface or at least the surface region. Among these tools, we shall devote our attention only to the methods of photoelectron

122

spectroscopy which are, for practical and instrumental reasons, subdivided into UV (ultra-violet) and x-ray photoemission, with a short comment on inverse photoelectron emission spectroscopy (IPE).

There is a wealth of literature concerning the photoexcitation and -emission process, and we only list monographs and review articles by Plummer [132, 133], Roberts [134], Siegbahn [135], Spicer [136], and Feuerbacher and Fitton [137, 138]. Figure 4.36 illustrates the probing characteristics of photoelectron spectroscopy, i.e., the energy dependence of photoexcitation and, hence, the difference between UPS and XPS. In the lower portion we display the potential energy situation of a (metallic) solid, and in the upper portion the respective photoelectron spectra. The only important variable here is the photon energy, which can be spanned from wavelengths of, say, 400 nm (corresponding to ~3.1 eV energy, a wavelength just sufficient to excite photoelectrons from metals with low work function) up to the (soft) x-ray region, i.e., 0.8 nm (1500 eV), where all sorts of valence and core orbitals can be excited. Although the instrumentation will be dealt with in the subchapters, we may premise here that there are light sources with fixed frequencies emitting line spectra (resonance lamps and x-ray tubes with filters) and sources emitting practically a radiation continuum (unfiltered x-ray (Bremsstrahlung) tubes and synchrotron storage rings). Returning to Fig. 4.36, it immediately can be seen that low-energy photons can only excite the outermost orbitals (shells) of atoms with relatively

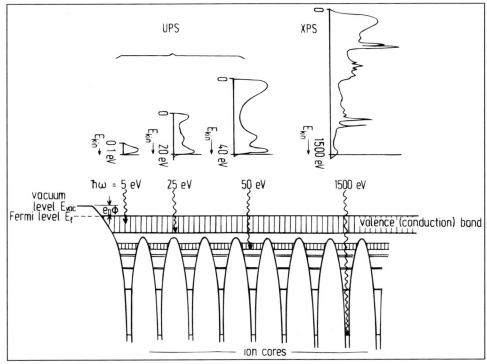

Fig. 4.36. Schematic illustration of the probing characteristics of a photoelectron excitation experiment (UPS and XPS). Shown is a representation of the valence/conduction band region as well as the core-level electronic structure of a metallic solid, whereby the electron bands are indicated by hatching. Depending on the photon energy (5 eV on the left, 1500 eV on the right) electrons with low or high binding energies can be excited. The corresponding electron-energy distribution curves (EDCs) may be, at first approximation, regarded as replicas of the electronic structure of the solid.

small electron-binding energies (the valence or conduction band region), whereas radiation of much higher energy can also ionize and probe inner atomic shells (the so-called core levels) with their relatively high electron-binding energies. The first (lower-energy) photoelectron spectroscopy is called UPS, because ultraviolet radiation is used; the high-energy regime is probed by XPS, because x-ray photons are utilized for excitation. Another name for this technique is ESCA (electron spectroscopy for surface analysis), which underlines the fact that the method is very useful in chemical analysis of surface species (as will be explained later). Before we enter the photoemission spectroscopies with a description of the basic physics and instrumentation, a few historical and organizing remarks may be worthwhile.

The use of UPS as a surface spectroscopy dates back to Spicer's work [139, 140, 140a]. In addition, the investigations conducted in the early 1970s by Eastman and his group deserve much attention. Eastman and Cashion were the first to introduce a UHV-compatible windowless-discharge photon source and were thus able to expand the accessible photon energy range appreciably [141, 142]. Among others, they irradiated clean and CO-covered Ni single-crystal surfaces with He I photons of 21.2 eV photon energy and observed not only an image of the Ni-electron density of states, but also CO-induced electronic levels, which they attributed correctly to the chemisorptive interaction between CO and Ni. These studies were still performed in an angle-integrated mode, that means the electrons were collected over a whole range of emission angles, without any spatial resolution. Today, it is an increasing trend to conduct angle-resolved UPS which accounts for the fact that the spatial distribution of the emitted photoelectrons carries important information about the spatial symmetry of the excited orbitals. Preferably, synchrotron radiation sources, together with rotable energy analyzers, are used for this purpose.

As far as XPS is concerned, it was Siegbahn and his crew at the university of Uppsala who developed in the 1950s all the instrumentation for generating soft x-ray photon sources and photoelectron energy analyzers to probe the kinetic energy of emitted electrons with sufficient resolution [135]. They were able to detect even small chemical shifts of core orbitals as a function of the excited atoms' chemical environment. It soon turned out that ESCA represented an extremely powerful and versatile technique in gas- and bulk-material analysis. Somewhat later (1965 or so), it was also realized that ESCA had a pronounced surface sensitivity and could provide a lot more information in the field of elementary surface-electronic excitations, surface core-level shifts, chemical-surface analysis, and practical catalytic material characterization.

In the following, we will comment on UPS and XPS photoemission techniques.

4.2.1 UV-Photoelectron Spectroscopy (UPS)

This method (as well as XPS, see the following section) is entirely based on the outer photoelectric effect. An ionizing radiation (UV light or vacuum UV light VUV) is directed onto a surface, and electrons in the conduction or valence band are excited. If they gain sufficient energy, they can overcome the surface potential barrier and escape into the vacuum, where they are monitored by an electron detector or energy analyzer. Very useful here is the naive representation of the electronic structure of a metal by means of the so-called Fermi sea (Fig. 4.37a), which we had already used when illustrating the field-emission process (cf., Fig. 4.14) and in which three important energy levels must be distinguished: the bottom of the sea, which is arbitrarily defined as zero; the filling level of the sea, which is called Fermi energy E_f; and the height of the outer barrier,

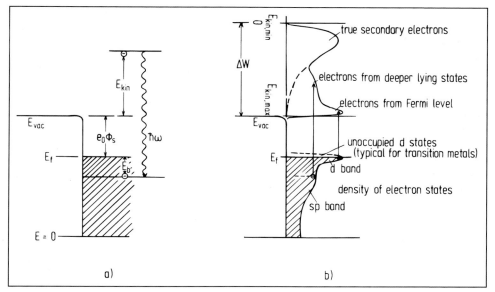

Fig. 4.37. Sketch of the energy balance of the UV photoexcitation process (UPS): **a)** Simple electron-sea model accounting for the energy balance, (Eq. 4.32); **b)** Refined electron-sea model considering the density of electron states within the conduction band, whereby a typical transition metal is chosen as an example. Also shown is the corresponding photoemission spectrum of width ΔW.

the so-called vacuum level E_{vac}. It is certainly known to the reader that the work function of the sample Φ_S practically corresponds to the difference $E_{vac} - E_f$, whereby this work function (cf., Sect. 4.4) is a very revealing and characteristic quantity of the surface-electronic structure. In this very simple picture, one can nevertheless conveniently illustrate the photoemission process. Figure 4.38 describes the energy level situation pertinent to photoelectron spectroscopy. This figure shows that actually the work function of the electron-energy analyzer Φ_A enters the formulae as a decisive quantity, because the electron kinetic energy is measured in and referenced with respect to the spectrometer. A photon having an energy $\hbar\omega$ is incident on the metal and ionizes an electron with binding energy E_b (which is normally referenced to E_f). If it can leave the surface with energy E_{kin} (measured by means of the analyzer), the balance reads:

$$\hbar\omega - E_b - e_0\Phi_A = E_{kin} \ . \hspace{3cm} 4.32$$

Therefore, for a given photon energy, and work function of the spectrometer, a measurement of a photoelectron's kinetic energy immediately yields its binding energy in the metal. The observable maximum kinetic energy is carried by those electrons excited directly at the Fermi edge. They give rise to a typical threshold behavior, and with a given analyzer the position of E_f coincides for all samples. The maximum kinetic energy then amounts to ($E_b = 0$):

$$E_{kin \, (max)} = \hbar\omega - e_0\Phi_A \ . \hspace{3cm} 4.33$$

Of course, electrons in deeper lying states, i.e., with larger binding energies, can and will be also excited. They would leave the surface, according to Eq. 4.32, with greatly reduced kinetic energy. Among the electrons leaving the surface with low kinetic energy are all those that have suffered energy exchange processes with the solid, so-called true-secondary electrons. The cut-off emission of a photoelectron spectrum then is determined by

125

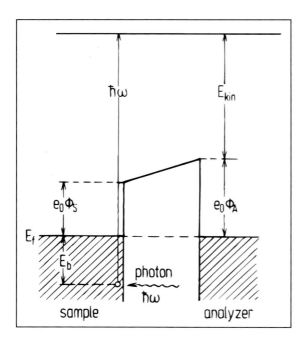

Fig. 4.38. Energy-level diagram applying to UV photoelectron spectroscopy. A photon of energy $\hbar\omega$ excites an electron of binding energy $E_{\rm b}$ (referred to as the Fermi level $E_{\rm F}$). The measured kinetic energy of the photoelectron $E_{\rm kin}$ is referenced to the electron spectrometer with work function $e_0\Phi_{\rm A}$.

those electrons (cf., Fig. 4.37) that had just sufficient energy $e_0\Phi_S$ to overcome the surface potential of the sample.

At the spectrometer they appear with a kinetic energy

$$E_{\rm kin\,(min)} = e_0(\Phi_S - \Phi_A) \ . \tag{4.34}$$

Hence, between the electrons with a maximum and a minimum kinetic energy a characteristic so-called electron energy distribution curve (EDC) is obtained, which is the typical photoemission spectrum. The work function of the sample Φ_S as a quantity of interest can then be calculated from the width ΔW of a photoelectron spectrum:

$$\Delta W = E_{\rm kin\,(max)} - E_{\rm kin\,(min)}$$

$$= \hbar\omega - e_0\Phi_A - e_0\Phi_S + e_0\Phi_A = \hbar\omega - e_0\Phi_S \ . \tag{4.35}$$

Looking again at Fig. 4.36, it is immediately realized that it contains nothing but a collection of EDCs obtained after excitation with ionizing radiation of increasing photon energy $\hbar\omega$. Evidently, there is a great deal of relevant fine structure seen in these curves, and this is why we must leave the naive electron sea model of Fig. 4.37a and turn to a more realistic band-structure description of the irradiated solid's electronic states. In other words, we must take into account the *density of states* of a metal. Again, a typical transition metal, such as nickel, can serve as a good example (Fig. 4.37b). It is well known that Ni exhibits a high density of d states right at E_f which is superimposed on a broad sp band with large dispersion. We repeat that for constant photon energy, the electrons excited right at the Fermi level will exhibit the highest kinetic energy, and the number of electrons in a given energy interval will be, in a first approximation, directly proportional to the density of states inside the metal. For the sake of simplicity, we shall not submit matrix-element (excitation-probability) effects here. We simply state that the actual shape of a UV photoemission spectrum near E_f is determined by the electronic

126

states of the metal, near the cut-off edge, however, by true secondary electron emission. Details here can be taken from the literature [136–140].

A very important application of UPS is derived from the fact that the photo-excitation and emission process is by no means restricted to a solid's electron states. Any adsorbate particle at the surface can be ionized as well, and will contribute density of states to the metal either by its own occupied molecular orbitals or by altering the charge distribution of the underlying substrate surface. These changes are most pronounced for chemical interactions (chemisorption), and in turn, the analysis of adsorbate-induced spectral features in the EDCs allows frequently valuable conclusions on the kind and strength of the particle-substrate interaction, particularly in comparison with quantum chemical calculations. We shall return to this point later.

Another potential of UPS lies in its angular dependence. The photoionization process occurs on such a short time scale ($\sim 10^{-16}$ s) that the emitted electrons cannot really equilibrate and still carry the spatial information of their parent orbitals. Therefore, angle-resolved measurements are increasingly more carried out, although the additional instrumental effort is appreciable. In addition, the use of polarized and tunable light at a synchrotron storage ring enables surface scientists to even map the electronic band structure of solids in great detail. These ARUPS studies have greatly contributed to our improved understanding of the electronic structure of the solids [133].

We add here a remark on the surface sensitivity of the UPS method. Although photons can enter a solid to appreciable depths (in the μm range) UV photoemission nevertheless is a relatively surface-sensitive spectroscopy, because the photoelectrons (which carry the relevant information) obey the universal mean free-path-energy relation of Fig. 4.1. With He I or He II resonance radiation ($\hbar\omega = 21.2$ eV and 40.8 eV, respectively) we are well in the minimum of this relation and hence probe between one and at most three atomic layers.

The principal instrumentation of UPS is relatively simple (Fig. 4.39). One needs a (monochromatic or tunable) photon source spanning the range from VUV to the soft x-ray region, a sample mounted on a moveable manipulator, and an electron energy analyzer of sufficient resolution equipped with a sensitive detector (SEV or channeltron). The point of light-incidence, photoelectron ejection, and focus of the analyzer must coincide. The operating pressure must be lower than, say, 10^{-5} mbar in order to meet the requirement for unperturbed electron trajectories. It is, however, attempted to keep it below 10^{-10} mbar to avoid surface contaminations while taking UP spectra.

In a surface analytical laboratory, most frequently the aforementioned gas-discharge resonance lamps, which have been commercially available for a number of years, are utilized for photon generation. Depending on the kind of noble gas used in the discharge, fairly intense line spectra, ranging from 11.62 eV (argon I) to 40.8 eV (helium II) radiation, are obtained. For normal lamps, photon fluxes of up to 10^{11} photons/s can be provided with the line widths in the range of 5 to 20 eV. A minor problem is that VUV radiation in this spectral range is absorbed by almost any material, and one must introduce the photons to the UHV system in a windowless manner. Because the operating pressure in the discharge can reach several hundred millibars powerful differential pumping stages must be mounted between the UHV system and the photon source, whereby long glass or metal capillaries help to conduct and focus the light onto the sample and provide sufficient flux resistance so that the background pressure in the vacuum system only increases by several percent. Figure 4.39 gives a typical example for the set-up of a He-resonance lamp.

The other important ingredient of a photoelectron spectrometer is the electron-energy

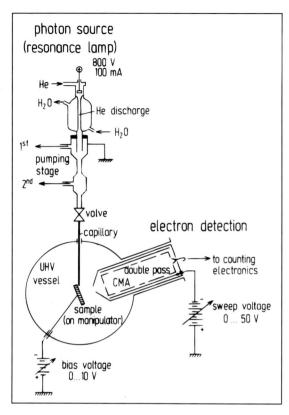

photon source
(resonance lamp)
800 V
100 mA

He →
H_2O ←
He discharge
← H_2O

1st
pumping
stage
2nd

valve
capillary

electron detection

UHV
vessel

double pass
CMA

→ to counting
electronics

sample
(on manipulator)

sweep voltage
0 ... 50 V

bias voltage
0...10 V

Fig. 4.39. Principal instrumentation required for UPS, consisting of a photon source (differentially pumped and water-cooled He discharge lamp) and the analyzer and detector part (CMA, channeltron, and counting electronics, in this case). The UV photons are conducted and focused onto the sample by means of a 1.5 mm diameter glass capillary. The focal point at the sample surface (from where the photoelectrons originate) must be well aligned with respect to the cylindrical mirror analyzer in order to obtain the best possible resolution and intensity (signal-to-noise ratio).

analyzer with a resolution of 100 meV or better. Principally, the analyzer part of the HREEL spectrometer described in Sect. 4.1.5 could be used. Usually however, somewhat larger and commercially available hemispherical analyzers are common, whereby the operating principle is completely the same. In some cases, a cylindrical mirror analyzer (CMA) also is utilized. It is characterized by good transmission characteristics, especially the double-pass CMA. We regret that we cannot describe the interesting physics of electron-energy analyzers, and instead refer to the literature. Useful brief excurses here are given in the books of Ertl and Küppers [10], Ibach and Mills [109], and Woodruff and Delchar [8]. For the sake of convenience, surface-analysis laboratories are very often equipped with a combination UPS/XPS apparatus, where a common analyzer and detector part counts the respective UV- or x-ray excited photoelectrons. The considerably higher electron energies involved in an XPS experiment, however, make great demands upon analyzer voltage stabilization. The typical resolution of an x-ray experiment is somewhat lower, about 500 meV or so.

Finally, we present some selected examples to demonstrate the usefulness of UV photoemission in surface analysis application. It is appropriate to distinguish clean surfaces and adsorbate-covered systems here. With regard to clean surfaces (we shall consider primarily metal surfaces), ultra-violet photoemission is easily capable of unraveling the typical electronic structure of a material. Thus, it is possible to observe the position of the metal's d band and to determine the position of the Fermi level in a single experiment. Thereby, the characteristic differences between a transition (d-band) metal and a free

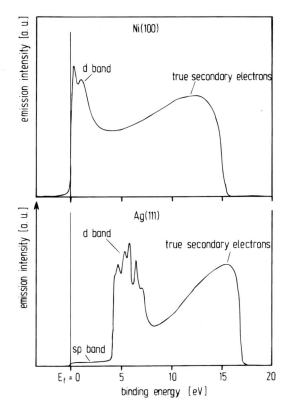

Fig. 4.40. Comparison of photoelectron spectra of a typical d electron metal (nikkel(100), top curve) and a free-electron-like metal (silver(111), bottom curve). Note the different position of the d band with respect to the Fermi level E_F and the different density of states at E_F.

electron-like (alkali, noble) metal become immediately apparent. As pointed out earlier, a *d* metal is characterized by a high density of states (DOS) right at the Fermi energy (maximum in the EDC), whereas an *sp* electron metal has a comparatively low DOS at E_f, as seen by a low onset of the electron emission. This is illustrated in Fig. 4.40, where EDCs of Ni (as a typical *d* metal) and Ag (as an exponent of a free electron-like metal) are contrasted with each other [143].

Owing to its surface sensitivity, it is convenient to follow changes of surface structure by means of UPS. As mentioned earlier, there are various metals whose surfaces tend to reconstruct under certain conditions. Usually, the geometrical changes are accompanied by alterations of the surface electronic structure (e.g., quenching of surface states), which in turn show up in the density of states and hence in the EDCs obtained in a photo-emission experiment. Examples here are Ir(100) [144] and Pt(100) surfaces [145].

Of much more chemical relevance is the photoemission in the field of adsorption and surface interaction in general. Although a UPS experiment probing electronic states of adsorbed particles cannot resolve (due to inherent broadening and relaxation effects) the respective vibrational fine structure as is possible in the gas phase, it is, nevertheless, a very useful method for probing and characterizing the nature of electronic adsorbate–substrate interaction, as pointed out before. Chemisorbed and physisorbed species are easily distinguishable; often also those molecular orbitals of an adsorbed particle can be detected, which are involved in and therefore responsible for the chemisorptive binding. According to chemisorption theory, particularly those orbitals which interact most strongly with the surface, broaden and exhibit an electron-binding energy shift. This phenomenon was emphasized among others by Demuth and Eastman [146] and called chemical bonding shift $\Delta\varepsilon_b$. $\Delta\varepsilon_b$ can reach several eV. While this is, as an inherent system property, clearly an initial-state effect, there always occurs another shift of all the molecular orbitals of the adsorbate as a final state effect, which is caused by relaxation processes of the electronic environment of the hole created by the photoionization process. The relaxation (including fluctuations of charge due to screening the photoelectron hole) changes the apparent binding energy of the emitted photoelectron somewhat, due to image-charge effects in the solid, thus resulting in some additional kinetic energy of the emitted electron. This phenomenon is known as extra-atomic relaxation shift $\Delta\varepsilon_r$ and is of the order of several electron volts. Usually, $\Delta\varepsilon_r$ depends on the electronic structure, i.e., polarizability of the environment surrounding the adsorbate. Upon completion of an adsorbed monolayer the free valencies of a surface become saturated, and relative abrupt changes occur in the overall charge distribution at the surface. Accordingly, the quantity $\Delta\varepsilon_r$ can be used to determine the transition from monolayer to multilayer coverages fairly accurately. If we take into account these initial- and final-state effects, we may relate the (often tabulated) ionization energy of the gaseous particle I_g with the actual electron-binding energy of the adsorbed particle via

$$E_b = I_g - (\Phi_s + \Delta\Phi) + \Delta\varepsilon_r + \Delta\varepsilon_b , \qquad\qquad 4.36$$

where $\Delta\Phi$ denotes the change of the surface potential induced by the adsorbate. In Fig. 4.41 we present, from our own work on pyridine adsorption on silver surfaces [147, 148], a typical application of this aspect of UV photoemission. Not only can the pyridine orbitals be identified (and hence a pyridine dissociation be ruled out), but also the type of bonding and the monolayer sorption capacity can be inferred from the

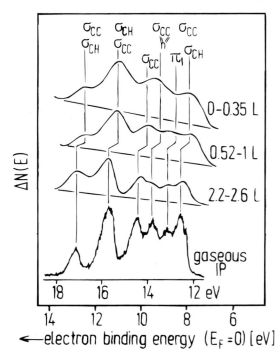

Fig. 4.41. He I photoelectron (difference) spectra of pyridine adsorbed on a silver (111) surface. Displayed are pairs of successive pyridine exposures, as indicated to the right. The ionization features (pyridine orbitals are marked at the top) demonstrate a pronounced energy shift starting at exposures above 1.2 L. This increase in binding energy for physisorbed pyridine as compared with the chemisorbed species is attributed to a reduction of relaxation effects due to molecular polarization, charge transfer, and final-state image-charge screening for the physisorbed layer. In the bottom part of the figure we present a photoemission spectrum of pyridine in the gas phase. After Demuth et al. [148].

sequence of spectra. This points to another important application of UPS in surface chemistry, namely, the identification of surface species and reaction intermediates. Rubloff and Demuth [149] investigated the adsorption and reaction of methanol on a Ni(111) surface as a function of temperature. They showed that below 200 K most of the adsorbed methanol was present in its molecular (undissociated) form as evident from the four typical methanol orbitals. Upon heating to beyond room temperature, however, two characteristic orbital features disappeared, leaving behind merely two orbitals, which could be attributed to a methoxide CH_3O species. The results of this work were confirmed later in a HREELS investigation by Demuth and Ibach [150]. A related study concerned the interaction of methanol with Pd(100), and arrived at similar conclusions [151,151a], although on Pd different and less stable forms of CH_3O species were found.

We are aware there are many more aspects of UV photoelectron spectroscopy that would deserve attention here. However, we must leave this interesting topic and turn to another.

4.2.2 X-ray Photoelectron Spectroscopy (XPS)

Because the XPS physical process is practically the same as in UPS, the description of basic physical principles and apparative methods can be kept very short. We once again refer to Fig. 4.36, showing curves obtained after excitation with photons of higher energy ($\hbar\omega \geq 1\,keV$). In order to describe the excitation process, the consideration of the electronic structure of a solid must be extended well to the proximity of the ion cores, i.e., to the inner shells with electron-binding energies of several hundred to 1000 eV. As compared to conduction and valence band structure, where several bands usually overlap, we must appreciate the fact that *atom-specific* electronic structure is retained more as a given orbital is located closer to the nucleus, i.e., the better it is screened against neighbors by outer electrons. Accordingly, the inner atomic shells are distinguished by sharp energy levels, and the respective electron-binding energy is a characteristic quantity for a given atomic species (atomic number Z). The well-known Aufbau principle (quantum numbers n and l) giving rise to K, L, M ... shells, determines the energetic position of the emitted photoelectrons, and an ESCA spectrum even shows the shell fine structure (K_α, K_β, L_α, L_β. . .) if it is measured with sufficient resolution. We present in Fig. 4.42 a typical spectrum of a stainless steel sample spanned over the appreciable energy range of 1000 eV. Clearly, excitations of Fe and Cr can be distinguished as well as carbon contaminations and the gold-binding energy reference mark. Because the relaxation of a photoionized atom can also occur via the Auger process, Auger electrons also usually appear to some extent (cf., Sect. 4.3.2) in an XP spectrum. While this kind of spectra provides a very useful overview of a solid's chemical composition at the surface (although it must be borne in mind that XPS is somewhat less surface sensitive than UPS, cf. Fig. 4.1) information about the chemical state of a given constituent atom is not so easy to obtain. For this purpose, high-resolution XPS is required. This application is based on the fact that, depending on their chemical environment and bonding status, atomic orbitals, even in inner shells, exhibit small so-called chemical shifts. An extremely high potential of XPS is analyzing such shifts, and, for example, the energetic position of Fe $2p$ or the splitting of $3s$ electronic levels differ slightly, depending on whether one deals with elementary clean iron metal or with Fe compounds [153]. These chemical shifts and fine structures of the spectra can help distinguish various valence states or even electronic environments of atoms. It is hence possible to delineate graphitic carbon from carbidic carbon or to identify carbon in organic compounds of various complexities [152]. This is compiled in

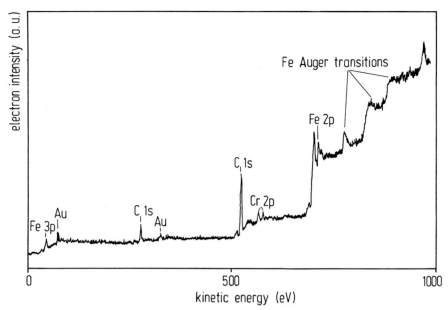

Fig. 4.42. ESCA spectrum of stainless steel after heating to 450°C in UHV as obtained after irradiation with Al K$_\alpha$ x-ray photons. The Fe and Cr levels show up as well as carbonaceous contaminations and the Au reference mark. After Ertl and Küppers [113].

Fig. 4.43a. In the field of heterogeneous catalysis, XPS is widely used to map the valence states of the atoms under consideration, which may depend on the degree of reduction in the course of a heterogeneous reaction. An example here may be taken from the work of Ertl and Thiele [154], where XP spectra of the Fe $2p_{3/2}$ were recorded for an industrial ammonia synthesis catalyst. While, with the oxidized catalyst, the Fe $2p_{3/2}$ peak appeared around 711.4 eV, it shifted with increasing reduction down to 706.9 eV (Fig. 4.43b).

In order to fully understand these chemical shifts, we must expand somewhat on the physical background of XPS (which resembles that of UPS). As pointed out before, the photoionization event occurs at inner atomic shells and leaves behind extremely reactive systems, which relax in a very short time scale. Let us first consider the evaluation of electronic-binding energies in general. The ionization process with photon energy $\hbar\omega$ obeys the energy balance

$$E_{in(N)} + \hbar\omega = E_{fin(N-1,n)} + E_{kin(A)} ,$$
4.37

where $E_{in(N)}$ represents the initial energy of the atom with N electrons *before* ionization, $E_{fin(N-1,n)}$ the final state energy where one electron (excited at level n) has been removed, and $E_{kin(A)}$ is the kinetic energy of the photoelectron detected in the analyzer A. Equation 4.37 may be rearranged to yield the original core binding energy of the emitted photoelectron:

$$E_{b(n)} = E_{fin(N-1,n)} - E_{in(N)} ,$$
4.38

which, in this case and contrary to UPS, is referenced to the vacuum level. This is why the work function does not appear in Eq. 4.38. Usually, because the final (ionic) state of the atom has a very short lifetime, there is an unavoidable line-width broadening. Evaluation of the binding energy most frequently makes use of the approximation of the

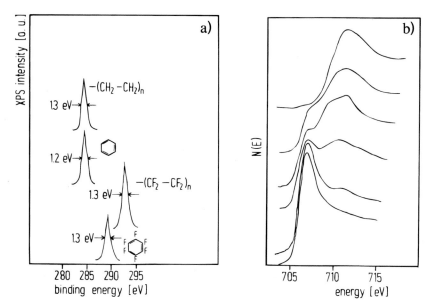

Fig. 4.43. High-resolution XP spectra showing the chemical shift for different environments: **a)** C 1s electron-binding energy as a function of the bonding situation of the C atom (hydrocarbon polymers, benzene, fluorocarbon polymers, and fluorobenzene) [152]; **b)** energy position of the Fe $2p_{3/2}$ level of an industrial ammonia catalyst at various states of reduction. Top curve: fully-oxidized, bottom curve: fully-reduced sample. After Ertl and Thiele [154].

so-called Koopmans' theorem [155], which corresponds to a one-electron description of the excitation process. It is therein assumed that the energy situation, including the spatial distribution of the ionized (N–1) system, is the same as in the initial state prior to the ejection of the electron (approximation of frozen orbitals) with the consequence that the electron-binding energy simply equals the negative orbital energy E_n of the emitted electron:

$$E_{b(n)} \approx -E_n.$$ 4.39

As in UV photoemission there occur, of course, relaxation effects ΔE_{relax} which alter the kinetic energy of the photoelectrons. Likewise, relativistic and electron correlation effects $\Delta\varepsilon_{\text{rel}}$, and $\Delta\varepsilon_{\text{corr}}$, respectively, have some influence on the exact value of E_b. Considering these terms, $E_{b(n)}$ is more precisely defined as

$$E_{b(n)} = -E_n - \Delta\varepsilon_{\text{relax}} + \Delta\varepsilon_{\text{rev}} + \Delta\varepsilon_{\text{corr}}.$$ 4.40

It may be added in brief that the intensity of an ESCA peak depends on the probability of a respective transition, which is proportional to the square of the transition matrix element and can be expressed in terms of the dipole approximation [156]. As with UV photoemission, one must consider the ESCA process as an optical transition, where electrons are excited from an occupied to an unoccupied electronic level. Accordingly, the shape of the spectra is governed by initial and final state effects. For example, the well-known spin-orbit coupling shows up in the final-state effects and leads to a splitting of the respective core-emission lines. In a fully occupied core subshell, e.g., the $2p$ orbital, the excitation can occur from two possible spin states ($m_2 = \pm 1/2$). The respective spin momentum vectors couple to the orbital angular momentum vectors in a different way,

133

giving rise to states with different total momentum vectors j. The respective energy difference clearly appears as fine structure in the XP spectra. In detail, p subshells split up into two lines corresponding to $p_{3/2}$ and $p_{1/2}$ states, d subshells into $d_{5/2}$ and $d_{3/2}$, and f subshells into $f_{7/2}$ and $f_{5/2}$ states. The degree of splitting is thereby proportional to the charge of the nucleus [157].

Finally, we make a short remark on the possibility of using XPS for quantitative surface analysis. For a given element, the line intensity usually is proportional to the number of excitable atoms and hence to its concentration, because the excitation probability of a core level is independent of the valence state of the atom. This allows exploiting XPS for quantitative determination of surface-adsorbate coverages.

Let us now return to the chemical shift problem by starting with the definition of two different binding states of a given element, $E_{b(1)}$ and $E_{b(2)}$. The chemical shift ΔE_b, again tied to the vacuum level, can then simply be expressed as:

$$\Delta E_b = E_{b(2)} - E_{b(1)} = E_{\text{kin}(2)} - E_{\text{kin}(1)} \,. \qquad 4.41$$

If one refers to $E_F = 0$, eventual work-function effects of the analyzer must be additionally considered, as well as charging effects in case insulating samples are used. We recall Fig. 4.38, which was discussed in the context of UV photoemission. One can, however, circumvent these problems by always including an internal reference standard element, e.g., the $C1s$ level, to which the orbital binding energies are referred. Taking advantage of the considerations leading to Eq. 4.40, the chemical shift can now accurately be written as:

$$\Delta E_b(n) = -\Delta E_n - \Delta(\Delta\varepsilon_{\text{relax}}) + \Delta(\Delta\varepsilon_{\text{rev}}) + \Delta(\Delta\varepsilon_{\text{corr}}) \,, \qquad 4.42$$

whereby in most cases the last two terms will be almost negligible. The problem here is unraveling the relaxation contribution, which may be split up into two parts: the *intra-atomic* relaxation represents the relaxation contribution of the individual atom, regardless of its environment; the *extra-atomic* part stems from additional screening of the core hole by the neighboring atoms of the condensed phase. For metals, the extra-atomic relaxation may be appreciable ($\Delta\varepsilon_{\text{extra}} \approx 5-10\,\text{eV}$). Many chemical shifts of core level binding energies are tabulated. For further details, [158] is recommended.

A special application of core level shift spectroscopy concerns the so-called *surface core level shifts*. In order to understand their physical origin, the *charge potential* model is helpful [159, 160]. It relates the binding energy $E_{b,i}$ of a certain core level of atom i with a reference energy $E_{b,i}^0$, its own charge q_i and the point charges located at the neighboring atoms j via (α being a proportionality factor and r_{ij} the distance between atoms i and j):

$$E_{b,i} = E_{b,i}^0 + \alpha q_i + \sum_j \frac{q_i}{r_{ij}} \,. \qquad 4.43$$

For any valence state, the last two terms have a specific value. The energy difference between the same level of a given atom being in two different valence states 1, 2 can be written ($V_i = \sum_j q_i / r_{ij}$ and $V_{i(2)} - V_{i(1)} = \Delta V_i$)

$$\Delta E_b = E_{b(2)} - E_{b(1)} = \alpha(q_{i(2)} - q_{i(1)}) + \Delta V_i \,. \qquad 4.44$$

V_i is actually closely interrelated with the Madelung potential of an ionic crystal, and Eqs. 4.43 and 4.44 can best be rationalized by assuming the photoionized level being located in the center of a hollow sphere with radius r, and by the electrostatic action of the surrounding potential inside the sphere, given everywhere by q_i / r. Hence, the binding energies of all core levels will be subjected to this field and more or less shifted in energy, whereby normal chemical shifts can reach appreciable magnitudes, up to 10 eV.

Surface core level shifts are now much smaller – up to several tenth of an eV – and the superior resolution of a synchrotron-radiation experiment is required for their observation. These shifts are solely caused by the smaller coordination of surface atoms as compared to the bulk, i.e., the potential terms of Eqs. 4.43 and 4.44 are affected. Likewise, altered periodicities of the surface, induced by relaxation and reconstruction phenomena, show up in the core level energy position. Examples for semiconductors as well as for metal surfaces have been provided among others by Eastman's group. It is even possible, by careful curve deconvolution procedures, to disentangle the XPS signals and determine quantitatively the relative concentrations of the displaced or altered surface atoms [161].

Although the essential parts of a photoelectron spectroscopy experiment have been described in the UPS section, a few remarks on ESCA equipment are nevertheless worthwhile, particularly with regard to the excitation source. Soft x-ray radiation is usually provided in the laboratory by an x-ray tube, in which an electron beam of several keV is incident on a (water-cooled) anode (anti-cathode) where Bremsstrahlung and characteristic x-ray radiation are generated, depending on the anode material. Most frequently, the Al-K$_\alpha$ ($\hbar\omega = 1486.6$ eV) and Mg-K$_\alpha$ ($\hbar\omega = 1253.6$ eV) radiation is used for XPS, whereby thin Al or Be foils transparent for the radiation separate the x-ray gun from the ESCA apparatus. In a few cases, also yttrium anode coatings are used with the Y-Mζ radiation emitted at the considerably lower energy of 132.3 eV [162,162a]. A problem encountered in this kind of excitation sources is the non-negligible line width of the x-rays, which amounts to 500 meV to 800 meV and limits the resolution of an XPS experiment, regardless of the performance of the spectrometer. In some cases, the additional use of monochromators can help improve the line width, albeit, at the cost of intensity. Outside the lab, of course, a synchrotron storage ring provides a whole spectrum of soft x-ray radiation, which can beneficially be used for XPS experiments.

Today, in XPS, as in UPS, apparatuses are generally equipped with electrostatical deflection energy analyzers; commercially, hemispherical (180°) or 127° spherical analyzers are often available. The efficiency of this type of (dispersive) analyzer can be improved if the ejected photoelectrons are preretarded. The small electron current is amplified by a channeltron and fed into an ordinary counting electronics device. As usual, the energy sweep is obtained by varying the retarding voltage between the sample and the entrance slit (in the case of preretardation) or of the electric potentials applied to the hemispherical condensor plates. Modern computer facilities have been widely exploited in XPS instrumentation; commercial spectrometers equipped with carousel sample holders allow almost automatic data acquisition for a wide variety of samples, which is of great importance for routine analyses in industrial application.

4.2.3 Inverse Photoemission Spectroscopy (IPE)

While XPS is not so often used to determine the surface electronic structure, but rather as a fingerprint technique for chemical analysis and state of oxidation or reduction of catalysts, another spectroscopy has been developed primarily by Dose and his group [163], which is almost ideally suited for complementing the information obtained by UV photoemission. As we remember, UPS detects the *occupied* electron states of a solid's valence or conduction band located below the Fermi level E_f. Bremsstrahlung isochromat spectroscopy (BIS), or inverse photoelectron spectroscopy (IPE), has the invaluable potential of probing the density of states of *unoccupied* electronic levels *above* E_f and hence allows, by combining ARUPS and angle-resolved IPE, a full mapping of the band structure of a solid. We, therefore, conclude this chapter about surface-electronic structure by some (short) remarks on IPE.

This technique is based on the well-known fact that a surface bombarded by electrons can emit photons within a certain spectral range, depending on the kinetic energy of the impact electrons (the so-called Bremsstrahlung). The emitted photons carry information about the density of states (DOS) above E_f [164–166]. This method has been adopted for analysis of surface-electronic structure, and more details are communicated in various review articles [163, 167, 168].

The physical mechanism of IPE can be understood based on the idea that the inverse photoemission is the *time reversed process* of ordinary photoelectron emission. Pendry [169, 170] laid the theoretical basis for this view. In principle, electron waves incident on a surface are de-excited and the corresponding energy is released in the form of photons. In close analogy to the three-step-model developed for UV photoemission by Berglund and Spicer [140] (consisting of i) photoionization, ii) transport of the photoelectron to the surface, and iii) the escape of the electron from the surface), a similar model was proposed by Dose [163]. The three steps are: i) capture and radiative decay of an incident electron with energy E in an unoccupied state at $E - \hbar\omega$, ii) transport of the created photon to the surface, and iii) escape of the photon from the surface to the detector.

Therefore, in order to perform an IPE experiment, essentially three ingredients are necessary: there must be an electron source (hot cathode) in the first place; a sample that emits the photons; and, as a more delicate part, a photon detector. Before we describe a typical experimental set-up, we present, in Fig. 4.44, the energy-level diagram pertinent to IPE.

The threshold for the emitted radiation of energy $\hbar\omega_0$ is given by the expression

$$\hbar\omega_0 = e_0\Phi_c + \tfrac{3}{2}kT + e_0 V_{\mathrm{acc}} , \qquad\qquad 4.45$$

where Φ_c represents the work function of the cathode, T the filament temperature, and V_{acc} the accelerating potential between sample and cathode. According to Fig. 4.44 the electrons incident on the sample have the maximum energy $E = e_0 V_{\mathrm{acc}} + e_0\Phi_c + {}^3/_2\,kT$,

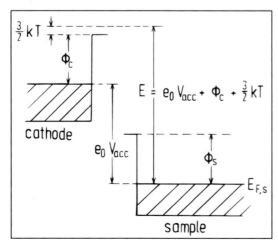

Fig. 4.44. Potential energy diagram for inverse photoemission IPE. The energy difference between cathode and sample is composed of the thermal energy part of the cathode electrons plus cathode work function Φ_c plus acceleration voltage V_{ac}. After Dose [163].

which is referred to the Fermi level of the sample. The work function of the sample Φ_s does not enter the balance equation Eq. 4.45!

Returning to the description of the IPE experiment, it is in principle required to detect the emitted photons with energy resolution. Because the photon yields are generally small, there is a problem of obtaining sufficient signal-to-noise ratio if the radiation is made passing a monochromator [171]. A more convenient way was proposed and developed by Dose [172]. He explicitly abandons analyzing the spectral distribution of the emitted photons, but instead uses a simple photon detector based on the principle of a Geiger-Müller counter, which consists of a tube filled with a gas mixture of He and iodine vapor and with a CaF_2 or SrF_2 entrance window for the radiation. This counter represents kind of a band-pass filter, limited on the low-energy side by the photodissociation threshold of the iodine vapor (~9.23 eV), and on the high-energy side by the transparency of the window, which has a cut-off around 10 eV. Accordingly, the pass energy is around $\hbar\omega = 9.4$ to 9.6 eV. Using a hot tungsten filament (2200 K) as a primary energy source, there is some thermal broadening to approximately 250 meV, at a primary current of 1 to 10 μA. The accelerating voltage is swept and the pulses of the counter detector are measured. A typical experimental set-up as proposed by Dose [163], is reproduced in Fig. 4.45.

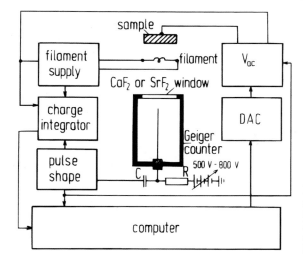

Fig. 4.45. Schematic set-up of a Bremsstrahlung isochromat spectrometer employing the energy-selective Geiger-Müller counter. DAC = digital-analogue converter; V_{ac} accelerating voltage. After Dose [163].

With IPE, clean surfaces as well as adsorbate-covered systems can be analyzed. Useful band-structure information is, however, only obtained if the experiment is performed in an angle-resolved mode. Examples of band-structure determinations can be found, among others, in [173]. In view of surface chemistry, however, the detection of unoccupied electronic states of surface complexes are more important. This renders a full comparison with quantum-chemical calculations of the adsorbate-surface interaction possible, including the location of empty levels. Various examples, which were mainly obtained by Dose's group [163], could be listed here. Among others, measurements of the O/Ni(100) system's unoccupied states [174] made clear the onset of oxidation showed up in the DOS above E_f. Also, the hydrogen interaction with Ni(110), which leads to surface reconstruction at a certain threshold coverage, was successfully investigated by Rangelov et al. [175].

137

4.3 Surface Chemical Composition

Probably the greatest interest of surface chemists and physicists concerns the exploration of the kind and number of atoms present at the surface or in the surface region. We have already learned that there are a variety of predominantly structure-sensitive spectroscopies, which also bear valuable additional chemical information. Because we shall not return to these methods again, as a reminder we will repeat their names. HREELS analyzes surface vibrations, which are often characteristic of certain chemical compounds. UV photoemission can detect characteristic emissions arising from excitation of specific adsorbate orbitals. In some cases, even complex organic adsorbates can be identified when guided by an analysis of their additional photoemission levels. Most important as a chemical probe seems to be XPS, as we have pointed out before, because each element has characteristic core orbital energies and leaves behind its finger print in an ESCA spectrum.

There are, however, several experimental techniques, whose power lies almost exclusively in the field of chemical-surface analysis. We consider Auger electron spectroscopy (AES) and secondary ion-mass spectrometry to be among these tools. Hence, we devote our attention in this chapter to these two methods.

4.3.1 Auger Electron Spectroscopy (AES)

The experimental appearance of Auger electrons was for the first time reported in 1925 when Pierre Auger irradiated photoplates with x-ray radiation [176]. However, 30 years passed before the Auger electrons regained interest [177], and it was not until 1968 that Auger electron spectroscopy was discovered as a surface-analysis tool. In that year, reports by Harris [178, 178a, 179] using a 127° electron analyzer appeared, and independently by Weber and Peria [180] and Palmberg and Rhodin [181] utilizing conventional LEED optics in conjunction with electronic-differentiation techniques as a retarding field analyzer. Another milestone was the use of a cylindrical-mirror analyzer (CMA) in AES by Palmberg [182], resulting in a higher recording speed and an improvement of the signal-to-noise ratio.

Today, Auger electron spectroscopy has developed to one of the most useful analytical tools providing access to surface-chemical composition. Despite its great importance, we shall deal with AES here only briefly; the interested reader is referred to the vast amount of literature devoted to AES [8–10, 183–190]. In the following discussion, we will first provide a concise introduction into the basic physics of the Auger process, describe a typical Auger electron spectrometer, and present of some selected examples of Auger spectra to elucidate various aspects of AES application.

If a solid is irradiated with a beam of electrons of medium or higher energy ($1 \, \mathrm{keV} < E_p < 10 \, \mathrm{keV}$), outer and inner electronic shells of the atoms become ionized and relaxation can occur in two different ways, which are illustrated in Fig. 4.46:

i) The core hole becomes filled by an energetically higher electron of the same atom, and the resulting energy is emitted via electromagnetic (x-ray) radiation, according to $\Delta E = h\nu$.

ii) As before, the core hole is filled by an outer electron, but the energy equivalent is, in a radiationless manner, transferred to a second electron of the atom, which is ejected and leaves the atom with a characteristic kinetic energy E_{kin}.

One can understand this as an internal photoelectric effect – the originally formed x-ray photon (process i) is absorbed within the atom; its energy is then used to emit a photoelec-

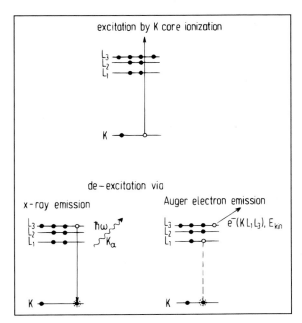

excitation by K core ionization

L_3
L_2
L_1

K

de−excitation via

x−ray emission

L_3
L_2
L_1

$\hbar\omega$
K_α

K

Auger electron emission

L_3
L_2
L_1

$e^-(KL_1L_3), E_{kin}$

K

Fig. 4.46. Energy-level diagram illustrating the two possible filling mechanisms of a K-shell core hole generated, for example, by electron-impact ionization. On the left, an x-ray photon is emitted; on the right, the radiationless Auger process occurs, giving rise to emission of an electron (the Auger electron) with well-defined and characteristic kinetic energy (Eq. 4.46).

tron, namely, the Auger electron (process ii). However, this naive view is not quite correct. The Auger process is actually radiationless, as an inspection of the selection rules for optical processes i) and for electronic transitions ii) demonstrate. In the Auger transition, electrostatic forces caused by the interaction of the surrounding electron cloud with the core hole dominate. The most prominent property of an Auger electron is its kinetic energy, which is characteristic for a given atom. This can be clarified by explaining the three processes that occur. The first process is the ionization of an inner shell, say, a K shell ($1s$ electron). The second process is the internal transition of an outer electron, for example, an L_2 electron ($2p_{1/2}$), to the K shell to fill the hole. Apparently, a second electronic level becomes involved here. The third process is the energy transfer to a third electron (the Auger electron), often of the same shell (L_2 or $L_3 = 2p_{1/2}, 2p_{3/2}$), but of course, also from an outer (M) shell. No matter from where the Auger electron is emitted, three electronic states participate in the process, and with relaxation phenomena neglected, the kinetic energy of the Auger electron can be written

$$E_{kin} = E_1 - E_2 - E_3 , \qquad\qquad 4.46$$

where E_1 denotes the binding energy of the initial core electron prior to ionization, E_2 that of the electron that fills the core hole, and E_3 the binding energy of the ejected electron. The essence of using Auger electron spectroscopy as an element-specific analytical tool is that in each case the emitted electron carries a *characteristic energy*, which arises from the combination of energetically well-defined atomic levels unique for a given atom. In this respect, AES highly resembles the XPS method, which is also (even more directly) a core level spectroscopy.

For atoms with many electron states (high Z atoms), there are many different Auger transitions possible. According to the above process, Auger transitions are assigned by capital letters denoting the shells, whereby sub-figures indicate the participating subshells. The sequence of these capitals is chosen according to Eq. 4.46. A KL_2L_3 transition then means that an electron hole produced in the K shell is filled by an L_2 electron, and the energy surplus is transferred to kick out an L_3 Auger electron. Some other possible transitions within the K and L shells are $K\,L_1L_1$, $K\,L_1L_2$, $K\,L_1L_3$, $K\,L_2L_2$, and $K\,L_3L_3$. Likewise, KLM, LMM, MNN, and NOO transitions can occur and lead to a wealth of Auger emission features for high Z elements. If the Auger process is subjected to a detailed consideration, the so-called Coster-Kronig transitions [191], among others, must be mentioned. These are of the L_2L_3M type and involve the energy transfer within the same subshell having the same principal quantum number. Such processes are extremely fast and lead to significant lifetime broadening effects. Another experimentally important fact is that the primary step, viz., the ionization by electron impact, is also very fast ($t < 10^{-16}$ s) as compared to the lifetime of the core hole ($t \geq 10^{-15}$ s). Hence, the energetic width of the ionizing electron beam does not affect the line widths of the Auger transitions. By the way, the primary ionization can also be provided by any other ionizing radiation, for example, x-rays; however, with reduced efficiency.

The exact correlation between the kinetic energy of an Auger electron and the electron-binding energies of an atom is actually more complicated than the simple Eq. 4.46 predicts. The main reason is that E_1 and E_2 refer to the neutral state of the atom, whereas the level E_3 corresponds to a slightly increased energy compared with the neutral state, because the E_3 electron moves in an orbital of increased positive charge, which in turn is caused by the missing electron of energy E_1. For detailed considerations of this effect, see [192]. For practical application of Auger electron spectroscopy, particularly for quantitative use, it is essential to regard the excitation probability of a certain Auger transition, which depends on several parameters, including the atomic number of the involved atom. The interesting quantity here is the so-called Auger yield. According to Fig. 4.46, there are two competing processes (Auger electron and x-ray quanta emission). If we denote the probability of ejecting an Auger electron as P_A, $1 - P_A$ remains for the probability P_x of x-ray emission. Then the Auger yield Y_A is

$$Y_A = P_A/(P_A + P_x) \, . \tag{4.47}$$

Transitions involving the K shell (KLL and K_α, K_β, ..., respectively) lead to x-ray emission proportional to Z^4, based on dipole interaction, while a calculation of the Auger transition probability should consider the electrostatic interaction between the electrons involved in the process. For hydrogen-like wave functions (a reasonable approximation for inner-shell electrons), it can be shown [157] that P_A is independent from the number of the nuclear charge, and one has:

$$Y_A = \frac{1}{1 + \beta Z^4} \, , \tag{4.48}$$

and the x-ray emission yield is

$$Y_x = \frac{\beta Z^4}{1 + \beta Z^4} \, , \tag{4.49}$$

where β is a parameter and must be fitted according to experimental data. Burhop [193] derived a semi-empirical function:

$$Y_A(Z) = \left\{ 1 + (3.4 \cdot 10^{-2}Z - 6.4 \cdot 10^{-2} - 1.03 \cdot 10^{-6}Z^3)^4 \right\}^{-1} \tag{4.50}$$

which is the physical basis for Fig. 4.47 [159].

140

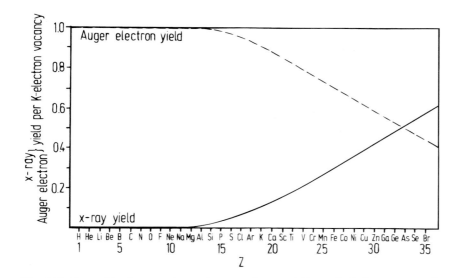

Fig. 4.47. Competition between Auger electron and x-ray yields per K-shell vacancy as a function of the atomic number Z. After Siegbahn et al. [159].

Obviously, up to the element with $Z = 20$ (calcium), almost 90% of the emission after K shell excitation is due to the Auger process. Only for germanium ($Z = 32$) are Auger and x-ray yield comparable; for higher Z-elements the x-ray deexcitation clearly predominates. (This is a lucky circumstance, which enables surface spectrocopists to monitor abundant non-metallic impurity elements, such as carbon or sulfur, with preferential sensitivity!)

When discussing the cross section Q for Auger emission, another important factor concerns the dependence of Q on the primary energy E_p of the ionizing electron beam. It was shown theoretically [194] that this cross section is roughly proportional to the product $3 \cdot E_1$, when E_1 refers to the core-electron energy. Q decreases strongly for lower, but only slightly for higher E_p values. Because most of the analytically relevant Auger lines appear between 50 eV and 1000 eV, this means that $3 \, \text{keV} \leq E_p \leq 5 \, \text{keV}$ is an appropriate choice.

We have seen so far that there are many different kinds of possible Auger excitations, all of which involve more or less discrete electronic states. This actually holds for dilute, i.e., gaseous material consisting of free atoms or molecules. However, condensed matter requires a separate treatment, because of the electronic bands that are formed in the outer electronic shells. In these valence or conduction bands, electronic states are delocalized and have a finite energy width. This situation gives rise to the so-called LVV Auger transitions, which are depicted in Fig. 4.48 for silicon, according to Chang [183]. With solid Si, the M electron states form a valence band of about 10 eV in width, and this represents an electron reservoir which allows filling the L core vacancy and emitting the Auger electron into the vacuum. It is apparent that the band structure will have an effect on the lineshape of the Auger emission peak. Furthermore, there is always a work-function contribution for condensed matter, which has to be surmounted by the ejected electron. Because the Auger electrons are detected and sampled with respect to their kinetic energy in an energy analyzer with a work function Φ_A, we face similar problems as in photoemission (cf., Fig. 4.38), and Φ_A must be taken into account in practical energy measurements.

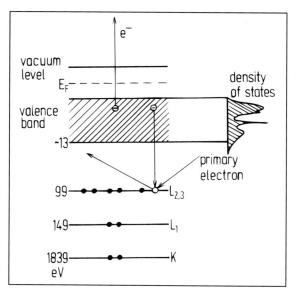

Fig. 4.48. Example for an LVV Auger transition as illustrated by means of the electronic structure of solid silicon. The primary impact electron creates a hole in the $L_{2,3}$ shell, which is filled by an electron from the valence band. The energy is then transferred to a second electron from the valence band with a probability that depends on the electron density of Si states. After Chang [183].

However, this is only important in quantitative AES when the exact energy position of an Auger line of a solid is to be monitored. As will be shown below, some chemical and even structural information similar to XPS is contained in these energy positions. These effects are discussed in great detail by Weissmann and Müller [188]. In most of the Auger applications, however, the absolute-peak position on the energy scale is not explicitly examined; rather the Auger spectrum as a whole and the relative-peak positions are considered as a fingerprint of the surface region's chemical composition. Here, of course, the *intensity* of a given Auger line deserves most attention, because – as in XPS – there is a direct proportionality between the number of excited atoms and the Auger intensity. Unfortunately, the actual situation in quantitative determination of elements present in the surface is rather more complicated, because Auger electrons also can be excited in deeper layers of the solid. According to Fig. 4.1, at least three, sometimes up to five, atomic layers can contribute to an Auger electron spectrum, and any attempt to perform quantitative surface analysis must take into account the sampling depth of an Auger experiment. This being one of the most important properties of AES, investigators tried to obtain information on the detected volume soon after AES was introduced as a surface analytic tool. Palmberg and Rhodin [181] deposited silver on Au surfaces and recorded Auger spectra as a function of Ag surface deposition. They found that AES was sensitive to amounts of ~10% of a complete monolayer and that deposition of more than five monolayers made Auger emission of the substrate practically disappear. Within a homogeneous attenuation model, the Auger electron current of the deposit $i_{A,dep}$ which may be thought of as a homogeneous metal layer of thickness d, should exhibit a simple exponential increase (λ_0 being the escape depth of Auger electrons and $i_{\infty,dep}$ the signal of the pure deposit) according to:

$$i_{A,dep} = i_{\infty,dep} \left(1 - e^{-d/(\lambda_0 \cos \alpha)}\right) .$$

4.51

142

The decay of the substrate Auger signal $i_{A,s}$ should obey the relation

$$i_{A,s} = i_{A,s,\infty} \cdot e^{-d/(\lambda_0 \cos \alpha)} , \qquad\qquad 4.52$$

where $\cos \alpha$ is a correction term due to the electron analyzer used in the experiment ($\alpha =$ emission angle) [195].

In a solid, however, not only the primary impact electrons can and will excite Auger transitions, but also all the scattered electrons (elastically *and* inelastically scattered ones), and hence also Auger electrons with a kinetic energy larger than the ionization potential of a certain shell. These effects comprise the backscattering process; consequently, there is an increase in the excitation cross section leading to an enhancement (up to 25%) of the Auger signals [196, 197]. Gallon, therefore, introduced a layered model [198] and proposed a relation, whereby I_n, the intensity originating from a slab of thickness n (monolayers), is related to the bulk emission intensity I_∞ and that of a single monolayer I_1 via

$$I_n = I_\infty \left\{ 1 - \left(1 - \frac{I_1}{I_\infty} \right)^n \right\} . \qquad\qquad 4.53$$

Using formulae of this kind, Auger electron-escape depths were determined for Ag layers on an Au substrate on the order of 4 Å for Auger electrons of 72 eV and of 8 Å for electrons of 365 eV energy [181].

Before we present some actual experimental examples for Auger electron spectra, a few words on the AES instrumentation are appropriate. From the foregoing, it has become apparent that there are two essential ingredients (as in XPS). In the simplest case, the excitation source typically consists of an electron gun with up to 10 keV adjustable-beam energy and currents on the order of microamps. In a few cases, an x-ray tube also can be utilized here along with an electron analyzer/detector, which provides the energy separation of the emitted electrons. Note that AES, in contrast to LEED, probes *inelastically* scattered electrons, whereby (and this discriminates AES from HREELS) the inelastic processes involve energies up to 1000 eV and more, with a typical energy resolution of 1 eV to 2 eV. A surface irradiated with medium-energy electrons gives the energy distribution of backscattered electrons displayed in Fig. 4.49. At very low kinetic energies, the true secondary electrons dominate. At the high-energy end, there is a sharp (asymmetric) maximum due to elastically reflected electrons. In the intermediate range, there is a smooth, but strongly varying background with small wiggles, and these wiggles actually represent the Auger electrons. It would be possible, of course, to amplify the electron current by several orders of magnitude to obtain larger signals, but the steep background would cause problems. A much more elegant way to separate the Auger electrons is an electronic-differentiation modulation technique used in conjunction with a lock-in amplifier. For this purpose, a small alternating voltage is superimposed on the measured detector current. Details of this method have been communicated first by Leder and Simpson [199] and lead to an experimental set-up, which is shown schematically in Fig. 4.50. A standard 4-grid LEED optics device can be chosen as a retarding field-energy analyzer (Fig. 4.50a) with a (negative) sweep potential applied to the second and third grid [200] and the LEED screen acting as an electron collector. The function principle corresponds to that of a high-pass filter. Even with electronic-differentiation techniques [199], where the second derivative of the collector-electron current (corresponding to the first derivative of the energy distribution curve $dN(E)/dE$) is actually measured, only a relatively poor signal-to-noise ratio is achieved, because all electrons with energies higher than a

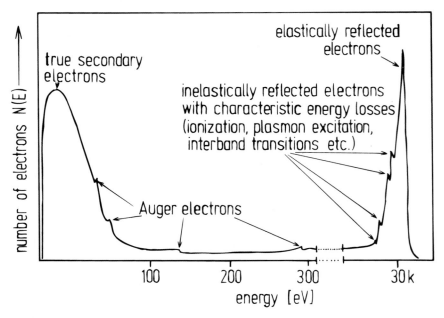

Fig. 4.49. Typical electron-energy distribution curve ($N(E)$ vs E) obtained after electron impact with 30 keV primary electrons. The sharp peak on the right indicates the elastically scattered primary electrons, with small inelastic contributions on the low-energy side, owing to so-called characteristic energy losses (mainly plasmon excitations), whose energy positions depend on and shift with the primary beam energy. In the medium-energy range the Auger excitations are visible as small bumps superimposed on the smooth background. The low-energy region (left part of the spectrum) is entirely dominated by the true secondary electrons, which have lost most of their initial energy by the cascade-excitation processes.

momentarily adjusted value are collected and contribute to the noise level. Much more convenient is the use of a cylindrical mirror analyzer (CMA), whose operation is comparable to a band-pass filter [182] (Fig. 4.50b).

The Auger electron gun required for excitation emits a focused beam of medium-energy electrons incident on the sample with an energy width of $0.5-1\,\text{eV}$ and a diameter of ca. $0.5\,\text{mm}^2$. Usually, the sample is mounted on a manipulator (as in a LEED experiment) and can be rotated with respect to the electron beam so that a whole range of impact angles from grazing to normal incidence can be adjusted. Thereby, a flat angle between electron beam and sample assures a particularly high surface sensitivity [201, 202], because surface atoms are struck preferentially. Often mounted on commercial Auger spectrometers containing a CMA as an energy-dispersive element are integral electron guns, which are suited only for perpendicular electron impact. The spatial distribution of the ejected secondary electrons usually varies with the cosine of the polar angle (cosine distribution), and in order to obtain Auger spectra with best performance the focal point of the electron-energy analyzer and the point of electron impact must be carefully aligned. Sometimes, significant deviations from a cosine distribution can occur, and one also can use AES as a structure-sensitive tool [203]. Experimental data regarding Auger intensities as a function of polar and azimuthal angle for fixed electron-beam incidence show fine structure correlated with the orbital symmetry of the surface atoms. Furthermore, diffraction of Auger electrons plays a role [204, 205], and can, in some cases, significantly influence the shape of Auger electron spectra.

144

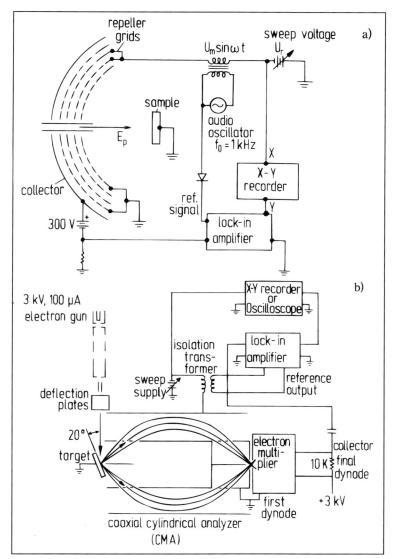

Fig. 4.50. Experimental set-up for Auger electron spectroscopy by means of electron-impact excitation: **a)** employment of a retarding-field analyzer (RFA), often realized by a conventional 4-grid LEED optics in conjunction with lock-in technique to obtain the differentiated electron-energy distribution d$N(E)$/dE; **b)** use of a cylindrical mirror analyzer as an energy-dispersive element.

A crucial property of the excitation-electron source is the spatial width of the electron beam or the beam diameter, which determines largely the lateral resolution of the AES experiment. In many technological applications, there is need to analyze lateral concentration profiles, for example, in semiconductor fabrication. Here, a resolution in the μm range is often required, along with beam scanning over the sample. The use of lanthanum hexaboride (LaB$_6$) cathodes or even field-emission electron guns has largely improved the brilliance and lateral beam-divergence properties.

145

During recent years, various experimental developments and improvements have been made in automatizing Auger analyses. By scanning AES it is possible to form an image of the surface using Auger electrons of a particular energy and hence of a selected chemical element, whereby lateral chemical distributions can be conveniently monitored. Even Auger microscopes have been developed and are progressively utilized for industrial surface analysis [10].

So far, Auger electron spectroscopy could appear as an almost ideal instrument for probing elemental surface concentrations. However, one must be aware that a finely focused electron beam of several keV energy and current densities of up to $50\,\mu A$ per mm^2 ($\triangleq 5\,mA/cm^2$) can and often will induce severe damage effects on the illuminated area of the surface, especially on insulating or semiconducting crystals with small heat conductivity. As a consequence of the local heating effect, there may occur melting, desorption, decomposition, and hence depletion or segregation effects particularly with adsorbed layers. These conditions prevent AES from analyzing the original surface composition. In many cases, it will help to reduce the electron-beam current density by minimizing the emission current and/or defocusing the beam. However, this is done at the expense of lateral resolution and signal-to-noise ratio. At any rate, for these reasons, AES can be reckoned among the *destructive* surface-analysis techniques, in contrast to photoelectron spectroscopy.

Finally, we shall present some examples for typical applications of Auger electron spectroscopy.

As emphasized above, AES serves, in almost any surface laboratory, as a fingerprint technique to identify chemical elements in the surface region. A typical application here is the control of surface cleanliness in catalytic model studies using metal single crystals. As an example, we display in Fig. 4.51 an Auger spectrum of a nickel (111) crystal (measured by the retarding field analyzer) directly after mounting in vacuo (where large concentrations of impurity elements (C, S) are visible) and after final cleaning [206]. The sensitivity of Auger electron spectroscopy has meanwhile been improved to an extent that (depending on the element) about 1 % of a monolayer can be detected.

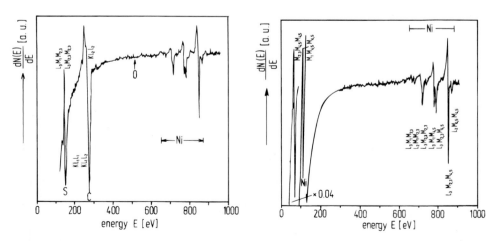

Fig. 4.51. Examples for use of Auger electron spectroscopy in surface analysis: **a)** Ni(111) surface heavily contaminated with sulfur and carbon impurities; **b)** same surface after cleaning by prolonged argon-ion sputtering. Respective Auger transitions are indicated. After Christmann and Schober [206].

In order to identify a certain element, the respective Auger transition lines must be correctly assigned. To facilitate this task, a huge data body of Auger spectra exists for practically all chemical elements (except, of course, H and He, where for physical reasons Auger transitions cannot occur). We mention here the collection of Auger spectra in the Auger Handbook, in which, beginning with lithium ($Z = 3$) and ending with uranium ($Z = 92$) Auger spectra of each chemical element are reproduced [207].

In some cases, AES can also be exploited for quantitative analysis of adsorbate concentrations if certain precautions are taken (low-beam densities to avoid thermal or electron-induced desorption). In this respect, for instance, xenon-surface concentrations on Pd(100) were determined by Palmberg [208]. Other applications are determining the growth mode of a given deposit material, whereby basically two mechanisms are possible – small three-dimensional crystallites on a complete first monolayer (Stranski-Krastanov growth) or strictly layer-by-layer growth (Frank-van-der Merwe growth). Particularly the layer-by-layer growth mode can be identified by characteristic breaks in a plot of the deposit Auger signal vs the overall deposited amount, which occur whenever a monolayer is completed. This has been exploited in many cases, where metal or carbon vapors were condensed onto metallic substrates [210].

An example, in Fig. 4.52, is presented for silver growth on a Cu(111) surface, taken from Bauer's work [211]. Furthermore, chemical surface reactions, in particular decomposition reactions in which split-off particles desorb in a certain temperature range, can be followed by AES. Glycine H_2N-CH_2-COOH adsorbed on Pt(111) at low temperatures, for example, decomposes at elevated temperatures into nitrogen- and oxygen-containing fragments. The disappearance of the O Auger signal and the persistence of the N signal around 450 K indicates the desorptive removal of a formic-acid-like fragment from the surface [209].

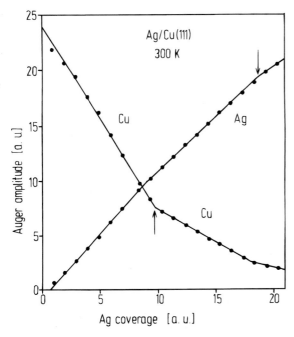

Fig. 4.52. Auger-electron spectroscopy as a monitor of thin-film growth. In the experiment by Bauer [211], silver was deposited onto a Cu(111) surface by vacuum evaporation. At 300 K, layer-by-layer growth (Frank-van der Merwe mechanism) was observed as indicated by the breaks in the slope of the declining Cu and increasing Ag Auger transition intensities, respectively, which are marked by arrows.

A very significant application of AES is in the field of determining alloy-surface composition. Here, vertical concentration gradients need to be detected as they arise from segregation, particularly, if the catalytic activity of a chemically modified surface is to be correlated with the number of atoms of the additive in the surface. A vast number of studies has been and still is being devoted to this issue, and we can only refer to some selected publications [202, 212–214] in this field.

Closely connected with determining alloy composition is a field known as depth profiling, which is of great practical importance in thin-film and semiconductor technology. Here, perpendicular surface concentration gradients in materials are analyzed. The procedure is actually relatively simple – the material is gradually removed (preferentially in a layer-by layer mode) by ion sputtering, and Auger electron spectra are simultaneously recorded. More details can be found in a recent communication by Hofmann [215].

The line shape and line widths of Auger transition peaks is also a source of rich chemical and physical information. Here, the so-called high resolution AES, in which small modulation voltages lead to energy resolutions of less than 0.5–1 eV, must be performed. Similar to XPS, the line shape characteristic of those transitions involving outer electronic shells react sensitively to the chemical bonding state and the environment of an atom and hence indicate the chemical nature of the excited atom. Thus, in a certain respect, AES can at least for low Z elements also probe valence states. Striking examples here are the silicon LVV transition around 93 eV or the carbon KLL transition around 275 eV. Depending on whether elemental silicon, silicon monoxide, or dioxide are present, characteristic line-shape and intensity changes of the Si LVV peak group occur [216], and graphitic or carbidic carbon can well be distinguished by the C KLL group peak shape [217]. Quite generally, oxidation processes, for example, lead to a red shift of the Auger transition; the naive, but straightforward explanation is that the electronegative O atom withdraws charge from the adjacent atoms.

All in all, and this has made the foregoing discussion perhaps a little lengthy, AES represents one of the most important methods for surface analysis. We regret that we could only touch on some of its applications here.

4.3.2 Secondary-Ion-Mass-Spectroscopy (SIMS)

Besides AES which has found its way into practically all the surface analysis laboratories even in the chemical industry, there is a second method, secondary ion mass spectroscopy (SIMS) [218], which has become quite popular particularly for routinely analyzing the chemical composition of solid materials. It is based on the mass spectrometric detection and identification of particles formed when the solid surface is bombarded with medium-energy ions. The following discourse is based on several articles by Benninghoven and coworkers [219, 220], who first realized the power of secondary ion emission for surface analysis. In contrast to AES, the SIMS method has the additional potential to identify chemical *compounds* present in the surface and obtain information about the local configuration of adsorption complexes [221, 222]. Furthermore, *all* elements are detected (including hydrogen and helium), and an extremely high sensitivity (0.1% of a monolayer) can be achieved. Also, surface reactions can be followed by means of SIMS [219]. For further details of the method and its applications in surface science, we refer to the literature [218–224].

The experimental equipment for secondary-ion mass spectrometry is actually very simple. One needs an ion gun, which produces noble gas ions of ca. 3 keV energy (the so-called primary ions) – often argon being used –, the sample target to be investigated,

148

which is exposed to the primary ion beam, and a mass spectrometer with sufficient resolution and mass range. Depending on the current density of the primary ions (which can be varied within wide limits) one distinguishes static (SSIMS) and dynamic SIMS. In the first case, the current density is maintained around 10^{-10} A/cm^2, which ensures that it takes several hours until a complete monolayer of material is removed by sputtering. Under these static conditions, it is therefore guaranteed that the incident ion really probes the original composition of the surface, in contrast to dynamic SIMS, where much higher primary ion-current densities of typically 10^{-2} A/cm^2 provide a rapid removal of material (10^{-2} s lifetime per layer). This mode plays an important role in etching processes and depth profiling; it is less suited for surface analysis in the strictest sense. Therefore, our following considerations will primarily concern static SIMS. As with the descriptions of the other methods, we will present the basic physical principles first; then some examples are chosen to illustrate the advantages of SIMS.

Ion impact on a solid surface is a relatively complicated overall process. It may be viewed as a kind of a local earthquake induced at and near the collision point in the surface. For simplicity, it can be subdivided into several single processes. The incident primary ion collides with the surface and accommodates, if it is not reflected back into the gas phase. The kinetic energy of the ion is dissipated in the solid-surface region, whereby several mechanisms – chemical bonding effects and, most importantly, transfer of momentum to particles at the surface via phonon coupling to the lattice – can play a role. If a critical energy is thus accumulated, for example, in the adsorbate bonding, the adparticle can leave the surface as a neutral particle or in the ionic state. Secondary ions can, in principle, carry positive or negative charge(s). This is the decisive process in SIMS, and we shall subject it to closer consideration. Secondary-ion emission consists of i) momentum transfer of the impact ion to the surface, ii) ionization, and iii) emission of the secondary ion into the vacuum. Particularly high ion yields are obtained when chemical bonds are dissociated by impact processes. Following Benninghoven [219], one can write for the secondary-ion current of a certain species, $I_{s,i}$

$$I_{s,i} = \varkappa \cdot I_p \cdot \gamma_i \cdot \beta_i \cdot \Theta_i \,, \qquad\qquad 4.54$$

where $\varkappa$ = transmission factor of the analyzer, β_i = ionization probability of species i, Θ_i = (relative) coverage of surface with the component leading to emission of species i, I_p = primary ion current, and γ = average sputter rate of a solid i. The surface coverage Θ_i of a certain species at time t depends exponentially on the lifetime $\hat{t}$ of a monolayer according to

$$\Theta_{i(t)} = \Theta_{i,0} \cdot \exp\left(-\frac{t}{\hat{t}}\right) \qquad\qquad 4.55$$

with

$$\hat{t} = \frac{\vartheta_0 e_0}{\gamma \cdot j} \,, \qquad\qquad 4.56$$

where ϑ_0 stands for the particle density within the complete monolayer and j for the current density of the primary ions. In turn, j is connected with the primary-ion current I_p and the impact area A via

$$j = \frac{I_p}{A} \, . \qquad\qquad\qquad\qquad\qquad\qquad\qquad\qquad\qquad\qquad\qquad\qquad 4.57$$

Although there are certain prerequisites for the validity of Eq. 4.56, i.e., that molecular ion emission not be included, the quantity of interest $\hat{\imath}$ can be relatively accurately predicted.

In chemical-analysis application, it must be born in mind that the cross sections for sputtering and secondary-ion formation, respectively, depend strongly on the kind and nature of the chemical element. For quantitative analyses these cross sections must be known.

In the following discussion, we present an example of a typical secondary-ion mass spectrum of a molybdenum surface (Fig. 4.53), monitored at a pressure of 10^{-9} mbar [219]. Before mounting in vacuo, the Mo surface had been cleaned using trichloro-ethylene and distilled water. It is well known that Mo targets subjected to this procedure are covered with several monolayers of surface oxide and other contaminants, and the corresponding ion fragments along with Mo^+ and MoO^+ ions appear in the positive SIM spectrum (Fig. 4.53a). Likewise, the negative ion spectrum (Fig. 4.53b) also can be recorded to provide additional evidence of the presence of anions such as nitrate (NO_3^-) hydroxide (OH^-), or molybdate (MoO_4^-). It is remarkable that merely 1% of a monolayer was removed during the measurement of the spectra. The further procedure is straightforward – the Mo surface can be heated in vacuo to remove the volatile surface contaminants and then heated in an oxygen atmosphere to oxidize and vaporize the carbonaceous contaminants. During these cleaning cycles, SIM spectra can be taken continuously to control the success of the cleaning procedure.

Although we are concerned with chemical analysis in this paragraph, we will briefly comment on an aspect of SIMS, which allows conclusions to be made about possible dissociation reactions, structure, or local chemisorption geometry. This application of static SIMS has been pursued by Vickerman [221, 222] and may be illustrated by means of carbon monoxide adsorption on various metal surfaces. On Cu, Pd, Ni, and Ru, CO adsorbs molecularly below 300 K, which is reflected by the molecular ion yield, i.e., the formation of Me_xCO^+ ions (as compared to the Me_xC^+ and Me_xO^+ ions, which would arise if CO would dissociate). Furthermore, the relative abundance of metal-CO-cluster ions yields valuable information about the local CO coordination. A terminal metal-CO bond is suggested if $MeCO^+$ ions are preferentially detected, whereas the dominance of Me_2CO^+ ions (Pd(100)!) indicates a two-fold (bridge) coordination. The mixed appearance of Me_3CO^+, Me_2CO^+, and Me_3CO^+ cluster ions is compatible with the simultaneous occupation of linear and higher coordinated CO adsorption sites. In his article [222], Vickerman convincingly showed that the respective information derived from SIM spectra agrees completely with the results of other methods that probe the local coordination of CO (HREELS, LEED).

Returning to the surface-analysis application of SIMS, we want to emphasize once again the advantages and disadvantages of this method, which is, by the way, widely used in industrial laboratories. SIMS is extremely sensitive and by no means restricted to single-crystal surfaces. Rather, all kinds of material, such as real catalysts and complicated solid organic compounds, including amino acids or proteins, can be probed [225, 226]. A limiting factor here is only the mass spectral range and the resolution of the mass filter detector. Even compounds can well be analyzed, whereby the option of whether positive or negative ions are recorded represents an additional advantage. One must, however, keep in mind that even in the static mode, SIMS is a destructive method,

150

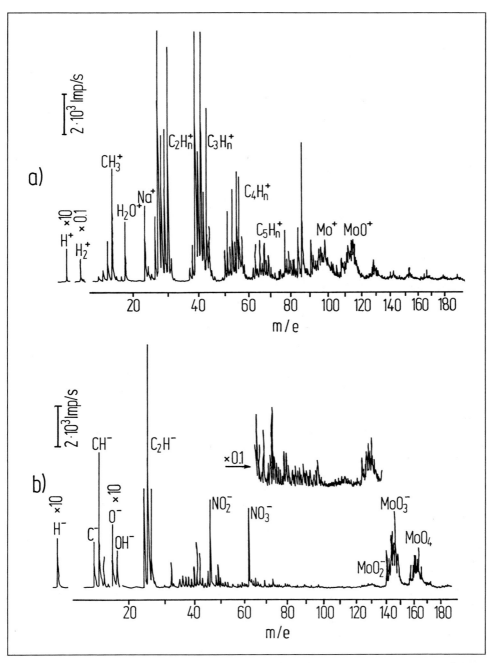

Fig. 4.53. Example of typical secondary-ion-mass spectra. Shown are a positive-ion spectrum **a)** and a corresponding negative-ion spectrum **b)** of a molybdenum surface, which was contaminated with various organic and inorganic impurities. The primary ion-current density was as low as $10^{-9}\,\mathrm{A\,cm^{-2}}$, and less than 1% of a monolayer was removed by sputtering while the spectra were recorded. From Benninghoven [219].

which sooner or later alters the properties of the surface. Another disadvantage is certainly the still persisting lack of complete understanding of the detailed physical processes during and after the ion-impact event, which, at present, prohibits a fully quantitative exploitation of SIMS.

4.4 Miscellaneous Methods

We move on to a group of methods, which are relatively easy to perform and, therefore, quite common in surface and adsorbate characterization. Among these are thermal-desorption mass spectroscopy (TDMS or simply TDS) and work-function change ($\Delta\Phi$) measurements, frequently referred to as contact-potential-difference (CPD) measurements. Particularly profitable is a combination of the two, and it is really surprising how detailed information about adsorbate coverages, surface-binding energies, kinetic coefficients, and surface-electronic interaction can be reached with relatively little experimental effort. In this section, we will confine ourselves to TDS and $\Delta\Phi$, whereby a variety of practical examples will be given.

4.4.1 Thermal-Desorption Spectroscopy (TDS)

Historically, TDS is one of the earliest methods used in investigating the state of adsorbates on surfaces. In the form of the so-called flash-filament desorption device, it consisted simply of a wire of the surface material of interest (most widely, tungsten was used), which was exposed to the adsorbing gas for a certain time interval and then rapidly heated, whereby the desorbing products were detected either by an ion gauge (total-pressure measurement) or a mass spectrometer tuned to the respective *e/m* ratio (partial-pressure measurement). Depending on whether the vacuum chamber was pumped or separated from the pump, the desorption, at the maximum rate of gas evolution from the surface, led either to pressure maxima or to step-like pressure increases. This will be illustrated below. Owing to the remarkably simple experimental set-up required for this kind of desorption investigations, it is no wonder that particularly the early work in surface science was almost dominated by thermal-desorption experiments.

As a consequence, a vast amount of literature on thermal-desorption spectroscopy has been accumulated over the years, and for the sake of brevity we refer the interested reader to several classical review papers by Ehrlich [227], Redhead [228], King [229], Menzel [230], and Péterman [231]. As in the previous descriptions, we shall first present the (simple) physical principles of TDS, and then provide some examples taken from relevant papers and our own work.

The ultimate goal of a TDS study is accurately determining the desorption parameters, such as activation energy for desorption ΔE_{des}^{*} (which is often coverage-dependent), frequency factor v, and reaction-order coefficient. Some of the respective equations can be found in Chapters 2 and 3. We want to stress here that – although the standard TDS experiment may be regarded as simple – the accurate data evaluation is by no means trivial, and kinetic models should always be checked by other methods.

Turning to some basic TDS physics, we remind the reader that one can describe the desorption reaction of species A as a kinetic process. the rate of which is expressed as

$$-\frac{d[A]}{dt} = k[A]^x ,$$

4.58

152

where the brackets denote (surface) concentrations and x the order of reaction. As in ordinary reaction kinetics, the rate constant k contains temperature dependence according to the Arrhenius equation

$$k = k_0 \exp\left(-\frac{\Delta E_{des}^*}{RT}\right) . \qquad 4.59$$

Using the conventional symbols ($k_0 \triangleq \nu$), and replacing surface concentrations σ_s by coverages Θ_σ, we arrive at the already mentioned Wigner-Polanyi equation (cf., Eq. 2.45).

$$-\frac{d\sigma}{dt} = -\sigma_{max} \cdot \frac{d\Theta_\sigma}{dt} = \nu_{(x)} \exp\left(-\frac{\Delta E_{des}^*}{RT}\right) \cdot \Theta_\sigma^x , \qquad 4.60$$

which is the basis for all further considerations and data evaluation of TDS.

We do not reiterate the discussion of the physical meaning and interpretation of the above listed quantities, which was presented in Chapters 2 and 3, we simply want to recall that the activation energy for desorption ΔE_{des}^* can frequently be identified with the binding energy of a molecular adsorbate on the surface. The naive view behind a thermal-desorption experiment is that a desorption maximum appears right at that temperature where the most adsorbate-substrate bonds per time interval are thermally dissociated, and the correct evaluation of the maximum temperature T_{max} is one of the tasks of TDS. In order to determine ΔE_{des}^* and the other desorption parameters correctly, one must consider a comprehensive expression for the rate of pressure variation in the UHV reaction chamber. A variety of reaction channels governs the net flow rate in the system and must be taken into account.

The basic equation

$$j = \frac{dN}{dt} = -j_{pu} + j_{des} - j_{ads} \qquad 4.61$$

is the mass (or flow) balance, that is to say, the net flow rate (or flux) is given by the difference of input flow and output flow (N = number of particles per cm^3 in the gas phase).

Positive fluxes are particle sources, negative fluxes represent a loss of molecules. While the particles desorbing from the surface represent the only source ($+ j_{des}$), withdrawal of particles can come about by i) removal by the vacuum pump ($-j_{pu}$) and ii) pumping effects by adsorption on the walls of the vacuum system, the gauge or mass spectrometer (or eventually by readsorption on the sample) which are combined as j_{ads}. To simplify our considerations, we regard j_{ads} as being approximately zero, because it is often a second-order effect compared to j_{pu}. The desorption flux j_{des}, on the other hand, is proportional to the desorbing sample area A and the rate of desorption $d\sigma_{(t)}/dt$, in which $\sigma_{(t)}$ represents the number of molecules on the surface at time t (cf., Eq. 4.60), and the pumping speed is accounted for by the "effective" pumping speed S_{eff} measured at the entrance of the pump.

In order to obtain j_{pu}, S_{eff} must be multiplied by the momentary gas concentration in the reaction vessel $N_{(t)}/V$, whereby V is the volume of the vessel. We can then write

$$\frac{dN}{dt} = A \cdot \frac{d\sigma_{(t)}}{dt} - S_{eff} \cdot \frac{N_{(t)}}{V} . \qquad 4.62$$

The number of particles in the gas phase can be expressed via the ideal gas law:

$$P \cdot V = \frac{N}{N_{\mathrm{L}}} \cdot RT ,$$ 4.63

where P stands for the system pressure, N_{L} for Avogadro's constant $= 6.023 \cdot 10^{23}$ [Mol^{-1}], and R for the gas constant $= 8.314$ [Mol^{-1} K^{-1}]. Hence, by substituting

$$N = \frac{PVN_{\mathrm{L}}}{RT}$$ 4.64

into Eq. 4.62, we arrive at

$$\frac{dP}{dt} = \dot{P} = \frac{ART}{N_{\mathrm{L}}V} \frac{d\sigma_{(t)}}{dt} - \frac{S_{\mathrm{eff}}}{V} \cdot P_{(t)} = \frac{ART}{N_{\mathrm{L}}V} \frac{d\sigma_{(t)}}{dt} - \frac{P_{(t)}}{\tau} ,$$ 4.65

if the characteristic pumping time τ

$$\tau = \frac{V}{S_{\mathrm{eff}}}$$ 4.66

is introduced. Integration yields for the area of a thermal-desorption trace $\int P dt$ the expression:

$$\int_{t=0}^{t=\infty} P dt = \frac{ART\tau}{N_{\mathrm{L}}V} \cdot \sigma_{(t)} ,$$ 4.67

from which σ, the absolute number of surface particles prior to the desorption, can be obtained:

$$\sigma_{(t)} = \frac{N_{\mathrm{L}} \cdot V \int P dt}{ART\tau} .$$ 4.68

This equation is the general basis for absolute determinations of surface coverages and kinetic parameters and requires, besides a numerical integration of a TDS peak area $\int P dt$, the accurate knowledge of the volume of the reaction vessel V, the area A from which desorption takes place, and the characteristic pumping time τ of the vacuum system. τ can be determined using, e.g., the pressure-drop method [2], where the system is exposed to a stationary gas pressure of the component (which is to be investigated in the TD experiment). At time t_0, P_i is suddenly (step-like) decreased by closing the gas-inlet valve and the pressure drop is monitored as a function of time on a storage oscilloscope. If we denote the base pressure of the system as $P_{i,\infty}$, we have the relaxation equation

$$\frac{dP_{(t)}}{dt} = -\frac{1}{\tau} \left(P_{i(t)} - P_{i,\infty} \right) ,$$ 4.69

which yields, upon integration

$$\frac{P_{i(t)} - P_{i,\infty}}{P_{i(0)} - P_{i,\infty}} = \frac{\Delta P_{i(t)}}{\Delta P_{i(0)}} = \exp\left(-\frac{t}{\tau} \right) .$$ 4.70

A semi-logarithmic plot of $\Delta P_i(t)$ vs time t thus gives a straight line, the slope of which allows determination of the characteristic pumping time τ.

As regards the evaluation of energetic and kinetic parameters from TDS experiments, we organize our presentation in two parts: First, we deal with the simplest case, namely the assumption of coverage-independent desorption parameters, which means that the adsorption process is entirely described by the Langmuir model (cf., Chapter 2), where

154

no particle-particle interactions occur, and we present kinetic expressions, which allow the determination of ΔE^*_{des} and v under the assumption of constant desorption order x. Second, we (briefly) enter the problem of coverage dependences ($\Delta E^*_{\text{des}}(\Theta)$, $v(\Theta)$) and the population of different binding states. Many of the following considerations make use of the equations presented in Chapters 2 and 3, where the fundamental physics behind the kinetic models is also explained.

In all our cases, the thermal desorption experiment is carried out in the following way. At temperature T_{ad}, the (single-crystal) surface of area A is exposed to a certain amount of chemisorbing gas i and becomes covered with σ_i particles/cm^2 or a coverage Θ_i. Then the sample is linearly and uniformly heated with a temperature program (heating rate $\beta = dT/dt$), while the vacuum chamber is effectively pumped (pumping speed S_{eff}) in order to avoid readsorption. During the heating process, the desorbing particles leave the surface with increasing rate until the adsorbate is exhausted, and the pressure returns to the initial value P_0, where the TDS experiment was started. Clearly, the partial pressure of the desorbing gas p_i, which is measured in a mass spectrometer, runs through a maximum at temperature T_{max}. After time t, the final temperature $T_f = T_{\text{ad}} + \beta t$ is reached, and the experiment is stopped when all particles have left the surface.

The desorption process is described by the Wigner-Polanyi equation in its general form:

$$-\frac{d\sigma_i}{dt} = v_x \cdot \sigma_i^x \exp\left(-\frac{\Delta E^*_{\text{des}}}{RT}\right).$$ (4.71)

At the desorption maximum T_{max}, differentiation leads to the condition:

$$\frac{d\left(-\frac{d\sigma_i}{dt}\right)}{dT} = 0\bigg|_{T=T_{\text{max}}},$$ (4.72)

and by making use of

$$\frac{d\sigma_i}{dt} = \beta \cdot \frac{d\sigma_i}{dT} = -\frac{d}{dT}\left(\frac{d\sigma_i}{dT}\right)_i = \frac{d}{dT}\left[\frac{v_x}{\beta}\sigma_i^x \cdot \exp\left(-\frac{\Delta E^*_{\text{des}}}{RT}\right)\right]$$ (4.73)

one obtains for the temperature derivation of the rate

$$\frac{d\left(-\frac{d\sigma_i}{dt}\right)}{dT}$$

$$= \frac{x \cdot v_x}{\beta}\sigma_i^{x-1} \cdot \exp\left(-\frac{\Delta E^*_{\text{des}}}{RT}\right) \cdot \frac{d\sigma_i}{dT} + \frac{v_x}{\beta}\sigma_i^x \left(\frac{\Delta E^*_{\text{des}}}{RT^2}\right)\exp\left(-\frac{\Delta E^*_{\text{des}}}{RT}\right).$$ (4.74)

This can be simplified using Eq. 4.72 to yield

$$\frac{\Delta E^*_{\text{des}}}{RT^2_{\text{max}}} = \frac{x \cdot v_x}{\beta}\sigma_{i,\text{max}}^{x-1} \cdot \exp\left(-\frac{\Delta E^*_{\text{des}}}{RT_{\text{max}}}\right),$$ (4.75)

where $\sigma_{i,\text{max}}$ stands for the surface-particle concentration still present at the desorption maximum. Remember that all decisive quantities (v_x, ΔE^*_{des}) are assumed coverage-independent!

In practice, four reaction types occur more frequently than any others, namely $x = 0$ (zero order), $x = 1/2$ (half order), $x = 1$ (first order), and $x = 2$ (second order). The most abundant case is first-order desorption which is characteristic for molecular (associative) desorption. For $x = 1$, Eq. 4.75 reads

$$\frac{\Delta E^*_{des}}{RT^2_{max}} = \frac{\nu_1}{\beta} \cdot \exp \left(-\frac{\Delta E^*_{des}}{RT_{max}} \right) , \qquad\qquad 4.76$$

and upon applying logarithms one has

$$\ln \left(\frac{\beta \Delta E^*_{des}}{\nu_1 RT^2_{max}} \right) = -\frac{\Delta E^*_{des}}{RT_{max}} , \qquad\qquad 4.77$$

which allows ΔE^*_{des} to be determined from the experimentally accessible T_{max} (which is apparently independent of σ_i) if the frequency factor ν_1 is known. A first-order desorption process includes the breaking of the adsorbate-substrate bond as the rate-limiting step, and as a rough approximation ν_1 may be assumed as $\approx 10^{13}\,\mathrm{s^{-1}}$. Redhead [228] has given an equation, based upon Eq. 4.77, which is often used to derive ΔE^*_{des} values for first-order desorption processes:

$$\Delta E^*_{des} = RT_{max} \left(\ln \frac{\nu_1 T_{max}}{\beta} - 3.64 \right) . \qquad\qquad 4.78$$

One can, however, circumvent any assumption about ν_1, if the heating rate β is varied according to Eq. 4.77, which can be rearranged to

$$\ln \left(\frac{T^2_{max}}{\beta} \right) = \frac{\Delta E^*_{des}}{RT_{max}} + \ln \left(\frac{\Delta E^*_{des}}{\nu_1 \cdot R} \right) , \qquad\qquad 4.79$$

and ΔE^*_{des} can be found from the slope of a plot of the expression $\ln(T^2_{max}/\beta)$ vs T^{-1}_{max}, whereas ν_1 is obtained, by inserting ΔE^*_{des} in Eq. 4.79, from the intercept. It should be noted that, for the first-order process, the shape of desorption peaks is asymmetric and the temperature maximum depends on the heating rate β. A good example for first-order desorption is the system CO/Pd(100) [232], from which we have chosen Fig. 4.54, which shows – in a coverage regime where ΔE^*_{des} does not depend on coverage – a series of TDS traces.

Second-order processes also occur quite frequently, namely, whenever recombination of two surface fragments becomes rate-determining. This is the case in many dissociative adsorption reactions (hydrogen, oxygen, nitrogen). Second-order desorptions can be identified from their symmetrical desorption peaks, which shift with increasing coverage σ_i to lower temperatures [233].

Referring again to Eq. 4.75, we obtain

$$\frac{\Delta E^*_{des}}{RT^2_{max}} = \frac{2\sigma_{i,max} \cdot \nu_2}{\beta} \cdot \exp \left(-\frac{\Delta E^*_{des}}{RT_{max}} \right) . \qquad\qquad 4.80$$

From the symmetrical peak shape, it follows that $2\,\sigma_{i,\,max}$ may be taken as equal to $\sigma_{i,0}$, the surface concentration prior to the application of the temperature program (which equals the total TDS peak area $\int P dt$) and accordingly to

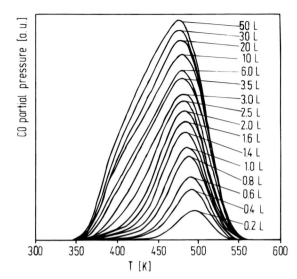

Fig. 4.54. Example for first-order thermal-desorption spectra: CO desorption from a Pd(100) surface. The heating rate β was 14 K/s; CO adsorption was performed at 350 K. The desorption maximum appears, for small coverages, around 490 K and shifts only by ~15 K to lower temperatures as coverage increases, indicating first-order kinetic behavior and (almost) constant activation energy for desorption. After Behm et al. [232].

$$\ln\left(\sigma_{i,0}\cdot T_{\max}^2\right) = \frac{\Delta E_{\mathrm{des}}^*}{RT_{\max}} + \ln\left(\frac{\nu_2 R}{\beta\Delta E_{\mathrm{des}}^*}\right) \ . \tag{4.81}$$

A plot of $\ln(\sigma_{i,0}\cdot T_{\max}^2)$ against $1/T_{\max}$ should yield a straight line with positive slope, from which $\Delta E_{\mathrm{des}}^*$ is easily calculated. The intercept, in turn, can be used to evaluate ν_2, if β and $\Delta E_{\mathrm{des}}^*$ are known. A typical example for a second-order process is presented in Fig. 4.55 and concerns hydrogen desorption from a Ni(100) surface [234]. Up to about 20 L exposure, there is only a single desorption state visible and application of Eq. 4.80 is possible.

Sometimes, zero-order desorption processes occur. This is the case whenever the concentration of the adsorbed particles is not rate-limiting for the desorption reaction, for example, if condensed multilayers of adsorbate grown on a substrate are removed. We then have from Eq. 4.71, by letting $x = 0$, the simple exponential relation for the rate of desorption

$$-\frac{d\sigma_i}{dt} = -\beta\cdot\frac{d\sigma_i}{dT} = \nu_0\exp\left(-\frac{\Delta E_{\mathrm{des}}^*}{RT}\right) \ , \tag{4.82}$$

which states that regardless of the initial coverage there is a single exponential function, which describes all desorption curves. Once the surface particle reservoir is exhausted, the rate simply returns to zero, which is seen experimentally as a cut-off on the high-temperature side of the spectra. $\Delta E_{\mathrm{des}}^*$ may be evaluated from a plot of $\ln(d\sigma_i/dT)$ vs $1/T$, whereby a single TD trace contains all the information:

$$\ln\left(\left|\frac{d\sigma_i}{dT}\right|\right) = -\frac{\Delta E_{\mathrm{des}}^*}{RT} + \ln\left(\frac{\nu_0}{\beta}\right) \ . \tag{4.83}$$

157

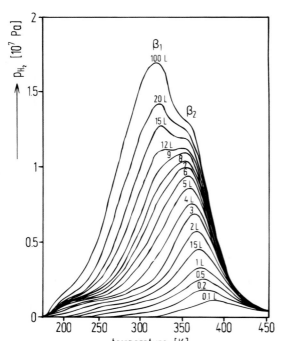

Fig. 4.55. Example for a second-order desorption process: hydrogen desorbing from a Ni(100) surface after increasing exposures at 120 K. The heating rate was $\beta = 10$ K/s. A second-order plot according to Eq. 4.81 reveals an activation energy for desorption of 96 kJ/mol [234].

The desorption of metal multilayers from refractory metal substrates often obeys the zero-order kinetics, and we provide, in Fig. 4.56, an example from our own work regarding Cu multilayer desorption from a Ru(0001) surface [235].

Finally, fractional-order desorption processes may deserve some attention, because they occasionally appear if the adsorbate forms two-dimensional islands on the substrate and the rate-determining step is the removal of a particle from the perimeters of these islands. It can be shown that in this case, for circular shaped islands of uniform diameter, the number of particles residing in perimeter positions N_p is proportional to the square root of the mean overall coverage σ_i times $4\pi z$, with z being the total number of islands on the surface divided by σ_{isl}, the local particle concentration in the interior of an island:

$$N_p \propto \sqrt{\frac{\sigma_i}{\sigma_{isl}} 4\pi z} \ . \qquad\qquad 4.84$$

Because the rate of desorption is proportional to N_p, we have a fractional-order process, which can approximately be described by

$$-\frac{d\sigma_i}{dt} = \nu_{1/2} \cdot \sigma^{1/2} \cdot \exp\left(-\frac{\Delta E^*_{des}}{RT}\right) \ . \qquad\qquad 4.85$$

Recalling Eq. 4.75, we arrive, with $x = 1/2$, at the expression

$$\frac{\Delta E^*_{des}}{RT^2_{max}} = \frac{\nu_{1/2}}{2\beta\sqrt{\sigma_{i\,max}}} \exp\left(-\frac{\Delta E^*_{des}}{RT_{max}}\right) \ , \qquad\qquad 4.86$$

and, after taking the logarithm, at the expression

158

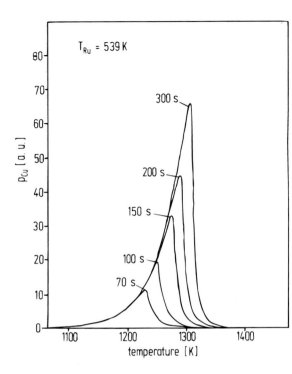

300 s

200 s

150 s

100 s

70 s

temperature [K]

p_{Cu} [a. u.]

Fig. 4.56. Thermal desorption of copper multilayers deposited onto a Ru(0001) substrate showing zero-order desorption kinetics. Deposition was performed at 539 K for different time intervals at a constant rate. Heating rates of $\beta = 10$ K/s were chosen in the desorption experiments [235].

$$\ln\left(\frac{T_{max}^2}{\sqrt{\sigma_{imax}}}\right) = -\frac{\Delta E^*}{RT_{max}} + \ln\left(\frac{2\beta\Delta E_{des}^*}{\nu_{1/2}R}\right) . \qquad 4.87$$

Accordingly, a plot of $\ln(T_{max}^2 \cdot 1/\sqrt{\sigma_{i,max}})$ against $1/T_{max}$ yields the activation energy for desorption ΔE_{des}^* for this kind of reaction. A characteristic property is that the desorption maxima shift to *higher* temperatures as the initial coverage is increased. Examples for this behavior have been reported for methanol desorption from Pd(100) [151], oxygen from Ag(110) [236], and copper from Ru(0001) in the sub-monolayer regime [235]. From this work, we present a typical example in Fig. 4.57.

Unfortunately, there are many cases in which the desorption processes are not so simple as might be suggested from the foregoing remarks. Complications can arise from i) multiple-peak structures, where particles are bound in different adsorption (binding) states, whose contributions cannot be separated in the experiment, and ii) coverage-dependent activation energies for desorption $(\Delta E_{des}^*(\sigma_i))$ and frequency factors $(\nu_x(\sigma_i))$. Both complications can have the same origin, namely the occurrence of lateral interaction forces between adsorbed particles, as was discussed in Sect. 3.3.3.

King [229], Bauer et al. [237], and Chan and Weinberg [238] have worked out data-evaluation procedures to determine both $\Delta E_{des}^*(\sigma_i)$ and $\nu_x(\sigma_i)$ as well as the reaction order x, based on the validity of the Wigner-Polanyi equation (Eq. 4.60) which is now formulated for all independent states as

$$-\frac{d\sigma_i}{dt} = \nu_{x,i}\sigma_i^{x_i} \exp\left(-\frac{\Delta E_{des,i}^*(\sigma_i)}{RT}\right) . \qquad 4.88$$

159

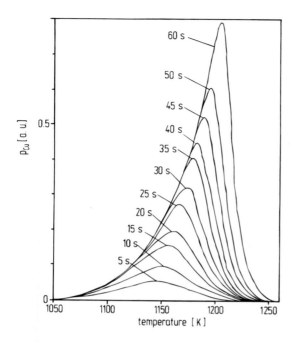

Fig. 4.57. Thermal desorption of Cu from Ru(0001) in the sub-monolayer regime. The deposition and desorption conditions were similar to those of Fig. 4.56, except for the observed reaction order, which is clearly fractional (1/2) [235].

In Chan and Weinberg's analysis [238], TDS peak widths also are taken into account. We do not want to enter this matter in great detail and simply present some ideas from King's work [229]. The procedure is illustrated in Fig. 4.58. Consider a family of thermal desorption curves, as they are reproduced in Figs. 4.54–57, where the x-axis is converted to a time scale (according to Eq. 4.73). Perpendicular cuts through the curves parallel to the desorption-rate axis after equally spaced time intervals and integration of the areas yield the remaining adsorbed amount (right-hand part of the TD traces) at time t or temperature T, σ_t, and σ_T, respectively.

The lengths of perpendicular cuts (Fig. 4.58a) correspond directly to the respective rate of desorption $d\sigma/dt$ at time t or temperature T. For each thermal-desorption curve, the amount σ_t is plotted vs temperature, and one obtains typical s-shaped descending curves (Fig. 4.58b), which reach zero when the adsorbed amount σ_t – for different curves reached after different times or at different temperatures T_1, T_2, T_3 – has been completely removed from the surface. Hence, isosteric-type conditions are established, and one obtains triples of values of coverage σ_t, temperatures T, and rate of desorption (from Fig. 4.58a). Considering the logarithm of Eq. 4.88

$$\ln\left(\left|\frac{d\sigma_i}{dt}\right|\right) = \ln \nu_{x,i} + x_i \ln \sigma_i - \frac{\Delta E^*_{\mathrm{des},i}(\sigma_i)}{RT} \; , \qquad\qquad 4.89$$

it is immediately evident that $\Delta E^*_{\mathrm{des},i}$ can be obtained as a function of particle concentration σ_i by plotting the logarithm of the desorption rate vs reciprocal temperature. Straight lines (isosteres) are associated with each surface concentration σ_i or coverage Θ. The intercept can be used to derive the pre-exponential factor $\nu_{x,i}$, provided the reaction order x is known. By checking the linearity of a plot of the intercept against $\ln x$, a possible coverage dependence of ν can be realized.

160

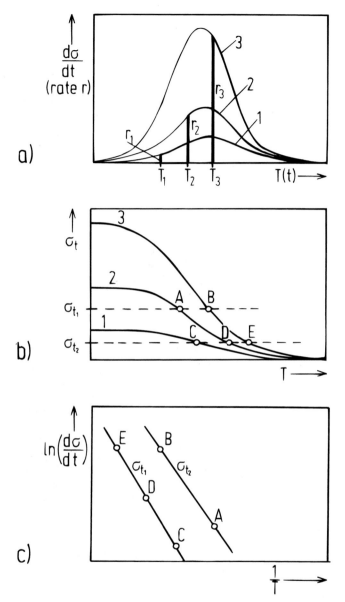

Fig. 4.58. Line-shape analysis of thermal desorption spectra according to King [229]. **a)** Family of three TD curves (desorption rate $d\sigma/dt$ vs time t). After various desorption times t different amounts of adsorbates σ_t remain on the surface. The times are linearly correlated with the surface temperatures T. At t_1, t_2, and t_3 different desorption rates $d\sigma/dt$ are obtained for each individual TD curve. **b)** For each desorption curve 1, 2, and 3, σ_t is plotted against time or surface temperature T. As can be seen, cuts parallel to the T axis at various σ_t values establish the condition that the same coverage at is obtained at different temperatures T_1, T_2, and T_3 for each thermal desorption curve. **c)** A plot of $\ln(d\sigma/dt)$ vs $1/T$ according to Eq. 4.89, yields isosteric straight lines whose slopes allow determination of ΔE^*_{des} for each coverage σ_t. If either the frequency factor v_x or the desorption order x is known, the respective other quantity can be evaluated from the intercept, as predicted by Eq. 4.89.

King [229] has described means of disentangling various overlapping desorption states, and we refer to his work for further information on this subject.

So far, thermal-desorption analysis has been exploited to yield the typical desorption parameters ΔE_{des}^*, v, and x. It is, however, also possible to derive sticking probabilities by means of TDS, simply by comparing the amount of gas offered to the sample for adsorption (exposure $P{\cdot}t$) with the actually adsorbed amount, which is equal to the desorption peak area $\int P dt$. In this way, either absolute integral sticking probabilities are accessible (by using Eq. 4.68) or relative peak areas can be compared, which leads to relative sticking coefficients. Of course, these latter quantities can be determined with much higher accuracy, which is only limited by the precision of the desorption peak area integration. A comprehensive employment of TDS to deduce desorption and surface kinetics parameters was performed, among others, in our study on hydrogen adsorption on a platinum (111) surface [30].

It is now interesting to compare the individual features of thermal-desorption spectra for various gases from different surfaces, for example, hydrogen and carbon monoxide, nitrogen and oxygen, organic vapors, and noble gases. Usually, a weakly interacting gas, such as xenon, on a smooth surface gives rise only to a single and extremely sharp desorption peak at relatively low temperatures, which hardly shifts with coverage [239]. Similar behavior is found with most organic molecules, which exhibit weak interaction and do not thermally decompose. Examples are acetonitrile CH_3CN on Au(100) [240] or methanol on Ag(111) [241]. If the interaction is stronger, the TD maximum appears at higher temperature, as manifested by desorption of carbon monoxide from Ni or Ru surfaces [242, 243]. As long as the adsorbate exists in a single binding state at the surface, there is only a single desorption maximum observed, which is coverage-invariant. For hydrogen and oxygen desorption, single desorption states occur, too, although for these molecules often second-order behavior is found, owing to the dissociative character of adsorption. However, if the surface coverage is raised to a limit where mutual particle interactions come into play, peak shifts, split-off states, and multiple peak structures develop in the TD spectra. A revealing example here is hydrogen desorption from a Rh(110) surface (Fig. 4.59). At coverages up to one monolayer, the H-H interactions are still small leading to a single second-order desorption maximum (β state). Beyond one monolayer coverage, the mutual H-H distance becomes noticeably smaller and the beginning repulsive interactions produce a second TD state around 220 K (α_2). At saturation, the hydrogens are so densely packed (cf., Fig. 3.15) that strong repulsions prevail. These reduce the overall binding energy and, hence, the desorption temperature considerably, resulting in a narrow low-temperature (α_1) state at 155 K [245]. It is well-known that crystallographically rough surfaces (high-index planes) usually show a greater multiplicity of TD states than flat surfaces. As an example, we present oxygen-desorption spectra from a Rh(110) surface, where not less than five individual states can be distinguished (Fig. 4.60) [244].

Besides adsorbate-induced (*a-posteriori*) heterogeneity of the adsorption energy, also *a-priori* heterogeneities may exist, for example with rough and heterogeneous samples that exhibit large concentrations of defect sites (i.e., steps and kinks). Of course, a TDS experiment is not able to delineate *a-priori* and *a-posteriori* heterogeneities, and the net results will lead to similar multiple-peak structure in the spectra. In case of an *a-priori* heterogeneity, additional high-temperature binding states are populated with an amount of adsorbate that is proportional to the defect-site concentration. A simple but suitable example in this context is desorption from artificially stepped surfaces, where a high concentration of step sites causes new peaks in a desorption experiment, such as our own investigation of H on Pt(997) [233].

162

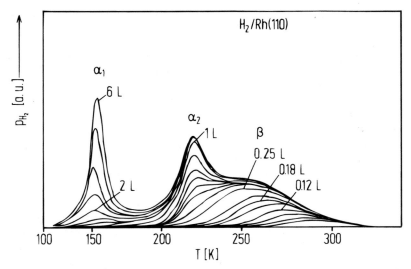

Fig. 4.59. Series of thermal-desorption spectra of hydrogen from a Rh(110) surface for increasing exposures [L]. Up to 0.30 L, only a single β state is populated; increasing coverages result in the formation of an α_2 state (~1L) and at saturation (1×1-2H phase) a split-off α_1 state appears. The heating rate β was 10 K/s; the adsorption temperature 85 K throughout. After Ehsasi and Christmann [245].

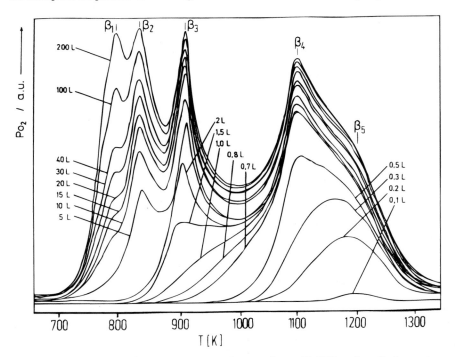

Fig. 4.60. Series of thermal-desorption spectra of oxygen from a Rh(110) surface, for increasing exposures (L). In order to achieve the good two-dimensional ordering of the adsorbate, O_2 adsorption was performed at 573 K. Up to saturation, five clear TD states ($\beta_1 - \beta_5$) are observed, reflecting different ordered oxygen phases on Rh(110). After Schwarz et al. [244].

Regardless of these complications, for carefully prepared clean single crystal surfaces and high-purity adsorbates the TD spectra should represent a characteristic replica or fingerprint of the chemical activity of the surface in question.

A compilation of many H_2, O_2, N_2, and CO desorption curves from metal surfaces is reproduced in the review article by Morris et al. [246], from which the reader can obtain further information. While normal desorption of surface species gives rise to TD spectra of a kind as shown in Figs. 4.54–4.57 and 4.59–4.60, sometimes broad peaks occur at relatively high temperatures and without a sharp maximum. Desorption features of this kind often indicate strongly thermally activated diffusion processes, where particles travel from regions deep in the bulk to the surface on which the desorption takes place. Hydrogen in Pd(110) may be quoted as an example. This system exhibits a broad desorption maximum around 500 K–700 K, far beyond the surface-hydrogen desorption maximum [247].

Another noteworthy property of thermal desorption spectroscopy is its principally destructive character. At the beginning, the surface is covered with adsorbate; after completion of the experiment all particles are removed. Nevertheless, this fact is often neglected when TDS results are discussed in terms of initial state properties of an adsorbate system. This concerns order-disorder or generally other phase-transition phenomena that may occur during a single temperature ramp, where the adsorbate exists in a phase with long-range order at low temperatures (T_{ad}) and changes to disorder upon heating prior to the desorption. Other examples are thermally activated reconstruction processes, which run on a similar time scale as the desorption temperature program. They may not be completed within a single temperature scan and hence leave behind a mixed surface consisting of patches already reconstructed (with adsorbate binding energy and thus activation energy of desorption $\Delta E_{des}^*(1)$) and patches not yet reconstructed, leading to desorption with activation energy $\Delta E_{des}^*(2)$. It can be seen that a multiple-peak structure results in these cases. A typical example is provided by the system H/Ni(110) [248], where a thermally activated streak-phase reconstruction provides a two-peak TDS structure, which depends on the heating rate.

Other complications that may arise from the application of a temperature program are, of course, all thermally activated surface-chemical-reaction steps, which lead to bond breaking (dissociation) or formation of new intermediate surface species. These finally desorb, too, but at a different temperature than the undecomposed molecule. Again, there is a wide range of (mostly organic) adsorbates where thermally induced fragmentation or isomerization processes obscur the normal desorptive removal of the adsorbate. Quite often, one can deduce valuable information about surface-decomposition paths of such molecules, and particularly in Madix's group there has been developed a special method, which has catalytical impact, the so-called temperature programmed reaction spectroscopy (TPRS) [249]. In this way, for example, all products evolved by thermal reaction from formic acid adsorbed initially onto a Cu(110) surface can be specified and analyzed [250]. Other examples are the decomposition of methanol on Pd(100) [151] or of glycine on Pt(111) [209].

We conclude our considerations of TDS by making some experimental remarks. Although the TD experiment seems to be very straightforward and simple, some experimental precautions must be taken as regards the state of the sample, its mounting on a sample holder, and its geometrical position with respect to the mass spectrometer. Furthermore, the way the sample is heated can have a great effect on the shape of the TD spectra. In the simplest case, the temperature program may be a linear ramp, i.e., $\beta = dT/dt = $ const, and a linear increase in the sample temperature with time is accomplished. For this most favorable situation, the aforementioned inter-conversion of time and temperature scale is easily possible. The surface temperature is usually measured by a thermocouple spot-welded to the sample, whereby the thermovoltage signal (which can span, if chromel-alumel is used as a thermocouple, from −5 to +50 mV) is not a linear function of the temperature, instead, the analogue output voltage is convoluted with the respective thermocouple characteristics. This property often exhibits appreciable deviations from linearity, particularly in the low-temperature range, which must be taken care of if, e.g., a linear sample heating is acquired by means of a PID regulator. Corresponding electronic temperature-control circuitry has been developed and tailored for a typical sample mounting design, where the single-crystal waver (often a disk of 1−2 mm thickness and 5−20 mm diameter) is spot-welded between two parallel-running 0.25 mm diameter tantalum wires, which in turn are attached to 2-mm diameter molybdenum support rods connected mechanically to the sample holder and electrically via feedthroughs to the dc heating power supply. This power supply is controlled by the electronic regulator circuitry to feed the sample with such a dc current so that a linear temperature rise is achieved [251].

Quite often, spurious contributions originating from the sample holder and the support and heating wires can simulate desorption maxima or give rise to background effects superimposed on the sample-desorption signal. Variation of the heating rate can frequently help to identify these unwanted contributions, which are particularly troublesome when desorption spectra are taken in the very-low-temperature range, far example in noble-gas-adsorption experiments.

Remedy against these spurions effects is provided by fading out the sample-holder contributions by means of a differentially pumped mass spectrometer mounted inside a hermetically close cone with a small (0.5−1 mm-diameter) orifice at the one end facing the sample, as is schematically sketched in Fig. 4.61. This figure also gives an impression about a typical set-up of a thermal-desorption experiment under UHV conditions.

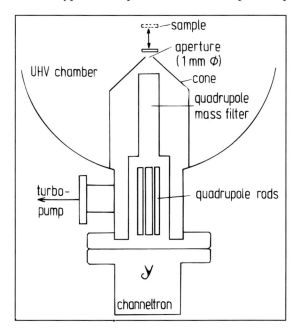

Fig. 4.61. Typical experimental set-up for thermal desorption spectroscopy using a differentially pumped mass spectrometer inside a cone with a small orifice at its tip in order to avoid spurious sample-holder contributions. The sample has direct-sight contact with the ionization source of the mass filter (the so-called line-of-sight conditions are established).

If other samples than metal single crystals are used, a TDS experiment becomes more difficult. First of all, heterogeneous and atomically rough samples usually possess a wide variety of different adsorption or binding sites that overlap and become sequentially emptied as the temperature is raised, thus leading to many separated desorption peaks in the most favorable case, but to a broad and uncharacteristic increase of the respective partial pressure in the worst case. An additional problem encountered mainly in investigations of practical catalyst material is the often very porous structure of these substances (for example, zeolites). Here, uniform heating is, if at all, only possible by selecting extremely small heating rates. These, however, decrease the sensitivity for detecting of a certain mass almost to zero when high pumping speeds are maintained. Therefore, other apparatus arrangements and designs are chosen along with different modes of detection i.e., flow tube reactors and gas chromatography). More details can be obtained from the special literature on catalysis [252].

As emphasized in the introduction to this chapter, a combination of thermal desorption spectroscopy and work-function change measurements is very useful, because both methods are quite complementary to one another. The next section, therefore, will be devoted to work-function measurements, in particular, contact potential difference (CPD) measurements.

Before we close our discourse on desorption spectroscopy, however, we should briefly mention that besides thermal desorption there are various other desorption spectroscopies used, where the energy to cleave the adsorbate-surface bond is not supplied thermally, but rather by electron impact (electron stimulated or electron impact desorption, ESD or EID), or by irradiation with photons (photon-stimulated desorption). By tuning the respective energies, a resonant bond-breaking or excitation of adsorbate-substrate bonds can be reached, resulting in large, energy-dependent cross sections for this kind of desorption. In this context we cannot, once again, enter this interesting matter, and must refer to the respective literature [253–256].

4.4.2 Work-Function ($\Delta\Phi$) Measurements

Similar to TDS, $\Delta\Phi$ measurements are very simple to perform, and yet can provide fairly detailed information about microscopic processes. Moreover, the work function is, by definition, a surface-sensitive property, because it contains above all the surface potential χ, which will be discussed below. It is therefore natural that even the earliest surface studies utilized work-function techniques, mainly based on photoelectric phenomena, after the discovery of the photoelectric effect by Hallwachs in 1888 [257], and its correct interpretation by Einstein in 1905 [258]. Fowler [259] (on the theoretical side) as well as Suhrmann and his group [260, 261], and Wedler [262] (on the experimental side) developed the photoelectric work-function determination to a level of sophistication, which has hardly been reached since then anymore. Ultraclean thin-metal films were deposited onto glass surfaces in an all-glass apparatus under ultra-high vacuum conditions (maintained by mercury-diffusion pumps and prolonged high-temperature bake out-cycles). After deposition, the films were irradiated through quartz windows with UV radiation provided by a Hg resonance lamp. The photoelectron current, or better the quantum yield I (= electrons emitted per absorbed quantum), is related to the frequency of the light via the so-called Fowler equation:

$$I = M \cdot T^2 \cdot f(\xi) , \qquad\qquad 4.90$$

where M is a yield constant and $f(\xi)$ the so-called Fowler function with $\xi = (h\nu - e_0\Phi)/kT$ (h = Planck's constant, ν = frequency of the UV light, and Φ = work function) :

$$f(\xi) = e^{\xi} - \frac{e^{2\xi}}{2^2} + \frac{e^{3\xi}}{3^2} \ldots \pm \frac{e^{n\xi}}{n^2} \quad \text{for} \quad \xi \leq 0 \qquad 4.91\text{a}$$

and

$$f(\xi) = \frac{\pi^2}{2^2} + \frac{\xi^2}{2} - \left(e^{-\xi} - \frac{e^{-2\xi}}{2^2} + \frac{e^{-3\xi}}{3^2} \ldots \pm \frac{e^{-n\xi}}{n^2} \right) \quad \text{for} \quad \xi \geq 0 . \qquad 4.91\text{b}$$

For frequencies ν, which are not too close to the threshold frequency ν_0, where photoelectrons can leave the surface for the first time, Eq. 4.90 can be simplified by the approximation

$$I \approx \frac{M \cdot h^2}{2k^2} (\nu - \nu_0)^2 . \qquad 4.92$$

The threshold frequency is defined via the energy balance

$$\nu_0 = \frac{e_0\Phi}{h} . \qquad 4.93$$

Accordingly, a plot of the square root of I against the frequency ν allows the determination of M and ν_0. From Eq. 4.93, ν_0 immediately yields Φ. Details about this kind of work-function evaluation can be found in the literature [260], and we just remind the reader that nowadays UV photoelectron spectroscopy allows Φ to be determined from the width of an energy distribution curve more conveniently, but much less accurately (cf., Sect. 4.2.1).

Besides the photoelectric measurement, there are two other related techniques, which are also based on electron excitation and emission: namely field emission (FEM) and thermionic emission. In both cases, the emitted electron current is exponentially related to the work function. The field emission current is given by the Fowler-Nordheim equation [263] via

$$i = 1.54 \cdot 10^{-6} \frac{F^2}{e_0\Phi t^2(y)} \exp \left(-6.83 \cdot 10^7 \frac{e_0^{3/2}\Phi^{3/2}}{E} f(y) \right) , \qquad 4.94$$

where F = electric field strength, E = potential energy of the electron between cathode and anode (i.e., $E = \beta \cdot U$, with β being the geometry factor given by the curvature of the FEM tip and U = electric voltage between cathode and anode), and $t(y)$ and $f(y)$ are tabulated functions of $y = e^{3/2} \cdot F^{1/2}/\Phi$. From a suitable log plot, a straight line, whose slope gives access to the work function Φ, can be obtained.

The thermionic-emission current obeys, if the emitting diode operates in the current-saturation mode, the well-known Richardson equation:

$$i = \frac{4\pi m_e k^2 e_0}{h^3} T^2 \exp \left(-\frac{e_0\Phi}{kT} \right) . \qquad 4.95$$

If the current is measured as a function of the emitter temperature, a logarithmic plot of i/T^2 vs $1/T$ allows the calculation of the work function Φ of the emitting material. More details are communicated in the review article by Hölzl and Schulte [266].

The aforementioned techniques can be applied to clean metal surfaces and the photo-electric and field-emission methods additionally to adsorbate covered surfaces in order to derive adsorbate-induced work-function changes. Here, the thermionic method only works if adsorbates with very high adsorption energy, which survive the high tempera-ture of the experiment, are used. The other noteworthy point is that all these emission methods yield the *absolute* value of the work function and the work-function change, respectively, which are difficult to obtain otherwise. As will be seen later, there are many relative methods, which just monitor work-function *differences*, such as the Kelvin or diode-contact potential difference measurement. However, before we turn to a short ana-lysis of these techniques, we must come back to the term *work function* and its physical meaning. There is a vast amount of literature published on this topic [265–267], and it suffices here to emphasize just some selected points. Figure 4.37a in Sect. 4.2.1 displays the simple electron-sea model and shows the potential barrier for electrons inside the metal for the field-free case as a step function. While one can easily define the energy quantities within this model, it requires some more effort to handle the state and physical properties of a charged particle, e.g., an electron, in the presence of the surface-electric fields under equilibrium conditions at the phase-boundary solid-vacuum. We follow the short guide to the problem given by Wedler [262], and stress in the beginning that the equilibrium between charged particles in two phases requires, in the first instance, the equality of their electrochemical potentials:

$$\eta_1 = \eta_2 \ . \tag{4.96}$$

The electrochemical potential of conduction electrons in a metal is composed of the chemical potential $\bar{\mu}$ of the electrons inside the metal and the electrostatical potential $e_0\varphi$:

$$\eta_i = \bar{\mu}_i - e_0\varphi_i \ , \tag{4.97}$$

where $\bar{\mu} = (\partial G/\partial n_e)_{T,P}$ and n_e = number of moles of electrons in the conduction band. φ_i represents the so-called inner potential and is defined as the electrical work necessary to transfer an electron from infinity right to the Fermi level of the metal substrate. It is important to recall that this inner potential contains two independent contributions, namely the *outer* electrical potential ψ of the solid surface (which deviates from zero only if extra charges are supplied to it, e.g., by an external voltage) and the so-called sur-face potential (SP) denoted by χ, which is the more interesting quantity in our context:

$$\varphi = \psi + \chi \ . \tag{4.98}$$

χ is caused by any asymmetries in the charge distribution at the surface as they come about by the spill-out effect of the electrons of the topmost layer. Furthermore, the sur-face potential is, of course, influenced by any adsorbed particle at the surface – these par-ticles are very frequently more or less polarized and may be regarded as kind of an elec-tric double layer, which is schematically illustrated in Fig. 4.62 for three different bind-ing situations: physisorption of polarizable atoms or molecules, ionic chemisorption, and covalent particle bonding at the surface. If the surface is in the neutral state, its inner potential φ equals the surface potential, i.e., $\varphi = \chi$, and a certain amount of work called the *work function* $e_0 \cdot \Phi$ must be performed on the system to transport an electron from the interior of the metal (that is to say, from the Fermi level E_f) to just outside the surface, where the image-charge forces have declined to zero. This work is composed of the chemical potential of the electron at E_f, $\bar{\mu}$, and the contribution necessary to overcome the surface potential χ_i

$$e_0\Phi_i = \bar{\mu}_i - e_0\chi_i \ . \tag{4.99}$$

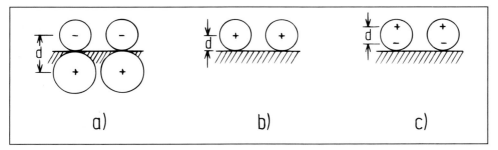

Fig. 4.62. Schematic illustration of the formation of surface dipoles within an adsorbed layer as a function of the bonding situation (d denotes the separation distance of the centers of gravity of charge): **a)** covalent (atomic) bonding. **b)** ionic, and **c)** physisorptive bonding. After Culver and Tompkins [265].

Remember that $\bar{\mu}_i$ is a bulk property of the respective metal and, hence, independent of the state of the surface (adsorbate) layer, and we emphasize again that it is only the surface potential that contributes to the work function and provides its surface sensitivity.

When discussing work functions and related phenomena, there is one important point frequently overlooked: namely, that Φ refers to the distance r_0 the electron must be moved in front of the surface. Usually, this distance is selected (somewhat arbitrarily) to be 10^{-6} m, an approximate value for the radius of action of the image-charge forces. The related image potential $V(r)$ obeys the relation

$$V(r) = -\frac{e_0}{4r} f^* , \qquad 4.100$$

where $f^* = (4\pi\varepsilon_0)^{-1}$. This potential corresponds, for a clean metal surface, to the surface potential χ defined above. Apparently, χ is distance-dependent (which implies the necessity to define a value for r_0). In the absence of external fields and for a given homogeneous lateral geometry (single-crystal surface) $V(r)$ tends to unity if r approaches infinity. In reality, this constant value is reached at about 10^{-6} m away from the surface. If there is an external field applied (cf., Fig. 4.14), the condition for r_0 is obtained by the potential energy maximum $|(dV(r))/dr|_{r=r_0} = 0$.

In a related way, surface potential effects caused by asymmetries of the charge distribution at the surface (electron spill-out) or adsorbed dipole layers may lead to a maximum (or minimum) of the overall electrostatic potential even in the absence of external fields and make an assessment of r_0 necessary. Here, we concentrate on adsorbate-induced work-function effects, and by returning to Eq. 4.99 we underline that work-function changes caused by adsorbates of kind i are totally equivalent to the (negative) change of the surface potential

$$-\Delta\chi_i = +\Delta\Phi . \qquad 4.101$$

To a great extent, this equation governs adsorption phenomena, because a change of the work function caused by adsorption can be directly correlated with the formation of a dipole layer of adsorbed molecules.

The term contact potential remains to be explained. Consider two conducting solids A and B characterized by the chemical potential of their electrons $\bar{\mu}_A$ and $\bar{\mu}_B$ with $\bar{\mu}_A > \bar{\mu}_B$. They form a plate capacitor, and the plates will be electrically connected with each other. According to Eq. 4.96, the electrochemical potentials η become equal, which can only be accomplished by a net flow of electrons from A to B. Applying Eq. 4.97, viz.

$$\eta_A = \bar{\mu}_A - e_0\Phi_A = \eta_B = \bar{\mu}_B - e_0\Phi_B$$

leads to the relation

$$\bar{\mu}_A - \bar{\mu}_B = e_0(\varphi_A - \varphi_B) \qquad\qquad 4.102$$

which is, if Eqs. 4.98 and 4.99 are considered, equivalent to

$$e_0(\Phi_A - \Phi_B) = e_0\Delta\Phi = e_0(\psi_A - \psi_B) . \qquad\qquad 4.103$$

Apparently, there is a distinct electrostatic potential difference built up if two different metals are electrically connected. This voltage is called contact potential and can also be observed if one of two initially identical metal surfaces becomes covered with an adsorbed dipole layer. The reason is that the surface potential of one metal changes and gives rise, according to Eqs. 4.98 and 4.99, to a contact-potential difference. This CPD can be measured by appropriate methods with a sensitivity of a few millivolts, for example by the Kelvin or the diode method, which will be explained below.

The adsorbate-induced surface-potential change is simply caused by the sum of the dipoles at the surface, and it is obvious that a given number of adsorbed particles (surface concentration σ_0) produces a work-function change, which is proportional to the magnitude of the dipole moment μ_0 of the individual adsorption complex, if depolarization effects are neglected. This leads directly to the Helmholtz equation, which we had dealt with before in Chapter 2 (Eq. 2.43) in the context of thermodynamical measurements, and it is straightforward to derive the initial dipole moment of an adsorbed particle μ_0 from this equation. Note that the dimension of μ is represented in SI units (As·m). In surface chemistry, the cm-g-s "Debye" [D] unit still is frequently used. The conversion factor is 1 [D] = $3.33 \cdot 10^{-30}$ [As·m]. Unfortunately, depolarization phenomena do play a role and must be considered in many cases, particularly at higher adsorbate concentrations and with strongly charged surface species, for example adsorbed alkali-metal atoms. Within a simple electrostatic model, depolarization effects can be accounted for, which leads to the well-known Topping formula [268]

$$\Delta\Phi = 4\pi\sigma\mu_0 f^*(1 + 9\alpha\sigma^{3/2})^{-1} , \qquad\qquad 4.104$$

where α stands for the polarizability of the adsorbed particle. Accordingly, polarizabilities can be determined from the coverage dependence of the adsorbate-induced work-function change, particularly for ionic adsorbates. Equation 4.104 can be rearranged, and a plot of the coverage Θ divided by $\Delta\Phi$ vs $\Theta^{3/2}$ should give, if the Topping model applies, a straight line, whose slope contains the polarizability α. (Remember that the coverage Θ has been defined by the ratio σ/σ_{max} (cf., Eq. 2.41.)) Using the aforementioned procedure, the polarizability of Xe atoms adsorbed onto a Ni(100) surface was evaluated to be $3.52 \cdot 10^{-24}$ cm^3 [239]. We remember that the polarizability reflects directly the size or volume of the respective adsorbed atom or molecule, which underlines the possibility of deducing atomic parameters from work-function measurements.

In Chapter 2, we also discussed a method of exploiting work-function data to receive thermodynamical information. It was based on the fact that $\Delta\Phi$ is often a very convenient and relatively precise monitor of the surface concentration of the adsorbate. On metal-single crystal surfaces $\Delta\Phi$ can frequently be correlated with the absolute coverage, for example by comparing a certain ordered LEED superstructure with the corresponding work-function change. Furthermore, thermal-desorption peak areas $\int P dt$ can be related to $\Delta\Phi$ values, and adsorption kinetics can easily be followed this way. The use of $\Delta\Phi$ as a coverage monitor dates back to Mignolet [269] and was further pursued and improved by Delchar and Ehrlich [270] as well as by Palmberg and Tracy [271].

Because adsorbate molecules usually tend to occupy surface sites with a high chemical coordination, $\Delta\Phi$ often reflects crystallographic imperfections of a surface (i.e., steps, kinks, and point defects) quite sensitively, for the following reason: the very first particles that arrive at a surface choose just these defect sites and adsorb therein. Only once these sites are filled, the adsorption sites characteristic of the nominal surface orientation become occupied. Because adsorption into a defect site is very often associated with a different charge transfer, the work function is different, too, and in some favorable cases even the sign of $\Delta\Phi$ is different compared with adsorption into non-defect sites. This is illustrated in Fig. 4.63 by an example taken from our work on H chemisorption on a stepped Pt(111) surface [233]: adsorption of H atoms in steps produces a work-function *increase* as opposed to adsorption on the terrace sites, which leads to a $\Delta\Phi$ *decrease*. Even two kinds of step sites can be distinguished, which are illustrated in the figure.

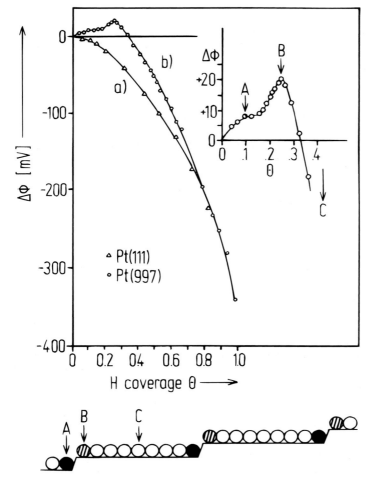

Fig. 4.63. Example demonstrating the influence of adsorption into step sites on the work-function change $\Delta\Phi$: H adsorption on a Pt(997) surface. Upper curve: $\Delta\Phi$ as a function of H coverage Θ for the flat Pt(111) surface **a)** and the stepped (997) surface **b)**, with the initial $\Delta\Phi - \Theta$ region magnified (inset). Three different kinds of adsorption sites leading to different work-function behavior (A, B, and C) are indicated. Lower part: schematic representation of the stepped surface (side view) showing the sites A, B, and C. After Christmann and Ertl [233].

We have seen that the overall work function of a sample is largely determined by the surface potential χ which reflects any asymmetries in the surface-charge distribution. This is the reason why different crystal-face orientations exhibit different work functions, and it also explains the fact that reconstructive phase transformations, such as pairing-row or missing-row reconstructions (which we dealt with in Chapter 3), are usually associated with a work-function change. In the case of a reconstruction, the atoms of the surface region change their lattice positions, and hence also the electron concentration or charge density near the surface is altered, resulting in an upward or downward shift of the surface potential χ. There are some metals that are reconstructed in the clean state and can be chemically prepared also in the unreconstructed (1×1) form, for example Ir and Pt(100) surfaces. The work functions of the reconstructed surfaces exceed those of the unreconstructed configurations by several hundred millivolts [144, 145]. This is clearly an effect of the different electronic-charge density at the surface.

Another complication must be discussed here. So far, we have tacitly assumed that the surfaces were always in a homogeneous state, that is to say, they had the same surface orientation everywhere. This is only fulfilled for defect-free, clean single-crystal surfaces. In practice, however, the relevant materials are often polycrystalline, and some short supplementary remarks on the work function behavior of these heterogeneous surfaces are worthwhile. A polycrystalline surface may be viewed as consisting of patches with a different local work function Φ_p, whereby this value again refers to a distance r_0 about 10^{-4} cm in front of the surface, as discussed before. It can be shown that at a distance from the surface, which is large compared to the dimensions of the patch, the surface potential χ_p will take a constant value χ_∞, which is kind of an algebraic average

$$\chi_\infty = \sum_i f_i \chi_{p,i} , \qquad\qquad\qquad 4.105$$

where f_i denotes the fractional area of the total surface occupied by the i-th patch. At closer distances to the surface, the surface potential exhibits lateral variations depending on the local work function of the patch and its extension. By contrast to χ, $\bar{\mu}$ does not vary locally with the patch orientation, which means that the overall work function Φ_{tot} of the polycrystalline surface is given by the expression:

$$e_0 \Phi_{tot} = e_0 \sum_i f_i \Phi_i = \bar{\mu}_i - e_0 \sum_i f_i \chi_{p,i} . \qquad\qquad 4.106$$

A consequence of the occurrence of patches is that there will be accelerating or retarding fields induced at the boundary of adjacent patches with different work function, so-called fringing fields, which may influence the electric and adsorptive properties of particles adsorbed in that local surface region.

In the following discussion, we will briefly describe some techniques for measuring work functions often used in adsorption studies of single-crystal surfaces: the Kelvin method and the (space-charge-limited) diode method. Both techniques have in common that only *relative* work functions and work-function changes are accessible, because virtually contact potential differences are measured. An absolute Φ or $\Delta\Phi$ determination is, however, possible, if a reference material with known absolute work function is available.

As mentioned above, the Kelvin method was successfully introduced by Mignolet [269] to follow work-function changes during gas adsorption. The physical principle

172

behind it is relatively simple. Actually, one measures the displacement current i_D, which flows inside the connecting wire of a charged-plate condenser as soon as the capacitance is periodically modulated. This can be achieved, for example, by vibrating one plate with respect to the other (fixed) plate, with frequency ω around a distance d_0 (Fig. 4.64), according to:

$$d(t) = d_0 + a \sin \omega t \ . \hspace{5cm} 4.107$$

A permanent alternating displacement current is thus obtained, which is given as

$$i_D(t) = dQ/dt = -\varepsilon\varepsilon_0 A V a\omega \cos \omega t (d_0 + a \sin \omega t)^{-2} \ , \hspace{1.5cm} 4.108$$

where Q= charge on the capacitor; ε, ε_0 = permittivity of the dielectricum and vacuum, respectively; A= plate area; V= voltage applied to the capacitor (V equals the contact potential difference, $\Delta\Phi$); d_0 = plate distance with plate at rest; ω = frequency of vibration, and a = amplitude of vibration. According to Eq. 4.108, i_D represents a time-dependent periodic function and differs from zero only if the contact potential difference V has a finite value. Because its magnitude is proportional to V, i_D can be utilized to monitor contact potential differences.

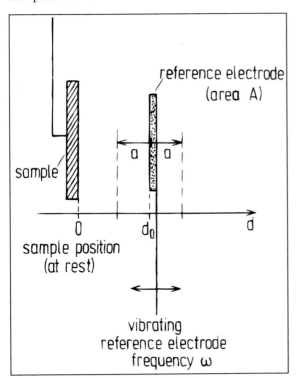

reference electrode
(area A)

a   a

sample

0   d_0 d

sample position
(at rest)

vibrating
reference electrode
frequency ω

Fig. 4.64. Schematic diagram elucidating the formation of a Kelvin vibrating condenser consisting of the (fixed) sample surface (left) and the (vibrating) reference plate electrode (right).

Experimentally, the sample is moved by means of a UHV manipulator in front of an inert reference electrode, which consists of a small gold plate or ring wire mounted to a glass arm. This arm is mechanically excited to up-down vibrations by an electromagnet, thus leading to the capacitor situation depicted in Fig. 4.64. The whole device is called Kelvin probe, in honour of Lord Kelvin, who first introduced the vibrating condenser method to measure small electrical currents [272]. The Kelvin method for monitoring

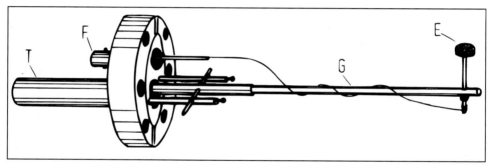

Fig. 4.65. Typical design of a Kelvin probe as developed in our laboratory. A vibrating glass arm G carries on its one end a gold-plate (4-mm diameter) reference electrode E and on its other end a permanent magnet, which sits inside a closed stainless steel tube T and is excited magnetically by an ac transformer (not shown in the photograph). The i_D signal is fed via a current feedthrough F welded into the 70-mm flange to a BNC jack. The mechanical vibration is achieved by clamping the glass arm via two stainless steel rods (1-mm diameter) onto two stainless steel stands welded onto the bottom of the flange. Under resonance conditions, the reference electrode vibrates with up-down amplitudes of more than 1 mm.

work-function changes was improved among others by Zisman [273] and Simon [274]. A typical design of a Kelvin probe mounted on a Conflat flange and used in our laboratory is shown in Fig. 4.65. Typical working frequencies are around $60-120\,\mathrm{s}^{-1}$, with amplitudes of about $0.5-1$ mm. In some cases it may be advantageous to work with the second harmonic mode in order to reduce the noise level [275, 276]. A high-impedance preamplifier converts the current signal to a voltage, which is then fed into a lock-in amplifier

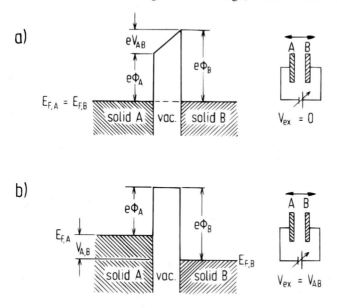

Fig. 4.66. Physical principle of the Kelvin method: electrical potential situation with two different metals A and B connected to form a condenser. Because the Fermi levels equilibrate, a contact potential difference V_{AB} is built up (**a**). If an adjustable external voltage V_{ex} is connected to the plates of the capacitor, the CPD V_{AB} can be compensated to zero. Then the condition $V_{AB} = V_{ex}$ holds (**b**). In a self-compensating circuitry device, this is achieved automatically using lock-in techniques [275].

tuned to the appropriate frequency, providing at its output jack a dc voltage directly proportional to the contact potential difference of the plate condenser.

When following adsorption-induced work-function changes, it is mandatory that the adsorption occurs *only at the sample surface*; gases must not interact with the reference electrode, which then consists of a chemically inert material (Au or oxidized Ta or W). The physical operation is as follows (Fig. 4.66). Because the two Fermi levels equilibrate to the same height, there exists an initial CPD between the clean-sample surface and the reference electrode. By applying an external voltage V_{ex} between sample and reference electrode the CPD can be compensated to zero, whereby the two Fermi levels are shifted with respect to each other accordingly. Any adsorption now changes the surface potential of the sample and, hence, its work function leading to the build-up of a renewed CPD. The external voltage necessary to compensate this CPD again to zero is then equal to the adsorbate-induced work-function change. There are various experimental solutions and set-ups how this can be accomplished automatically using the self-compensating lock-in technique; for details we refer the reader to the respective original communications [270, 271, 273–276].

A particularly intriguing solution to the problem of varying the capacitance of the condensor formed by the sample and the reference electrode is the so-called pendulum device developed by Hölzl and Schrammen [277]. Here, the reference electrode is made up by the (rectangular) end of a mechanical pendulum, which moves *parallel* to the sample with such amplitudes that there occurs a periodic plate area (A) rather than a distance variation. One advantage of this device is that work-function changes also can be followed during deposition of metal vapors, in that deposition is chopped with the pendulum frequency. Other electrical and mechanical solutions of the Kelvin probe as well as descriptions of electronic circuitries and precautions that must be taken against stray capacitance influences are collected in the monograph by Hölzl and Schulte [266].

Another way to measure CPDs, which requires even less effort than the Kelvin method and does not use any mechanical device, is the diode method. Its principle is very simple and is based on the fact that electrons emitted from a hot-cathode filament positioned in front of the sample surface are collected by the sample, which acts as an anode similar to a radio tube. It can be shown that for constant cathode parameters (work function Φ_C!), the anode current i_A of this diode device (which can be operated either in the space-charge-limited or in the retarding-field mode) depends only on the difference between the applied anode voltage V_A and the work function of the anode Φ_A [278]. In the space-charge operational mode we have then

$$i_A = B(U_A + \Phi_C - \Phi_A)^n ,$$ 4.109

where Φ_C and Φ_A are the work functions of the cathode and anode, respectively, and B is a constant involving the geometry of the diode device, the position of the space charge, and the filament temperature. The exponent n takes a value of approximately 1.5.

A variation of the anode's work function by the amount $\Delta\Phi$, for example, during gas adsorption, results in a parallel shift of the current-voltage curve $i_A(U_A)$, and provided that Φ_C, T, and B remain constant, Eq. 4.109 predicts that the change of the anode voltage U_A is equal to the change of the surface potential and hence the work function $\Delta\Phi$. A series of current-voltage curves is displayed in Fig. 4.67. For the sake of convenience, electronic devices have been developed to keep the anode current constant and record the corresponding voltage continuously, which corresponds in fact to a continuous $\Delta\Phi$ measurement. This type of operation was initially proposed by Klemperer and Snaith [279] and later improved in our laboratory to yield an accuracy of ~100 μV with a long-

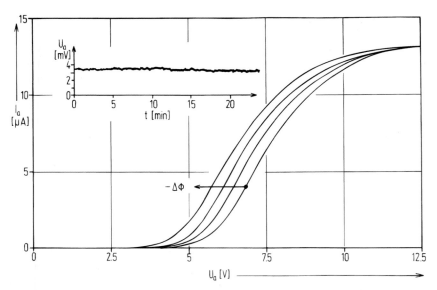

Fig. 4.67. Series of current-voltage curves obtained with the diode method following adsorption. A $\Delta\Phi$ increase causes a shift towards higher voltages, and by adjusting constant anode-current conditions the time dependence of the anode voltage U_A reflects directly the work-function change.

term drift of less than 1 mV/h [280]. However, there are two problems that limit a universal use of the diode method – the major difficulty is the stability of the space charge in front of the cathode, which depends extremely sensitively on its temperature and work function Φ_C, as predicted by the Richardson equation, cf., Eq. 4.95. Chemically active gases, such as oxygen, also can and will interact with the cathode filament material (often consisting of thoriated tungsten to obtain a good electron emission at comparatively low temperatures) and alter its electron-emission characteristics dramatically. Then Eq. 4.109 no longer applies, and the CPD measurement fails. The other complicating factor is that the cathode filament, in order to achieve a stable operation, must be carefully outgassed prior to the experiment and, furthermore, be placed relatively close to the sample (typical working distance around 3–5 mm). Therefore, a local heating of the sample surface is often inevitable, which can impair $\Delta\Phi$ measurements performed with weakly adsorbed gases at low temperatures. Nevertheless, there are many reports in the literature where the diode method was successfully employed, and we give some references for further information [278–282]. The advantages and shortcomings of the diode method in general as well as various experimental verifications have been complied in several articles, which are recommended for further reading [278–285].

To conclude this brief survey of the most important work-function measurement techniques, we refer again to the remarks made in the introduction to this section. Together with thermal desorption spectroscopy (which offers the potential of determining relative and/or absolute adsorbate coverages) the work-function measurement not only allows profound statements to be made about the electronic interaction between adsorbed particles and the surface (for example, dipole moments can be derived) but can also yield invaluable information about thermodynamical (energetic) and kinetic properties of adsorbate systems. This was documented by means of the example: Xe on Ni(100) in Chapter 2 as well as on several other occasions (i.e., H on Pt(111) and CO on Pd(100)).

4.5 Trends and Conclusions

In the foregoing sections, we have presented a selection of methods and techniques, which are – in our opinion – most versatile and particularly suited to characterizing not only clean-solid surfaces with regard to topography, morphology, or electronic structure, but also adsorbate systems, that is to say, the interface region between a solid surface and the adsorbed layer(s) on top. The lack of available space has made it necessary to confine this presentation to just a few such methods, most of them being well established for many years. For this reason, many modern surface analysis tools could only be mentioned, but hopefully our selection has included those techniques and spectroscopies that will also in the future provide the highest potential towards the ultimate goal, namely to completely characterize the physical state of a surface. At present, we are still relatively far away from this goal, although the invention of scanning tunneling microscopy (and spectroscopy) together with the extended use of *combined* methods have led to a situation where the static (equilibrium) properties of surfaces are relatively well known, at least as far as single-crystal surfaces are concerned. In spite of this, there is still an overwhelming number of open problems awaiting experimental (and theoretical) solution. Our data body concerning the geometrical and electronic structure of clean surfaces, including restructuring phenomena, is far from being complete. Here, an increasing number of structure analyses is required, whereby LEED, LEEM, EXAFS or STM investigations could provide the necessary information. Much of our attention during previous decades was devoted to metal (especially transition metal) surfaces, which is understandable in view of the catalytic significance of these materials. Of course, there were also numerous investigations directed at semiconducting surfaces, their geometrical and electronic structure, and their surface chemical composition. However, for technological and practical reasons, these studies were mainly focused on selected materials, such as elemental silicon and some II-VI and III-V componds (i.e., HgTe, GaAs, and InSb, to list only a few). In view of materials science and catalysis the surface studies should in the future be intensified and extended to oxide, nitride, and sulfide surfaces; metallic alloys; bimetallic materials; and metallic glasses which are believed will steadily become more interesting in the future, along with the (relatively complicated) High-T_c superconducting materials. As long as single crystals are available, there are no foreseeable major difficulties in applying the established techniques, unless charging or local decomposition problems occur with insulating or poorly conducting materials, which can sometimes impair the application of electron impact spectroscopies. However, most of the new and interesting materials will not be available in single-crystalline form; by contrast, they are often amorphous or highly dispersed, and new methodical developments are necessary to examine their properties, whereby in many cases researchers will aspire to atomic resolution.

As far as the experimental tools suited for analyzing surface geometries and electronic structures are concerned, the interest in the future will almost certainly be directed to the microscopic methods, STM in particular, that will sooner or later become standard analysis methods much like LEED or AES. Researchers will try to improve the performance so that atomic resolution will be obtained routinely. They also will attempt to solve the thermal-drift problems pertinent to STM and improve pattern-recognition techniques. Besides these rather sophisticated methods, the established techniques dominant in the industrial laboratories, such as Auger electron spectroscopy (Auger microprobe), scanning-electron microscopy (SEM), ESCA, SIMS, and depth profiling, will undoubtedly be further improved as far as lateral resolution, sensitivity, or ease of handling is concerned.

Automatic sample transfer and computer-controlled data acquisition are likewise major aims mainly in industrial application, as well as development of methods that do not make such great demands on the vacuum conditions. In addition, differentially pumped ESCA tubes working in the mbar regime are presently being developed. Optical methods, such as Fourier-transform infrared (FTIR), various kinds of laser spectroscopies, and improved x-ray techniques are capable of working at atmospheric pressures that would facilitate surface analysis studies considerably. A very noteworthy recent optical development is the use of nonlinear optics in surface analysis, for example, the second harmonic generation (SHG). This kind of optical spectroscopy can provide rich information regarding the structure and adsorption kinetics at an interfacial region [286–288], whereby the instrumental effort is relatively simple and straightforward. Laser light is directed onto a surface where a small fraction of it is frequency doubled, because of the second-order nonlinear polarizability of the crystal, which under the dipole approximation is nonzero only at the interface where the center of inversion is broken along the surface normal. These second harmonic photons are detected in the direction of the specularly scattered fundamental light beam. The aforementioned excitation mechanism brings about an intrinsic surface sensitivity of SHG, which makes this method a generally useful diagnostic tool. Without entering further details, we refer to original communications of Plummer's group, where the adsorption of oxygen and pyridine on silver (110) was studied [289], and by Tom et al., who examined CO adsorption on Rh surfaces [290]. A major advantage of SHG is, of course, that it does not require UHV conditions.

Because a careful investigation of an adsorption system will always require the best vacuum conditions possible, the atmospheric-pressure methods have a preferential impact for industrial applications, where the aforementioned type of sophisticated fundamental research is seldom acquired. It is worth mentioning in this context that the investigation of interfaces at the solid-liquid boundary by means of some formerly only UHV-compatible methods appears to move within the bounds of possibility. A particularly striking example is the use of the scanning tunneling microscope to study electrochemical processes, for instance etching or galvanic deposition of metals in situ which has recently been accomplished [291, 292].

A wealth of open questions and really challenging problems exist with regard to *surface dynamics*. Here, the body of experimental (and theoretical) instrumentation is still very poor, but the development of powerful laser systems and molecular-beam equipment leads one to suppose that in the near future time-resolved and state-selective methods will more and more enter the field and finally provide us with a much better understanding not only of diffusion and ordering phenomena (phase transitions) and surface vibrations, but also of surface-chemical *reactions* (the simplest being the dissociation of a diatomic molecule at the surface). Many promising initial steps have been made in the meantime; we are reminded of the pioneering work in Ertl's group, where a molecular beam of nitric oxide NO was scattered off a graphite [293] and a platinum(111) surface [294], whereby the population of the various NO rotational states could be analyzed before and after scattering by means of laser-induced fluorescence (LIF) An example is presented in Fig. 4.68, showing a comparison of rotational spectra of gaseous NO, beam of incoming NO, and NO backscattered from a graphite surface. Even at first glance, the redistribution of the rotational states due to surface scattering is evident. Similar experiments were performed in other groups around the same time, for example, by Auerbach and his crew on the system NO/Ag(111) [295, 296]. Parallel trajectory calculations were carried out by Kimman et al. [297], which gave much insight into the processes of energy and momentum exchange between diatomic molecules and solid

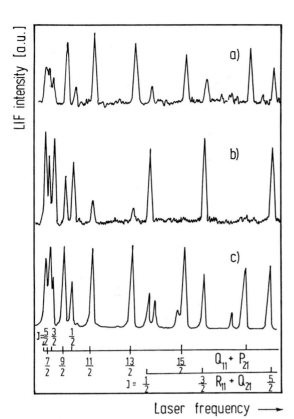

Fig. 4.68. Analysis of the rotational state distribution of nitric oxide NO using Laser-induced fluorescence: a) NO scattered off a graphite surface; b) LIF spectrum of the incoming NO beam; c) NO gas-phase-reference spectrum. On the x-axis, the various distinguishable rotational levels J_i are indicated. After Frenkel et al. [293].

surfaces. State-selective spectroscopy using supersonic molecular beams and time-of-flight mass spectrometry is currently performed in various laboratories with the aim of characterizing and disentangling translational, rotational and vibrational contributions of the internal energy of scattered molecules. A great deal of activity is still devoted to the NO molecule, as well as the hydrogen molecule [298–300]. Recent developments permit even the velocity distributions of molecules to be determined with internal quantum-state resolution. A comprehensive review of the present status of gas-surface interaction dynamics is presented in several recent articles [301–307]. Further developments in this area may comprise the excitation and state-selective spectroscopy of other molecules, such as CO, N_2, or H_2, where problems arise due to the relatively high photon energies required for excitation.

Another still-growing area is the *reactive scattering* of molecules at surfaces, where particles are detected that have a different mass than the molecules originally used in the molecular beam and are formed by a surface-chemical reaction. An important step forward has been made in this direction by investigations of the systems CO on Pd(111) [308] and Pt(111) [309] as well as NO_2 on germanium surfaces (in Ertl's group [310]); it is felt that this area is still in the beginning stages.

A somewhat related field, where dynamical processes are examined, concerns surface reactions in general, which may occur under steady-state conditions in a UHV chamber acting as chemical reactor. The catalytic oxidation of carbon monoxide on transition-metal surfaces is such a reaction that has been frequently investigated in the past, and it could be shown that it obeys the so-called Langmuir-Hinshelwood reaction mechanism, which will be explained in Chapter 5. Under most experimental conditions, bimolecular reactions of this type will run in the steady-state mode, which is characterized by constant input and output parameters. Under special circumstances, however, kinetic instabilities leading to oscillatory behavior (i.e., the occurrence of self-sustained kinetic oscillations in the product formation) can be observed. In some favorable cases, even lateral concentration gradients have been observed, for instance with CO oxidation over Pt(110) and (100) surfaces. Here, real reaction fronts, which travel like chemical waves across the surface [311], have been reported; it is one goal of these experiments to directly display the corresponding lateral inhomogeneities (e.g., CO-rich and CO-deficient surface areas). Here we are reminded of another beneficial role of the adsorbate-induced work-function change. Because $\Delta\Phi$ is strongly dependent on CO coverage, this quantity can be conveniently used to image CO-covered surface areas. A peculiar development in this context is the so-called scanning photoelectron microscope (SPM), which monitors the lateral intensity distribution of emitted photoelectrons (and which in turn is governed by the work function of the emitting patches, cf., Eqs. 4.90 and 4.92). Using SPM, reaction fronts (which sometimes really show a self-organization process) could be made visible in the form of spirals similar to the Belousov-Zhabotinsky reaction [312–314]. One could easily expand on this topic, however, we shall rather be concerned with more elementary features of surface reactions under the aspect of heterogeneous catalysis in the next chapter (Chapter 5).

References

1. Seah MP, Dench WA (1979) Quantitative Electron Spectroscopy of Surfaces: A Standard Data Base for Electron Inelastic Mean Free Paths in Solids. Surf Interf Anal 1:1–11
2. Trendelenburg EA (1963) Ultrahochvakuum. Braun, Karlsruhe
3. Dushman S, Lafferty JM (1962) Scientific Foundations of Vacuum Technique, 2nd edn. Wiley, New York
4. Edelmann C, Schneider HG (1978) Vakuumphysik und -technik. Akademische Verlagsgesellschaft Geest und Portig KG, Leipzig
5. Redhead PA, Hobson JP, Kornelsen EV (1968) The Physical Basis of Ultrahigh Vacuum. Chapman and Hall, London
6. Robinson NW (1968) The Physical Principles of Ultrahigh Vacuum Systems and Equipment. Chapman and Hall, London
7. Benninghoven A (1971) In: "Physik 1971", Plenarvorträge der 36. Physikertagung in Essen. Teubner, Stuttgart, p 331
8. Woodruff DP, Delchar TA (1986) Modern Techniques of Surface Science, Cambridge University Press, Cambridge
9. Brümmer O, Heydenreich J, Krebs KH, Schneider HG (eds) (1980) Handbuch der Festkörperanalyse mit Elektronen, Ionen und Röntgenstrahlen. Vieweg Braunschweig 1980
10. Ertl G, Küppers J (1985) Low Energy Electrons and Surface Chemistry, 2nd edn. Verlag Chemie, Weinheim
11. Prutton M (1983) Surface Physics; 2nd edn. Oxford University Press, Oxford
12. Estrup PJ, McRae EG (1971) Surface Studies by Electron Diffraction. Surface Sci 25:1–52
13. Bauer E (1969) Low–energy Electron Diffraction. In: Techniques of Metal Research, vol II. Wiley, New York, pp 559–639
14. Ertl G (1971) Verwendung niederenergetischer Elektronen zur Untersuchung von Oberflächenstrukturen und geordneten Adsorptionsphasen. Ber Bunsenges Phys Chem 75:967–979
15. Pendry JB (1974) Low Energy Electron Diffraction. Academic Press, New York–London

16. van Hove MA, Tong SY (1979) Surface Crystallography by LEED. Springer, Berlin Heidelberg New York

17. Somorjai GA, van Hove MA (1979) Adsorbed Monolayers on Solid Surfaces. Springer, Berlin Heidelberg New York

18. van Hove MA, Weinberg WH, Chan CM 1986) Low-Energy Electron Diffraction. Springer, Berlin Heidelberg New York

19. Davisson CJ, Germer LH (19271, The Scattering of Electrons by a Single Crystal of Nickel. Nature 119:558–560

20. Davisson CJ, Germer LH (1927) Diffraction of Electrons by a Crystal of Nickel. Phys Rev 30:705–740

21. Lang E, Heilmann P, Hanke G, Heinz K, Müller K (1979) Fast LEED Intensity Measurements with a Video Camera and a Video Tape Recorder. Appl Phys 19:287–293

22. Heilmann P, Lang E, Heinz K, Müller K (1984) The Necessity for Fast LEED Intensity Measurements. In: Marcus PM. Jona F (eds) Determination of Surface Structure by LEED Plenum Press, New York, p 463–481

23. Heinz K, Müller K (1982) LEED Intensities – Experimental Progress and New Possibilities of Surface Structure Determination. In: Tong SY, van Hove MA (eds) The Structure of Surfaces. Springer, Berlin Heidelberg New York, pp 1–54

24. Gronwald KD, Henzler M (1982) Epitaxy of Si(111) as Studied with a New High Resolving LEED System. Surface Sci 117:180–187

25. Scheithauer U, Meyer G, Henzler M (1986) A New LEED Instrument for Quantitative Spot Profile Analysis. Surface Sci 178:441–451

26. Ewald PP (1921)VII. Das reziproke Gitter in der Strukturtheorie. Z Krist 56:129–156

27. Marcus PM (1984) A Unified Formulation of Multiple Scattering Calculations. In: Marcus PM. Jona F (eds) Determination of Surface Structure by LEED. Plenum Press, New York, p 93–128

28. Clarke LJ (1985) Surface Crystallography – An Introduction to Low Energy Electron Diffraction, Wiley, Chichester Ch. 4

29. Woolfson MM (1978) An Introduction to X-ray Crystallography, Cambridge University Press, Cambridge

30. Christmann K, Ertl G, Pignet T (1976) Adsorption of Hydrogen on a Pt(111) Surface. Surface Sci 54:365–392

31. Telieps W, Bauer E (1985) An Analytical Reflection and Emission UHV Surface Electron Microscope. Ultramicroscopy 17:57–66

32. Telieps W, Bauer E (1985) The (7×7) – (1×1) Phase Transition on Si(111). Surface Sci 162:163–168

33. Telieps W (1983) Ein Ultrahochvakuum-Elektronenmikroskop zur Abbildung von Oberflächen mit langsamen reflektierten und emittierten Elektronen. PhD thesis, TU Clausthal

34. Siegel B, Menadue JF (1967) Quantitative Reflection Electron Diffraction in an Ultrahigh Vacuum Camera. Surface Sci 8:206–216

35. Nielsen PEH (1973) On the Investigation of Surface Structure by Reflection High Energy Electron Diffraction (RHEED). Surface Sci 35:194–210

36. Smith DJ (1986) High-Resolution Electron Microscopy in Surface Science. In: Vanselow R, Howe R (eds) Chemistry and Physics of Solid Surfaces IV. Springer, Berlin Heidelberg New York, pp 413–434

37. Urban K (1990) Hochauflösende Elektronenmikroskopie. Phys Bl 46:77–84

38. Smith DJ (1986) Atomic Imaging of Surfaces by Electron Microscopy. Surface Sci 178:462–474

39. van der Veen JF (1985) Ion Beam Crystallography of Surfaces and Interfaces. Surface Sci Repts 5:199–288

40. Engel T, Rieder KH (1982) Structural Studies of Surfaces with Atomic and Molecular Beam Diffraction. In: Höhler G (ed) Structural Studies of Surfaces. Springer, Berlin Heidelberg New York, pp. 55–180

41. Engel T (1984) Determination of Surface Structure Using Atomic Diffraction. In: Vanselow R, Howe R (eds) Chemistry and Physics of Solid Surfaces V. Springer, Berlin Heidelberg New York, pp. 257–282

42. Taglauer E, Heiland W (1976) Surface Analysis with Low-energy Ion Scattering. Appl Phys 9:261–275

42a. Heiland W (1982) Ion Scattering Studies of Surface Crystallography. Appl Surface Sci 13:282–291

43. Ernst HJ, Hulpke E, Toennies JP (1987) Observation of a Soft Surface Phonon Mode in the Reconstruction of Clean W(100). Phys Rev Lett 58:1941–1944

44. Niehus H, Comsa G (1988) Structure Analysis of Crystal Surfaces by Low Energy Ion Beams. Nucl Instr Methods B33:876–883

45. Müller EW (1937) Elektronenmikroskopische Beobachtungen von Feldkathoden. Z Phys 106:541–550

46. Müller EW, Tsong TT (1969) Field Ion Microscopy. Elsevier, Amsterdam

47. Müller EW (1971) Investigations of Surface Processes with the Atom-Probe Field Ion Microscope. Ber Bunsenges Phys Chem 75:979–987

48. Gomer R (1961) Field Emission and Field Ionization. Harvard University Press, Cambridge MA

49. Bowkett KM, Smith DA (1970) Field Ion Microscopy. North Holland, Amsterdam

50. Graham WR, Ehrlich G (1974) Direct Identification of Atomic Binding Sites on a Crystal. Surface Sci 45:530–552

51. Ehrlich G (1966) Chemisorption on Single Crystal Planes. Disc Faraday Soc 41:7–13

52. Ehrlich G (1977) Direct Observation of Individual Atoms on Metals. Surface Sci 63:422–447

53. Ayrault G, Ehrlich G (1974) Surface Self-Diffusion on an fcc–Crystal: An Atomic View. J Chem Phys 60:281–294

54. Fink HW, Ehrlich G (1984) Pair and Trio Interactions between Adatoms: Re on W(110). J Chem Phys 81:4657–4665

54a. Fink HW, Ehrlich G (1984) Lattice Steps and Adatom Binding on W(211). Surface Sci 143:125–144

55. Witt J, Müller K (1986) Direct Observation of Reconstructed Surfaces in the Field Ion Microscope. Appl Phys A41:103–106

56. Witt J, Müller K (1986) Evidence for the Ir(100) Surface Reconstruction by Field Ion Microscopy. Phys Rev Lett 57:1153–1156

57. Block JH (1989) Experimental Techniques in Evaluating the Dynamics of Catalytic Surface Reactions, Dechema Monographs 120:169

58. Binnig G, Rohrer H (1983) Scanning Tunneling Microscopy. Surface Sci 126:236–244

59. Binnig G, Rohrer H, Gerber C, Weibel E (1983) 7×7 Reconstruction on Si(111) Resolved in Real Space. Phys Rev Lett 50:120–123

60. Binnig G, Rohrer H (1984) Scanning Tunneling Microscopy. Physica 127B: 37–45

61. Binnig G, Rohrer H (1987) Scanning Tunneling Microscopy – From Birth to Adolescence. Rev Mod Phys 59:615–625

62. Behm RJ, Hösler W (1986) Scanning Tunneling Microscopy. In: Vanselow R, Howe R (eds) Chemistry and Physics of Solid Surfaces VI. Springer, Berlin Heidelberg New York, Ch 14, pp 361–412

63. Quate CF (1986) Vacuum Tunneling: A New Technique for Microscopy. Physics Today 39:26–33

64. Binnig G, Rohrer H (1985) The Scanning Tunneling Microscope. Scientific American 235, No. 2:40–46

65. Fuchs H (1989) Strukturen – Farben – Kräfte: Wanderjahre der Raster-Tunnelmikroskopie. Phys Bl 45:105–115

66. Kuk Y, Silverman PJ (1989) Scanning Tunneling Microscope Instrumentation. Rev Sci Instr 60:165–180

67. Wintterlin J (1989) Struktur und Reaktivität einer Metalloberfläche – eine Untersuchung mit dem Rastertunnelmikroskop am System Al(111)/Sauerstoff. PhD thesis, Freie Universität Berlin

68. Sommerfeld A, Bethe H (1933) Elektronentheorie der Metalle. In: Geiger H, Scheel K (eds) Handbuch der Physik Bd 24/2. Julius Springer, Berlin, pp. 333–622

69. Garcia N, Ocal C, Flores F (1983) Model Theory for Scanning Tunneling Microscopy: Application to Au(110) (1×2). Phys Rev Lett 50:2002–2005

70. Garcia N, Flores F (1984) Theoretical Studies for Scanning Tunneling Microscopy. Physica 127B:137–142

71. Stoll E (1984) Resolution of the Scanning Tunnel Microscope. Surface Sci 143:L411–L416

182

72. Tersoff J, Hamann DR (1983) Theory, and Application for the Scanning Tunneling Microscope. Phys Rev Lett 50:1998–2001

72a. Tersoff J, Hamann DR (1983) Theory of the Scanning Tunneling Microscope. Phys Rev B31:805–813

73. Baratoff A (1984) Theory of Scanning Tunneling Microscopy – Methods and Approximations. Physica 127B:143–150

74. Doyen G (1990) Theorie der Rastertunnelmikroskopie, Verh Dtsch Phys Ges 4:O–24.1 and to be published

75. Becker RS, Golovchenko JA, McRae EG, Swartzentruber BS (1985) Tunneling Images of Atomic Steps on the Si(111) 7×7 Surface. Phys Rev Lett 55:2028–2031

76. Becker RS, Golovchenko JA, Hamann DR, Swartzentruber BS (1985) Real Space Observation of Surface States on Si(111) 7×7 with the Tunneling Microscope. Phys Rev Lett 55:2032–2034

77 Gritsch T, Coulman D, Behm RJ, Ertl G (1989) A Scanning Tunneling Microscopy Investigation of the 1×2 – 1×1 Structural Transformation of the Pt(110)–Surface. Appl Phys A49:403–406

78. Binnig G, Rohrer H, Gerber C, Weibel E (1983) (111) Facets as the Origin of Reconstructed Au(110) Surface. Surface Sci 131:L379–L384

79. Moritz W, Wolf D (1985) Multilayer Distortion in the Reconstructed (110) Surface of Au. Surface Sci 153:L655–L665

80. Gritsch T (1990) Oberflächenstrukturen, Strukturumwandlungsprozesse und Adsorbate auf Pt(110) und Au(110) – eine STM-Untersuchung, PhD thesis, Freie Universität Berlin

81. Wintterlin J, Brune H, Höfer H, Behm RJ (1988) Atomic Scale Characterization of Oxygen Adsorbates on Al(111) by Scanning Tunneling Microscopy. Appl Phys A47:99–102

82. Schweizer EK, Eigler DM (1991) Untersuchung von adsorbiertem Xenon auf Ni(110) und Pt(111) mit einen 4K Rastertunnelmikroskop, Verh Dtsch Phys Ges 4:1157 (O.25.3.) and to be published

83. Gritsch T, Coulman D, Behm RJ, Ertl G (1989) Mechanism of the CO-induced 1×2 – 1×1 Structural Transformation of Pt(110) Phys Rev Lett 63:1086–1089

84. Hofmann P, Bare SR, King DA (1982) Surface Phase Transitions in CO Chemisorption on Pt(110). Surface Sci 117:245–255

85. Pötschke G, Behm RJ (1990) Struktur und Aufwachsverhalten von Cu auf Ru(0001) Verh Dtsch Phys Ges 4:1145 (O.19.13) and to be published

86. Magnussen O, Hotlos J, Behm RJ (1990) Atomare Struktur von Cu-Aufdampfschichten auf einkristallinen Au-Oberflächen im Elektrolyten. Verh Dtsch Phys Ges 4:1144 (O.19.22) and to be published

87. Sayers DE, Lytle FW, Stern CA (1971) Point Scattering Theory of x-ray K-Absorption Fine Structure. Adv X-Ray Anal 13:248–271

88. Lytle FW, Sayers DE, Stern EA (1975) Extended x-ray Absorption Fine Structure Techniques II: Experimental Practice and Selected Results. Phys Rev B11:4825–4835

89. Kincaid BM, Eisenberger P (1975) Synchrotron Radiation Studies of the K–edge Photoabsorption Spectra of Kr, Br_2, and $GeCl_4$: A Comparison of Theory and Experiment. Phys Rev Lett 34:1361–1364

90. Eisenberger P, Kincaid BM (1978) EXAFS: New Horizons in Structure Determination. Science 200:1441–1447

91. Lee PA, Citrin PH, Eisenberger P, Kincaid BM (1981) Extended x-ray Absorption Fine Structure – its Strengths and Limitations as a Structural Tool. Rev Mod Phys 53:769–806

92. Sandstrom DR, Lytle FW (1979) Developments in Extended x-ray Absorption Fine Structure Applied to Chemical Systems. Ann Rev Phys Chem 30:215–238

93. Teo BK, Toy DC ((eds) (1981) EXAFS Spectroscopy – Techniques and Applications, Plenum Press, New York

94. Citrin PH, Eisenberger P, Hewitt RC (1978) Extended x-ray Absorption Fine Structure of Surface Atoms on Single-Crystal Substrates: Iodine Adsorbed on Ag(111). Phys Rev Lett 41:309–313

94a. Stöhr I. Jaeger R, Brennan S (1982) Surface Crystallography by Means of Electron and Ion Yield SEXAFS. Surface Sci 117:503–524

95. Sinfelt JH, Via GH, Lytle FW, Greegor RB (1981) Structure of Bimetallic Clusters. Extended X-ray Absorption Fine Structure (EXAFS) Studies of Os–Cu Clusters. J Chem Phys 75:5527–5537

96. Lytle FW, Via GH, Sinfelt JH (1977) New Application of Extended X-ray Absorption Fine Structure (EXAFS) as a Surface-probe: Nature of Oxygen Interaction with a Ruthenium Catalyst. J Chem Phys 67:3831–3832
97. Via GH, Sinfelt JH, Lytle FW (1979) Extended x-ray Absorption Fine Structure (EXAFS) of Dispersed Metal Catalysts. J Chem Phys 71:690–699
98. Lytle FW, Wei PSP, Greegor RB, Via GH, Sinfelt JH (1979) Effect of Chemical Environment on Magnitude of x-ray Absorption Resonance at L_{III} Edges. Studies on Metallic Elements, Compounds, and Catalysts. J Chem Phys 70:4849–4855
99. Sinfelt JH, Via GH, Lytle FW (1982) Structure of Bimetallic Clusters. Extended x-ray Absorption Fine Structure (EXAFS) Studies on Pt–Ir Clusters. J Chem Phys 76:2779–2789
100. Sinfelt JH, Via GH, Lytle FW (1980) Structure of Bimetallic Clusters. Extended X-ray Absorption Fine Structure (EXAFS) Studies on Ru-Cu Clusters. J Chem Phys 72:4832–4844
101. Reed J, Eisenberger P, Teo BK, Kincaid BM (1977) Structure of the Catalytic Site of Polymer–Bound Wilkinson's Catalyst by x-ray Absorption Studies. J Am Chem Soc 99:5217–5218
102. Forstmann F, Berndt W, Büttner P (1973) Determination of the Adsorption Site by Low–energy Electron Diffraction for Iodine on Silver (111) Phys Rev Lett 30:17–19
102a. Maglietta M, Zanazzi E, Bardi U, Sondericker D. Jona F, Marcus PM (1982) LEED Structure Analysis of a $\sqrt{3} \times\sqrt{3}$ −30° Overlayer of Iodine on Ag(111) Surface Sci 123:141–151
103. Bianconi A, Incoccia L, Stipcich S (eds) (1983) EXAFS and Near Edge Structure. Springer, Berlin Heidelberg New York
104. Hogson KO, Hedman B, Penner-Hahn JE (eds)(1984) EXAFS and Near Edge Structure III. Springer, Berlin Heidelberg New York
105. Ibach H (1972) Low-Energy Electron Spectroscopy – a Tool for Studies of Surface Vibrations. J Vac Sci Technol 9:713–719
106. Ibach H (1971) Optical Surface Phonons in Zinc Oxide Detected by Slow-Energy Electron Spectroscopy. Phys Rev Lett 24:1416–1418
106a. Ibach H (1971) Surface Vibrations of Silicon Detected by Low-Energy Electron Spectroscopy. Phys Rev Lett 27:253–256
107. Froitzheim H, Ibach H, Lehwald S (1975) Reduction of Spurious Background Peaks in Electron Spectrometers. Rev Sci Instr 46:1325–1328
108. Froitzheim H, Ibach H, Lehwald S (1976) Surface Sites of H on W(100). Phys Rev Lett 36:1549–1551
108a. Froitzheim H, Ibach H, Lehwald S (1976) Surface Vibrations of Oxygen on W(100). Phys Rev B14:1362–1369
109. Ibach H, Mills DL (1982) Electron Energy Loss Spectroscopy and Surface Vibrations. Academic Press, New York
110. Froitzheim H (1977) Electron Energy Loss Spectroscopy. In: Ibach H (ed) Electron Spectroscopy for Surface Analysis, Springer, Berlin Heidelberg New York, pp 205–250
111. Willis RF (ed) (1980) Vibrational Spectroscopy of Adsorbates. Springer, Berlin Heidelberg New York
112. Newns DM (1980) Theory of Dipole Electron Scattering from Adsorbates. In: Willis RF (ed) Vibrational Spectroscopy of Adsorbates. Springer, Berlin Heidelberg New York, pp 23–54
113. Ertl G, Küppers J (1985) Low-Energy Electrons and Surface Chemistry, 2nd edn., Verlag Chemie, Weinheim, Ch. 11
114. Tong SY, Li CH, Mills DL (1981) Inelastic Scattering of Electrons by Adsorbate Vibrations in the Impact Scattering Regime: CO on Ni(100) as an Example. Phys Rev B24:806–816
115. Voigtländer B, Lehwald S, Ibach H (1989) Hydrogen Adsorption and the Adsorbate-induced Ni(110) Reconstruction – an EELS Study. Surface Sci 208:113–135
116. Lauth G, Schwarz E, Christmann K (1989) The Adsorption of Hydrogen on a Ruthenium (10$\bar{1}$0) Surface. J Chem Phys 91:3729–3743
117. Blyholder G (1964) Molecular Orbital View of Chemisorbed Carbon Monoxide. J Phys Chem 68:2722–2778

184

118. Blyholder G (1962) Infrared Spectrum of CO Chemisorbed on Iron. J Chem Phys 36:2036–2039
119. Blyholder G (1970) CNDO Model and Interpretation of the Photoelectron Spectrum of CO Chemisorbed on Ni. J Vac Sci Technol 11:865–868
120. Michalk G, Moritz W, Pfnür H, Menzel D (1983) A LEED Determination of the Structures of Ru(001) and of CO/Ru(001)$\sqrt{3} \times \sqrt{3}$ R 30°. Surface Sci 129:92–106
121. Thomas GE, Weinberg WH (1979) The Vibrational Spectrum and Adsorption Site of CO on the Ru(001) Surface. J Chem Phys 70:1437–1439
122. Ehsasi M, Wohlgemuth H, Christmann K (1991) Interaction of CO with a Rhodium(110) Surface, to be published
123. Pfnür H, Menzel D, Hofmann FM, Ortega A, Bradshaw AM (1980) High-Resolution Vibrational Spectroscopy of CO on Ru(001): The Importance of Lateral Interactions. Surface Sci 93:431–452
124. Bradshaw AM, Hoffmann FM (1978) The Chemisorption of Carbon Monoxide on Palladium Single Crystal Surfaces: IR Spectroscopic Evidence for Localized Site Adsorption. Surface Sci 72:513–535
125. Chesters MA, McDougell GS, Pemble ME, Sheppard N (1985) The Chemisorption of CO on Pd(110) at 110 and 300 K Studied by Electron Energy Loss Spectroscopy. Surface Sci 164:425–436
126. Eischens RP, Pliskin WA (1958) The Infrared Spectra of Adsorbed Molecules. Adv Catal Rel Subj 10:1–56
127. Benndorf C, Krüger B, Thieme F (1985) Unusually Low Stretching Frequency for CO Adsorbed on Fe(100) Surface Sci 163:L675–L680
128. Voigtländer B, Bruchmann D, Lehwald S, Ibach H (1990) Structure and Adsorbate–Adsorbate Interactions of the Compressed Ni(110)–(2×1) CO Structure. Surface Sci 225:151–161
129. Ibach H (1990) Auflösungsrekord in der Elektronenspektroskopie. Phys Bl 46:254
130. Ho W (1989) EELS of Molecular Beam and Temperature Induced Surface Processes: Implications on Time-dependent Surface Phenomena. Surface Sci 211/212:289–302
131. Froitzheim H, Köhler U (1987) Kinetics of the Adsorption of CO on Ni(111) Surface Sci 188:70–86
131a. Froitzheim H, Schulze M (1989) The Kinetics of the Adsorption and Desorption of the System CO/Pt(111) Derived from High-resolution TREELS. Surface Sci 211/212:837–843
132. Plummer EW (1975) Photoemission and Field Emission Spectroscopy. In: Gomer, R (ed) Interactions on Metal Surfaces. Springer, Berlin Heidelberg New York, pp 143–223
133. Plummer EW, Eberhardt W (1982) Angle-resolved Photoemission as a Tool for the Study of Surfaces. Adv Chem Phys 49:533–656
134. Roberts MW (1980) Photoelectron Spectroscopy and Surface Chemistry. Adv Catal Rel Subj 29:55–95
135. Siegbahn K, Nordling C, Fahlmann A, Nordberg H, Hamrin K, Hedman J. Johansson G. Bergmark T, Karlsson SE, Lindgren I, Lindberg B (1967) Electron Spectroscopy for Chemical Analysis. Atomic, Molecular and Solid State Structure Studies by Means of Electron Spectroscopy, Almquist and Wiksells, Stockholm
136. Spicer WE (1982) Development of Photoemission as a Tool for Surface Science: 1900–1980. In: Vanselow R, Howe R (eds) Chemistry and Physics of Solid Surfaces IV. Springer, Berlin Heidelberg New York
137. Feuerbacher B, Fitton B (1977) Photoemission Spectroscopy. In: Ibach H (ed) Electron Spectroscopy for Surface Analysis. Springer, Berlin Heidelberg New York
138. Feuerbacher B, Fitton B, Willis RF (1973) Photoemission and the Electronic Properties of Surfaces, Wiley, New York
139. Spicer WE (1958) Photoemissive, Photoconductive, and Optical Absorption Studies of Alkali-Antimony Compounds. Phys Rev 112:114–122
139a. Spicer WE (1962) Photoemission and Band Structure of the Semiconducting Compound CsAu. Phys Rev 125:1297–1299
140. Berglund CN, Spicer WE (1964) Photoemission Studies of Copper and Silver: Theory. Phys Rev 136:A1030–A1044
140a. Berglund CN, Spicer WE (1964) Photoemission Studies of Copper and Silver: Experiment. Phys Rev 136:A1044–A1064

141. Cashion JK, Mees JL, Eastman DE, Simpson JA, Kuyatt CE (1971) Windowless Photoelectron Spectrometer for High-resolution Studies of Solids and Surfaces. Rev Sci Instr 42:1670–1674
142. Eastman DE, Cashion JK (1971) Photoemission Energy-level Measurements of Chemisorbed CO and O on Ni. Phys Rev Lett 27:1520–1523
143. Christmann K, Demuth JE, (unpublished)
144. Küppers J, Michel H, (1979) Preparation of Ir(100)–(1×1) Surface Structures by Surface Reactions and its Reconstruction Kinetics as Determined with LEED, UPS, and Work Function Measurements. Appl Surface Sci 3:179–195
145. Bonzel HP, Helms CR, Kelemen S (1975) Observation of a Change in the Surface Electronic Structure of Pt(100) Induced by Reconstruction. Phys Rev Lett 35:1237–1240
146. Demuth JE, Eastman DE (1974) Photoemission Observations of π-d Bonding and Surface Reactions of Adsorbed Hydrocarbons on Ni(111) Phys Rev Lett 32:1123–1127
147. Sanda PN, Warlaumont JM, Demuth JE, Tsang JC, Christmann K, Bradley JA (1980) Surface–enhanced Raman Scattering from Pyridine on Ag(111) Phys Rev Lett 45:1519–1523
148. Demuth JE, Christmann K, Sanda PN (1980) The Vibrations and Structure of Pyridine Chemisorbed on Ag(111): The Occurrence of a Compressional Phase Transformation. Chem Phys Lett 76:201–206
149. Rubloff GW, Demuth JE (1977) Ultraviolet Photoemission and Flash-desorption Studies of the Chemisorption and Decomposition of Methanol on Ni(111) J Vac Sci Technol 14:419–423
150. Demuth JE, Ibach H (1979) Observation of a Methoxy Species on Ni(111) by High–resolution Electron Energy Loss Spectroscopy. Chem Phys Lett 60:395–399
151. Christmann K, Demuth JE (1982) The Adsorption and Reaction of Methanol on Pd(100). I. Adsorption and Condensation. J Chem Phys 76:6308–6317
151a. Christmann K, Demuth JE (1982) The Adsorption and Reaction of Methanol on Pd(100). II. Thermal Desorption and Decomposition. J Chem Phys 76:6318–6327
152. Clark DT, Feast WJ, Kilcast D, Musgrave WKR (1973) Applications of ESCA to Polymer Chemistry. III. Structures and Bonding in Homo-polymers of Ethylene and the Fluoroethylenes and Determination of the Compositions of Fluoro Co-polymers. J Polymer Sci 11:389–411
153. Fadley CS, Shirley DA (1971)): Multiplet Splitting of Metal–Atom Electron Binding Energies. Phys Rev A23:1109–1120
154. Ertl G, Thiele N (1979) XPS Studies with Ammonia Synthesis Catalysts. Appl Surface Sci 3:99–112
155. Koopmans T (1934) Über die Zuordnung von Wellenfunktionen und Eigenwerten zu den einzelnen Elektronen eines Atoms. Physica 1:104–113
156. Fadley CS (1978) Basic Concepts of X-ray Photoelectron Spectroscopy. In: Brundle CR, Baker AD (eds) Electron Spectroscopy: Theory, Techniques and Applications, Vol. 2, Academic Press, New York pp 1–157
157. Landau LD, Lifshitz EM (1985) Lehrbuch der Theoretischen Physik, Akademie-Verlag. Berlin
158. Muilenburg GE (ed) (1979) Handbook of X-ray Photoelectron Spectroscopy, Perkin–Elmer Corporation, Eden Prairie, Minnesota, USA
159. Siegbahn K, Nordling C. Johansson G, Hedman J, Heden PF, Hamrin K, Gelius U. Bergmark T, Werme LO, Manne R, Baer Y (1969) ESCA Applied to Free Molecules, North Holland, Amsterdam
160. Kowalczyk SP, Ley L, Martin RL, McFeely FR, Shirley DA (1975) Relaxation and Final–State Structure in XPS of Atoms, Molecules, and Metals. Disc Faraday Soc 60:7–17
161. van der Veen JF, Himpsel FJ, Eastman DE (1980) Structure–dependent 4f–Core Level Binding Energies for Surface Atoms on Ir(111) Ir(100)–(5x1) and Metastable Ir(100)–(1×1) Phys Rev Lett 44:189–192
162. Krause MO (1971) The Mζ-x-rays of Y to Rh in Photoelectron Spectroscopy. Chem Phys Lett 10:65–69
162a. Bonzel HP (1984)(personal communication)
163. Dose V (1983) Ultraviolet Bremsstrahlung Spectroscopy. Progr Surface Sci 13:225–284
164. Ohlin P (1942) Structure in the Short-wavelength Limit of the Continuous x-ray Spectrum. Ark Mat Astro Fys 29A:1–10

165. Nijboer BR (1946) On the Density Distribution of the Continuous x-ray Spectrum Near its Short-wavelength Limit. Physica (Utrecht) 12:461–465
166. Ulmer K (1959) New Method for the Evaluation of h/e from the Quantum Limit of the Continuous x-ray Spectrum. Phys Rev Lett 3:514–516
167. Scheidt H (1983) Neuere Entwicklungen auf dem Gebiet der Bremsstrahlungs-Isochromatenspektroskopie, Fortschr Phys 31:357–401
168. Woodruff DP. Johnson PD, Smith NV (1983) Inverse Photoemission. J Vac Sci Technol A1:1104–1110
169. Pendry JB (1980) New Probe for Unoccupied Bands at Surfaces. Phys Rev Lett 45:1356–1358
170. Pendry JB (1981) Theory of Inverse Photoemission. J Phys C14:1381–1391
171. Lang JK, Baer Y (1979) Bremsstrahlung Isochromat Spectroscopy Using a Modified XPS Apparatus. Rev Sci Instr 50:221–226
172. Dose V (1977) VUV Isochromat Spectroscopy. Appl Phys 14:117–118
173. Desinger K, Dose V, Glöbl M, Scheidt H (1984) Momentum-resolved Bremsstrahlung Isochromat Spectra from Ni(001) Solid State Commun 49:479–481
174. Scheidt H, Glöbl M, Dose V (1982) Unoccupied Electronic States on Ni(100) Induced by a c(2x2) Oxygen Overlayer. Surface Sci 123:L728–L732
175. Rangelov G, Memmel N. Bertel E, Dose V (1990) The Bonding of Hydrogen on Nickel Studied by Inverse Photoemssion. Surface Sci 236:250–258
176. Auger P (1925) Sur L'effet Photoélectrique Composé. J Phys Radium 6:205–208
177. Burhop EHS (1952) The Auger Effect and Other Radiationless Transitions, Cambridge University Press, London
178. Harris LA (1968) Analysis of Materials by Electron-excited Auger Electrons. J Appl Phys 39:1419–1427
178a. Harris LA (1968) Some Observations of Surface Segregation by Auger Electron Emission. J Appl Phys 39:1428–1431
179. Harris LA (1969) Angular Dependences in Electron-excited Auger Spectra. Surface Sci 15:77–93
180. Weber RE, Peria WT (1967) Use of LEED Apparatus for the Detection and Identification of Surface Contaminants. J Appl Phys 38:4355–4358
181. Palmberg PW, Rhodin TN (1968) Auger Electron Spectroscopy of fcc Metal Surfaces. J Appl Phys 39:2425–2432
182. Palmberg PW, Bohn GK, Tracy JC (1969) High-sensitivity Auger Electron Spectrometer. Appl Phys Lett 15:254–255
183. Chang CC (1971) Auger Electron Spectroscopy. Surface Sci 25:53–79
184. Sickafus EN (1974) Surface Characterization by Auger Electron Spectroscopy. J Vac Sci Technol 11:299–311
185. Grant JT (1982) Surface Analysis with Auger Electron Spectroscopy. Appl Surface Sci 13:35–62
186. Tracy JC (1973) In: Electron Emission Spectroscopy. Reidel D, Dordrecht, Boston, p 295
187. Müller K (1975) How Much can Auger Electrons Tell us About Solid Surfaces? In: Springer Tracts of Modern Physics, vol 77. Springer, Berlin Heidelberg New York, pp 97–125
188. Weißmann R, Müller K (1981) Auger Electron Spectroscopy – A Local Probe for Solid Surfaces. Surface Sci Repts 1:251–309
189. Fuggle JC (1982) High-resolution Auger Spectroscopy of Solids and Surfaces. In: Brundle CR, Baker AD (eds) Electron Spectroscopy: Theory, Techniques and Applications, vol. 4. Academic Press, New York
190. Larkins FP (1982) Theoretical Developments in High–resolution Auger Electron Studies of Solids. Appl Surface Sci 13:4–34
191. Coster D, Kronig RL (1935) A New Type of Auger Effect and its Influence on the x-ray Spectrum. Physica 2:13–24
192. Bergstrøm I, Hill RD (1954) Auger Electrons from the L-shell in Mercury. Arkiv Fysik 8:21–26
193. Burhop EHS (1955) Le Rendement de Fluoroscence. J Phys Radium 16:625–629
194. Drawin HW (1961) Zur formelmäßigen Darstellung der Ionisationsquerschnitte gegenüber Elektronenstoß. Z Phys 164:513–521

187

195. Seah MP (1972) Quantitative Auger Electron Spectroscopy and Electron Ranges. Surface Sci 32:703–728
196. Gallon TE (1972) The Estimation of Backscattering Effects in Electron-induced Auger Spectra. J Phys D56:822–832
197. Gerlach RL, DuCharme AR (1972) Total Electron Impact Ionization Cross Sections of K Shells of Surface Atoms. Surface Sci 32:329–340
198. Gallon TE (1969) A Simple Model for the Dependence of Auger Intensities on Specimen Thickness. Surface Sci 17:486–489
199. Leder LB, Simpson JA (1958) Improved Electrical Differentiation of Retarding Potential Measurements. Rev Sci Instr 29:571–574
200. Palmberg PW (1968) Optimization of Auger Electron Spectroscopy in LEED Systems. Appl Phys Lett 13:183–185
201. Ertl G, Küppers J (1971) Adsorption an Einkristall-Oberflächen von Cu/Ni–Legierungen I.. Surface Sci 24:104–124
202. Christmann K, Ertl G (1972) Adsorption of Carbon Monoxide on Silver/Palladium Alloys. Surface Sci 33:254–270
203. McDonell L, Woodruff DP, Holland BW (1975) Angular Dependence of Auger Electron Emission from Cu(111) and (100) Surfaces. Surface Sci 51:249–269
204. Aberdam D, Baudoing R, Blanc E, Gaubert C (1976) Theory of Angular Dependence in Electron Emission from Surfaces. Application to Auger Emission from Several Metal Surfaces. Surface Sci 57:306–322
205. Allié G, Blanc E, Dufayard D (1976) Angular Distribution of the Auger Emission from Aluminum and Nickel Surfaces. Surface Sci 57:293–305
206. Christmann K, Schober O, (unpublished)
207. Davis LE, MacDonald NC, Palmberg PW, Riach GE, Weber RE (1976) Auger Handbook 2nd edn.. Physical Electronics Industries, Inc., Eden Prairie, Minnesota, USA
208. Palmberg PW (1971) Physical Adsorption of Xenon on Pd(100) Surface Sci 25:598–608
209. Ernst KH, Christmann K (1989) The Interaction of Glycine with a Platinum (111) Surface. Surface Sci 224:277–310
210. Biberian JP, Somorjai GA (1979) On the Determination of Monolayer Coverage by Auger Electron Spectroscopy. Application to Carbon on Platinum. Appl Surface Sci 2:352–358
211. Bauer E (1982) Epitaxy of Metals on Metals. Appl Surface Sci 11/12:479–494
212. Vickerman JC, Christmann K, Ertl G, Heimann P, Himpsel FJ, Eastman DE (1983) Geometric Structure and Electronic States of Copper Films on a Ruthenium(0001) Surface. Surface Sci 134:367–388
213. Harendt C, Christmann K, Hirschwald W, Vickerman JC (1986) Model Bimetallic Catalysts: The Preparation and Characterisation of Au/Ru Surfaces and the Adsorption of Carbon Monoxide. Surface Sci 165:413–433
214. Bauer E, Poppa H, Todd G, Bonczek F (1974) Adsorption and Condensation of Cu on W Single-crystal Surfaces. J Appl Phys 45:5164–5175
215. Hofmann S (1983) In: Briggs O, Seah MP (eds) Practical Surface Analysis by Auger and X-ray Photoelectron Spectroscopy. Wiley, New York
216. Joyce BA, Neave JH (1971) An Investigation of Silicon-Oxygen Interactions using Auger Electron Spectroscopy. Surface Sci 27:499–515
217. Shelton JC, Patil HR, Blakely JM (1974) Equilibrium Segregation of Carbon to a Ni(111) Surface: A Structure Phase Transition. Surface Sci 43:493–520
218. Benninghoven A, Rüdenauer FG, Werner HW (1987) Secondary Ion Mass Spectrometry. In: Elving PJ, Winefordner JD, Kolthoff IM (eds) Chemical Analysis, vol. 86. Wiley, New York
219. Benninghoven A (1971) Beobachtung von Oberflächenreaktionen mit der statischen Methode der Sekundärionen-Massenspektroskopie. I. Die Methode. Surface Sci 28:541–556
220. Benninghoven A (1975) Developments in Secondary Ion Mass Spectroscopy and Applications to Surface Studies. Surface Sci 53:596–625

221. Barber M, Vickerman JC, Wolstenholme J (1977) The Application of SIMS to the Study of CO Adsorption on Polycrystalline Metal Surfaces. Surface Sci 68:130–137
222. Vickerman JC (1987) Static SIMS – A Technique for Surface Chemical Characterisation in Basic and Applied Surface Science. Surface Sci 189/190:7–14
223. Werner HW (1980) Quantitative Secondary Ion Mass Spectrometry: A Review. Surface Interf Anal 2:56–74
223a. Werner HW (1975) The Use of Secondary Ion Mass Spectrometry in Surface Analysis. Surface Sci 47:301–323
224. Wittmaack K (1979) Secondary Ion Mass Spectrometry as Means of Surface Analysis. Surface Sci 89:668–700
225. Benninghoven A, Sichtermann WK (1977) Secondary Ion Mass Spectrometry: A New Analytical Technique for Biologically Important Compounds, Org Mass Spectrom 12:595–598
225a. Benninghoven A, Sichtermann WK (1978) Detection, Identification, and Structural Investigation of Biologically Important Compounds by Secondary Ion Mass Spectrometry, Anal Chem 50:1180–1184
226. Holtkamp D, Lange W. Jirikowski M, Benninghoven A (1984) UHV Preparation of Organic Overlayers by a Molecular Beam Technique. Appl Surface Sci 17:296–308
227. Ehrlich G (1963) Modern Methods in Surface Kinetics. Flash Desorption, Field Emission Microscopy, and Ultra High Vacuum Techniques. Adv Catal Rel Subj 14:255–427
228. Redhead PA (1962) Thermal Desorption of Gases, Vacuum 12:203–211
229. King DA (1975) Thermal Desorption from Metal Surfaces: A Review. Surface Sci 47:384–402
230. Menzel D (1975) Desorption Phenomena. In: Gomer R (ed) Interactions on Metal Surfaces Topics in Applied Physics, vol. 4. Springer, Berlin Heidelberg New York, pp 101–142
231. Peterman LA (1972) Thermal Desorption Kinetics of Chemisorbed Gases. In: Progress in Surface Science, vol 1. Pergamon Press Oxford, pp 2–61
232. Behm RJ, Christmann K, Ertl G, van Hove MA (1980) Adsorption of CO on Pd(100) J Chem Phys 73:2984–2993
233. Christmann K, Ertl G (1976) Interaction of Hydrogen with Pt(111): The Role of Atomic Steps. Surface Sci 60:365–384
234. Christmann K (1979) Adsorption of Hydrogen on a Nickel(100) Surface. Z Naturf 34a:22–29
235. Christmann K, Ertl G, Shimizu H (1980) Model Studies on Bimetallic Cu/Ru Catalysts I., Cu on Ru(0001) J Catal 61:397–411
236. Engelhardt HA, Menzel D (1976) Adsorption of Oxygen on Silver Single Crystal Surfaces. Surface Sci 57:591–618
237. Bauer E, Bonzcek F, Poppa H, Todd G (1975) Thermal Desorption of Metals from Tungsten Single Crystal Surfaces. Surface Sci 53:87–109
238. Chan CM, Aris R, Weinberg WH (1978) An Analysis of Thermal Desorption Mass Spectra I.. Appl Surface Sci 1:360–376
238a. Chan CM, Weinberg WH (1978) An Analysis of Thermal Desorption Mass Spectra II. Appl Surface Sci 1:377–387
239. Christmann K, Demuth JE (1982) Interaction of Inert Gases with a Nickel (100) Surface I., Adsorption of Xenon. Surface Sci 120:291–318
240. Solomun T, Christmann K, Baumgärtel H (1989) Interaction of Acetonitrile and Benzonitrile with the Au(100) Surface. J Phys Chem 93:7199–7208
241. Christmann K, Rüstig J (1984) Interaction of Methanol with Nickel, Palladium and Silver Single Crystal Surfaces. Proc. 8th Intern Congr Catalysis. Berlin, Vol. IV, p 13–24
242. Christmann K, Ertl G, Schober O (1974) Adsorption of CO on a Ni(111) Surface. J Chem Phys 60:4719–4724
243. Lauth G, Solomun T, Hirschwald W, Christmann K (1989) The Interaction of Carbon Monoxide with a Ruthenium (1010) Surface. Surface Sci 210:201–224
244. Schwarz E, Lenz J, Wohlgemuth H, Christmann K (1990) The Interaction of Oxygen with a Rhodium(110) Surface, Vacuum 41:167–170

245. Ehsasi M, Christmann K (1988) The Interaction of Hydrogen with a Rhodium(110) Surface. Surface Sci 194:172–198

246. Morris MA, Bowker M, King DA (1984) Kinetics of Adsorption, Desorption, and Diffusion at Metal Surfaces. In: Bamford CH, Tipper CHF, Compton RG (eds) Simple Processes at the Gas-Solid Interface vol. 19. Elsevier, Amsterdam, pp 1–180

247. Conrad H, Ertl G, Latta EE (1974) Adsorption of Hydrogen on Palladium Single Crystal Surfaces. Surface Sci 41:435–446

248. Christmann K, Chehab F, Penka V, Ertl G (1985) Surface Reconstruction and Surface Explosion Phenomena in the Nickel (110)/Hydrogen System. Surface Sci 152/153:356–366

249. Stuve EM, Madix RJ, Brundle CR (1985) The Adsorption and Reaction of Ethylene on Clean and Oxygen-covered Pd(100) Surface Sci 152/153:532–542

250. Madix RJ (1979) The Adsorption and Reaction of Simple Molecules on Metal Surfaces. Surface Sci 89:540–553

251. Herz H, Conrad H, Küppers J (1979) Thermocouple-controlled Temperature Programmer for Flash Desorption Spectroscopy. J Phys E12:369–371

252. see, for example: Menon PG (1988) Hydrogen as a Tool for Characterization of Catalyst Surfaces by Chemisorption, Gas Titration, and Temperature-programmed Techniques. In: Paal Z, Menon PG (eds) Hydrogen Effects in Catalysis. Dekker, New York, pp 117–138

253. Menzel D, Gomer R (1964) Desorption of Metal Surfaces by Low–energy Electrons. J Chem Phys 41:3311–3328 (ESD)

254. Redhead PA (1964) Interactions of Slow Electrons with Chemisorbed Oxygen, Can J Phys 42:886–905 (ESD)

255. Knotek ML. Jones VO, Rehn V (1979) Photon–stimulated Desorption of Ions. Phys Rev Lett 43:300–303 (PSD)

256. Woodruff OP. Johnson PD, Traum MM, Farrell HH, Smith NV, Benbow RL, Huryzch Z (1981) Photon and Electron Stimulated Desorption of Adsorbates from W(100) Surface Sci 104:282–299 (PSD)

257. Hallwachs WI (1898) Wied Ann 38:157

258. Einstein A (1905) Zur Theorie der Lichterzeugung und Lichtabsorption. Ann Phys 20:199–206

259. Fowler RH (1931) The Analysis of Photoelectric Sensitivity Curves for Clean Metals at Various Temperatures. Phys Rev 38:45–56

260. Simon H, Suhrmann R (1958) Der lichtelektrische Effekt und seine Anwendungen, 2. Aufl.. Springer, Berlin

261. Suhrmann R (1956) Elektronische Wechselwirkungen bei der Chemisorption an elektrisch leitenden Oberflächen. Ber Bunsenges Phys Chem 60:804–815

262. Wedler G (1970) Adsorption, Verlag Chemie, Weinheim, p 107ff.

263. Fowler RH, Nordheim L (1928) Electron Emission in Intense Electric Fields. Proc Roy Soc (London) A119:173–181

264. Becker JA (1955) Adsorption on Metal Surfaces and its Bearing on Catalysis. Adv Catal Rel Subj 7:136–211

265. Herring C, Nichols MH (1949): Thermionic Emission. Rev Mod Phys 21:85–270

265a. Culver RV, Tompkins FC (1959) Surface Potentials and Adsorption Process on Metals. Adv Catal Rel Subj 11:67–13

266. Hölzl J, Schulte FK, (1979) Work Function of Metals. Springer Tracts of Modern Physics, vol. 85. Springer, Berlin Heidelberg New York

267. Wandelt K (1987) The Local Work Function of Thin Metal Films: Definition and Measurement. In: Wißmann P (ed) Thin Metal Films and Gas Chemisorption. Elsevier, Amsterdam, pp 280–368

268. Topping J (1927) On the Mutual Potential Energy of a Plane Network of Doublets. Proc Roy Soc London A114:67–72

269. Mignolet JCP (1950) Studies in Contat Potentials II., Vibrating Cells for the Vibrating Condenser Method. Disc Faraday Soc 8:326–329

270. Delchar TA, Ehrlich G (1965) Chemisorption on Single–crystal Planes: Nitrogen on Tungsten. J Chem Phys 42:2686–2702

271. Tracy JC, Palmberg PW (1969) Structural Influences on Adsorbate Binding Energy I. Carbon Monoxide on (100) Palladium. J Chem Phys 51:4852–4862

272. Thomson W (later: Lord Kelvin) (1889) Contact Electricity of Metals. Phil Mag 46:82–120

273. Zisman WA (1932) A New Method of Measuring Contact Potential Differences in Metals. Rev Sci Instr 3:367–370

274. Simon RE (1959) Work Function of Iron Surfaces Produced by Cleavage in Vacuum. Phys Rev B116:613–617

275. Ertl G, Küppers D (1971) Wechselwirkung von Wasserstoff mit einer Nickel (111)) Oberfläche. Ber Bunsenges Phys Chem 75:1017–1025

276. Engelhardt HA, Feulner P, Pfnür H, Menzel D (1977) An Accurate and Versatile Vibrating Capacitor for Surface and Adsorption Studies. J Phys E10:1133–1136

277. Hölzl J, Schrammen P (1974) A New Pendulum Device to Measure Contact Potential Differences. Appl Phys 3:353–357

278. Knapp AG (1973) Surface Potentials and their Measurement by the Diode Method. Surface Sci 34:289–316

279. Klemperer DF, Snaith JC (1971) A Recording Diode for the Measurement of the Surface Potential of Adsorbed Layers on Evaporated Metal Films. J Phys E4:860–864

280. Christmann K, Herz H (1979) A Precise Diode Method for Recording Contact Potential Changes Caused by Gas Adsorption. Rev Sci Instr 50:988–992

281. Rivière JC (1969) Work Function: Measurements and Results. In: Green M (ed) Solid State Surface Science. Dekker, New York, pp 180–289

282. Pritchard J (1963) Surface Potential Study of the Chemisorption of Hydrogen and Carbon Monoxide on Evaporated Copper and Gold Films. Transact Faraday Soc 59:437–452

283. Gundry PM, Tompkins FC (1968) In: Anderson RB (ed) Experimental Methods in Catalytic Research. Academic Press, London

284. Anderson JR (1960) The Adsorption of Halogens on Metal Films. I. Adsorption Measurements and Surface Potentials for Chlorine on Nickel. J Phys Chem Solids 16:291–301

285. Jones PL, Pethica BA (1960) The Chemisorption of Nitrogen on Polycrystalline Tungsten Ribbon: Kinetic and Contact Potential Studies. Proc Roy Soc A256:454–469

286. Shen YR (1984) The Principles of Nonlinear Optics, Wiley, New York

287. Shannon VL, Koos DA, Richmond GL (1987) The Observation of Rotational Anisotropy in the Second Harmonic Intensity from a Ag(111) Electrode. J Chem Phys 87:1440–1441

288. Driscoll TA, Guidotti D, Gerritsen HJ (1982) Generation of Synchronous, Continuously Tunable High-power Picosecond Pulses. Rev Sci Instr 53:1547–1549

288a. Driscoll TA, Guidotti D (1983) Symmetry Analysis of Second Harmonic Generation in Silicon. Phys Rev B28:1171–1173

289. Heskett D, Urbach LE, Kong KJ, Plummer EW, Dai HL (1988) Oxygen and Pyridine on Ag(110) Studied by Second Harmonic Generation: Coexistence of Two Phases within Monolayer Pyridine Coverage. Surface Sci 197:225–238

290. Tom HWK, Mate CM, Zhu XD, Crowell JE, Heinz TF, Somorjai GA, Shen YR (1984) Surface Studies by Optical Second Harmonic Generation: The Adsorption of O_2, CO, and Sodium on the Rh(111) Surface. Phys Rev Lett 52:348–351

291. Drake B, Sonnenfeld R, Schneir J, Hansma PK (1987): Scanning Tunneling Microscopy of Processes at Liquid-Solid Interfaces. Surface Sci 181:92–97

292. Wiechers J, Twomey T, Kolb DM, Behm RJ (1988) An In–situ Scanning Tunneling Microscopy Study of Au(111) with Atomic Scale Resolution. J Electroanal Chem 248:451–460

293. Frenkel F, Häger J, Krieger W, Walther H, Ertl G, Segner J, Vielhaber W (1982) Potential State Populations and Angular Distributions of Surface-scattered Molecules: NO on Graphite. Chem Phys Lett 90:225–229

294. Segner J, Robota H, Vielhaber W, Ertl G, Frenkel F, Häger J, Krieger W, Walther H (1983) Rotational State Populations of NO Molecules Scattered from Clean and Adsorbate-covered Pt(111) Surface Sci 131:273–289

191

295. Kleyn AW, Luntz AC, Auerbach DJ (1981) Rotational Energy Transfer in Direct Inelastic Scattering: NO on Ag(111) Phys Rev Lett 47:1169–1172

296. Kleyn AW, Luntz AC, Auerbach DJ (1985) Rotational Polarization in NO Scattering from Ag(111) Surface Sci 152/153:99–105

297. Kimman J, Rettner CT, Auerbach DJ, Becker JA, Tully JC (1986) Correlation Between Kinetic Energy Transfer to Rotation and to Phonons in Gas-Surface Collisions of NO with Ag(111). Phys Rev Lett 57:2053–2056

298. Kubiak GD, Sitz GO, Zare RN (1985) Recombinative Desorption Dynamics: Molecular Hydrogen from Cu(110) and Cu(111). Chem Phys 83:2538–2551

299. Schröter L, Zacharias H, David R (1989) Recombinative Desorption of Vibrationally Excited D_2 ($v'' = 1$) from Clean Pd(100). Phys Rev Lett 62:571–574

300. Schröter L, Ahlers G, Zacharias H, David R (1987) Internal State Selected Velocity and Population Distribution of D_2 Desorbing from Clean Pd(100). J Electr Spectr Rel Phen 45:403–411

301. Ceyer ST, Somorjai GA (1977) Surface Scattering. Ann Rev Phys Chem 28:477–499

302. Barker JA, Auerbach DJ (1985) Gas – Surface Interactions and Dynamics: Thermal Energy Atomic and Molecular Beam Studies. Surface Sci Repts 4:1–99

303. Lin MC, Ertl G (1986) Laser Probing of Molecules Desorbing and Scattering from Solid Surfaces. Ann Rev Phys Chem 37:587–615

304. Tully JC (1980) Theories of the Dynamics of Inelastic and Reactive Processes at Surfaces. Ann Rev Phys Chem 31:319–343

305. D'Evelyn MP, Madix RJ (1984) Reactive Scattering from Solid Surfaces. Surface Sci Repts 3:413–495

306. Houston PL, Merrill RP (1988) Gas – Surface Interactions with Vibrationally Excited Molecules. Chem Rev 88:657–671

307. Zacharias H (1990) Laser Spectroscopy of Dynamical Surface Processes, Intern J Mod Phys B4:45–91

308. Engel T, Ertl G (1978) A Molecular Beam Investigation of the Catalytic Oxidation of CO on Pd(111) J Chem Phys 69:1267–1281

309. Campbell CT, Ertl G, Kuipers H, Segner J (1980) A Molecular Beam Study of the Catalytic Oxidation of CO on a Pt(111) Surface. J Chem Phys 73:5862–5873

310. Mödl A, Robota H, Segner J, Vielhaber W, Lin MC, Ertl G (1986) Reaction Dynamics of NO_2 Decomposition on Ge: An Example of Dissociative Desorption. Surface Sci 169:L341–L347

311. Eiswirth M, Möller P, Wetzl K, Imbihl R, Ertl G (1989) Mechanisms of Spatial Self–organization in Isothermal Kinetic Oscillations During the Catalytic CO Oxidation on Pt Single Crystal Surfaces. J Chem Phys 90:510–521

312. Rotermund HH, Ertl G, Sesselmann WI (1989) Scanning Photoemission Microscopy of Surfaces. Surface Sci 217:L383–L390

313. Rotermund HH. Jakubith S, von Oertzen A, Ertl G (1989) Imaging of Spatial Pattern Formation in an Oscillating Surface Reaction by Scanning Photoemission Microscopy. J Chem Phys 91:4942–4948

314. Rotermund HH, Engel W, Kordesch M, Ertl G (1990) Imaging of Spatio-temporal Pattern Evolution During Carbon Monoxide Oxidation on Platinum. Nature 343:355–357

5 Surface Reactions and Model Catalysis

In the preceding chapters we have tried to lay a useful physical basis for an understanding of a phenomenon that has been known to chemists for several hundred years, namely, *heterogeneous catalysis.*

Its discovery dates back to the early 19th century when Sir Humphrey Davy and, independently, W. Döbereiner realized the peculiar reaction of highly dispersed platinum to spontaneously oxidize hydrogen even though the reaction mixture was far below the ignition temperature. Some time later the Swedish chemist J.J. Berzelius rediscovered this phenomenon and called it "catalysis", but without understanding the physical principles behind it. J. Liebig tried to explain catalysis as a process in which a substance being in "chemical motion" is able to communicate this motion to another substance of the reaction mixture, thus causing, for example, its decomposition. However, only Wilhelm Ostwald offered the correct interpretation of catalysis when he wrote *mutatis mutandis* in his famous textbook "Grundriß der allgemeinen Chemie" [1]: " . . . a catalyst is an agent that speeds up a chemical reaction without being consumed in it and that increases the rate at which a system tends towards thermodynamical equilibrium. Catalysis cannot increase or change any yields of a chemical reaction, since the position of the thermodynamical equilibrium is not altered by the catalyst".

The organization of this final chapter will be as follows: after a brief introductory explanation of the phenomenon *catalysis*, we will first consider the simultaneous adsorption of two chemically different species at a solid surface, a process which is called *coadsorption*, and we will consider the respective interaction phenomena. Thereafter, we will expand on surface reaction kinetics of unimolecular and bimolecular processes and the corresponding possible mechanisms, before we turn to some more practical problems of heterogeneous catalysis, such as catalyst poisoning, inhibition or promotion effects, and finally provide several selected examples for simple surface reactions of practical importance. However, we must emphasize that it is impossible within the framework of this chapter to elucidate all the facets of the subject heterogeneous catalysis, which has almost become an independent discipline in recent years. Rather, the interested reader is referred to the wealth of specific literature on catalysis, a selection of which was presented previously in Chapter 1, [4–45].

Catalysis can be relatively easily explained in terms of transition state theory (cf., Chapter 2.6). Some qualitative remarks will aid in understanding the basic kinetics: There are two fundamental principles that should always be kept in mind when dealing with catalysis, namely, the fact that a catalyst merely *accelerates* the adjustment of chemical equilibrium, but does not change its position. To state it differently, the Gibbs energy ΔG as a thermodynamic state function depends only on the initial and final states of the system, and not on how the change is accomplished. The other principle has to do with *microscopic reversibility* and can be illustrated in the following way: if a thermodynamic system has reached equilibrium and if there are transitions possible between different states of the system, then there must also be equilibrium between these states. One important consequence is that any molecular process and its reversal occur *at the same rate*. Thus, if a given reaction $A + B \rightarrow Pr$ can proceed via two channels, where, for example, 10% react directly to product Pr, and 90% react via a bypass towards Pr, also the reverse reaction, the decomposition of Pr into A and B, is split into the two channels with the

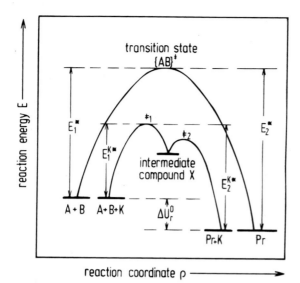

Fig. 5.1. Potential energy diagram (schematic) showing the uncatalyzed and the catalyzed reaction path of a bimolecular reaction A + B = Pr with reaction energy ΔU_r^0. The direct (uncatalyzed) path towards the transition state complex $\{AB\}^{\ddagger}$ is activated by an energy E_1^{K*} of appreciable height, whereas the catalyzed route (via formation of intermediate X with the activated complexes $^{\ddagger}_1$, and $^{\ddagger}_2$, respectively) requires the much smaller activation energy E_2^{K*} thus causing the much faster reaction rate

same ratio. This principle has an important consequence. It allows conclusions to be made about the path of the product *decomposition* reaction if just the product *formation* reaction is known, or vice versa. In the following kinetic scheme we again discuss the simple reaction in which a product Pr is formed from the reactants A and B. This reaction is assumed to proceed i) in a *direct way*, and ii) in a *catalyzed* way (catalyst symbol K, intermediate compound symbol X). The reaction energy diagram of Fig. 5.1 illustrates the situation. We have for the *direct* channel:

(i) $A + B \xrightarrow{k_1} \text{Pr}$ (slow) , 5.1

whose rate can be written

$$-\frac{d[A]}{dt} = +\frac{d[\text{Pr}]}{dt} = k_1[A][B] .$$ 5.2

This is an ordinary bimolecular reaction with an overall second-order kinetics. The *catalyzed* reaction must be formulated:

(ii) $A + K \xrightarrow{k_2} X$ (relatively slow) 5.3

$\qquad X + B \xrightarrow{k_3} \text{Pr} + K$ (fast) . 5.4

The reaction rate then reads

$$\frac{d[\text{Pr}]}{dt} = k_3[X][B] ,$$ 5.5

The unknown concentration of the intermediate species X can be obtained by applying the steady state approximation:

$$\frac{d[X]}{dt} \approx 0 = k_2[A][K] - k_3[X][B] ;$$ 5.6

194

$$[X] = \frac{k_2[A][K]}{k_3[B]} ,$$ 5.7

and insertion into Eq. 5.5 yields

$$\frac{d[Pr]}{dt} = k_2[A][K] .$$ 5.8

This means that (provided $k_3 \gg k_2$) the reaction rate can be greatly influenced by the concentration of the catalyst. It is worth mentioning that these considerations apply to homogeneous as well as to heterogeneous catalysis in quite a similar manner; note that in heterogeneous catalysis the catalyst is represented by a *surface* of active material, and the decisive quantity $[K]$ is the *number of active surface sites*. As repeatedly mentioned before, we are only interested here in *heterogeneous* catalysis, and we begin our discourse with a consideration of the physical and chemical interactions of atoms and/or molecules that are simultaneously present at a surface.

5.1 Coadsorption Phenomena

When we consider the simultaneous adsorption of two or more different particles on a surface, we must again distinguish (cf., Sect. 3.2) the limits of a sparsely covered surface (zero-coverage condition), where only the interaction between the individual particles *A, B, C* ... and the surface plays a role, and the situation where a whole ensemble of all these particles coexists on the surface, interacting with the surface *and* with each other (multiparticle interaction). As opposed to our preceding considerations with chemically identical particles, there is now a new component in the interactions, namely, the interaction between chemically *different* species. This may be repulsive or attractive, and clearly adsorption site and mutual orientational effects of the adparticles will become essential.

The interplay of the interactions between chemically identical and chemically different adparticles (which, of course, occur predominantly at elevated coverages) can lead to two different situations. We must distinguish the processes of *competitive* and *cooperative* adsorption. Consider, for example, the coadsorption of two kinds of particles *A* and *B* on a clean solid surface with their chemisorption properties being roughly similar, i.e., there exist no large differences with respect to their binding energies to the substrate. If we denote the mutual interaction energies, as before, by the symbol ω, the following (two-dimensional) interaction energy terms play a role: ω_{AA}, ω_{BB}, and ω_{AB}. It must be borne in mind that the overall distribution of the particles on the surface is practically entirely dominated by these interaction energies, whereby the following balance is decisive (provided that no surface diffusion barriers exist):

$$\Delta E_\omega = |\omega_{AA}| + |\omega_{BB}| - 2|\omega_{AB}| ;$$ 5.9

ΔE_ω is positive if the sum $\omega_{AA} + \omega_{BB}$ is greater than twice the interaction energy ω_{AB}. This means that equal particles tend to become neighbors, and a thorough mixture of species *A* and *B* is not expected, rather, equal particles will form islands with a structure characteristic of an *A* and *B* phase, respectively. In analogy to three-dimensional thermodynamics we have a miscibility gap, and the whole process will lead to *competitive* adsorption in that both species *A* and *B* will compete for a given adsorption site. Conversely, if different particles interact more strongly with each other than with equal ones or, in other words, if the term $2\omega_{AB}$ exceeds the sum $\omega_{AA} + \omega_{BB}$, ΔE_ω becomes negative. Given this situation, it is energetically more favorable for species *A* and *B* to become

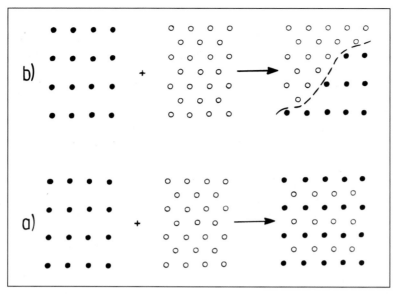

Fig. 5.2. Real space patterns of two kinds of adsorbates A and B, with different two-dimensional periodicities. After coadsorption on the same surface, A and B may form a composite phase with a new overall periodicity (cooperative adsorption), case a), or two phases with their inherent respective structure separated by a phase boundary (competitive adsorption), case b)

neighbors, and a pronounced miscibility between A and B leading to individual $A \cdots B$ complexes will result. The overall process is called *cooperative* adsorption and is schematically illustrated in Fig. 5.2. If both species form phases with long-range order with different periodicity (which leads to distinguishable LEED patterns) it is relatively easy to delineate competitive and cooperative adsorption. In the first case, both species will form large islands with their inherent order, in the second case the $A \cdots B$ complexes may very well exhibit a new single phase with a different periodicity. The LEED method is indeed often used to make a distinction between these possible types of coadsorption; it is, of course, restricted to single-crystalline systems. The advantage of LEED is that the actual diffraction pattern depends, among others, on the relation between the coherence width Δx of the electron beam (cf., Sect. 4.1.1) and the size (diameter d) of the respective island of uniform periodicity. Remember that, for standard LEED equipment, Δx is of the order of 50 to 200 Å. If $\Delta x < d$ the diffraction pattern consists of a superposition of the *intensities* (squared structure amplitude) of the individual scattering contributions, and with two differently ordered islands the actual LEED pattern will display a true superposition of the respective diffraction spots. The other possibility starts with the condition $\Delta x > d$, and there occurs a summation of the diffraction *amplitudes* of the electron wave originated at different small domains or $A \cdots B$ complexes, with a completely new diffraction pattern becoming possible. It is perhaps noteworthy that exactly the same considerations apply for *antiphase domains* of chemically identical species. Various examples are reported in the literature regarding competitive or cooperative adsorption and there exist several authoritative reports and discussions of coadsorption effects, among others by Ertl [2] and May [3,4].

May's work particularly appreciates the relation between coadsorption, surface morphology, and chemical reactivity. For the sake of brevity, we confine ourselves to a few

196

examples taken from the coadsorption studies of CO and O on Pd(111) by Ertl and Koch [5], and of CO and H on a Pd(110) surface by Conrad et al. [6]. On Pd(111), a CO coverage of $\Theta = 0.3$ produces a $\sqrt{3} \times\sqrt{3}\ R30°$ LEED structure, whereas adsorbed oxygen (0.25 ML coverage) forms a 2×2 pattern. The existence of individual CO and oxygen islands of appreciable size is documented by the fact that both aforementioned structures are simultaneously visible on the LEED screen; apparently, CO and O phases on Pd(111) are immiscible.

Other examples of similar phenomena are hydrogen and carbon monoxide on Ru(10$\bar{1}$0) surfaces [7,8], where, likewise, coexistence of the characteristic individual LEED phases is observed, thus pointing to competitive adsorption. From a chemical viewpoint, *cooperative* adsorption is certainly far more interesting, because this may be regarded as a very first step toward compound formation at surfaces. As mentioned before, coadsorption of CO and H on a Pd(110) surface is a good example for this type of process [6]. While CO alone forms a $c(2\times2)$ and hydrogen a (1×2) structure on Pd(110), the exposure of hydrogen on a CO-precovered surface gives rise to a novel (1×3) structure, indicating an intimate mutual interaction between CO and H (mixed adsorbate complex). It is not always possible to assess the actual number of molecules or atoms of the respective species participating in the complex, but in some cases LEED analyses, in conjunction with vibrational loss investigations, could reveal this important chemical information. An example is provided by the coadsorption of benzene and carbon monoxide on platinum (111) and rhodium(111) surfaces, which was extensively studied by Mate and Somorjai [9]. These authors reported on CO-induced ordering phenomena of benzene on Pt and Rh(111). On Pt(111), C_6H_6 alone does not form ordered overlayers; only in conjunction with coadsorbed CO are well-ordered phases of the coadsorbate [(2$\sqrt{3}$ ×4) and (2$\sqrt{3}$ ×5) LEED structures) formed. The coadsorption itself could be confirmed by concomitant HREELS studies. On Rh(111), ordered phases of benzene and CO alone could be observed, but also additional coadsorption structures (corresponding to a $c(2\sqrt{3}$ ×4) rect and a (3×3) LEED pattern). Real space structure models of the four aforementioned phases are reproduced in Fig. 5.3 and reveal evidence of interesting spatial and electronic mutual interactions between benzene and carbon monoxide. Very recently, Jacob and Menzel reported on quite similar cooperative coadsorption phenomena of benzene and oxygen on a Ru(0001) surface [10]. The coadsorption of hydrogen and carbon monoxide on a nickel(100) surface is another well-known prototype of cooperative adsorption. Among others, this system was studied by means of LEED, UV photoelectron spectroscopy, HREELS and thermal desorption spectroscopy by White and coworkers [11, 12], as well as by Goodman and his group [13], and interesting new LEED structures, shifts in UPS orbital energies of CO and, most importantly, a new TDS state, the so-called Σ state, appeared at the same temperature both in CO and H_2 desorption spectra. Unfortunately, HREELS measurements did *not* provide any evidence for new vibrations so that the idea of a methoxide or formaldehyde intermediate species being formed by coadsorption could not be supported. Nevertheless, the correlated CO/H Σ-desorption state points to peculiar intermolecular interactions between the two species and may thus be regarded as a precursor stage of a surface reaction.

It is quite obvious that the mutual arrangement of particles, which are to interact, will largely dominate the reaction kinetics and, hence, the rate of reaction. For a type of reaction that obeys the so-called *Langmuir-Hinshelwood* (LH) mechanism (implying a reaction of particles A and B, both in the adsorbed state; cf., Sect. 5.2) the mutual distance of these particles is decisive for the reaction rate. If one molecule A has to travel over a long distance until it meets a species B, the reaction will be slower than if both particles are

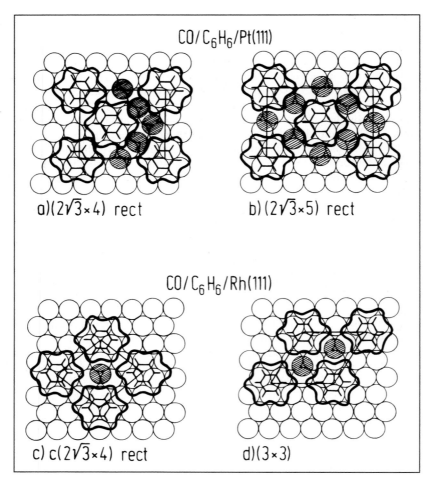

Fig. 5.3. Examples for real space structures of cooperative adsorption between benzene C_6H_6 and carbon monoxide CO (based on LEED and HREELS data and using van der Waals diameters). Structures a) and b) refer to coadsorption on Pt(111), structures c) and d) to coadsorption on Rh(111). The CO molecules are indicated by hatching, the metal atoms by open circles. After Mate and Somorjai [9]

neighbors on the surface. In this respect the question as to which phase is present at a given surface (non-miscible or miscible phases) is expected to have great implications on the kinetics of a Langmuir-Hinshelwood reaction. One can easily imagine that, with non-miscible phases, this type of reaction will only occur at the phase boundaries, i.e., at the perimeters of the domains or islands, and the majority of the species in the interior of an island cannot immediately participate in the reaction. Vice versa, if $A \cdots B$ complexes are preformed by cooperative adsorption, the LH reaction can take place immediately and no surface migration or diffusion is required. However, it is not that just the mutual arrangement of the reactive particles governs the reaction rate; equally important is whether or not the molecules really *compete* for a given adsorption site. Again, the non-miscibility of the competitive adsorption represents an obstacle for an efficient reaction, and complete coverage and thus, poisoning of the adsorption sites by a *single* species may result, which will extinguish any reaction activity for a bimolecular surface reaction. Examples for this behavior will be provided later (Sect. 5.4).

There are, however, cases feasible in which competitive adsorption may be *beneficial* for establishing certain favorable adsorbate coverage conditions or binding states. This shall be demonstrated with the aid of the Ru(10$\bar{1}$0)/CO + H coadsorption system [7,8]. From thermal desorption spectroscopy results it is well-known that CO is more strongly held on Ru than hydrogen; in other words, CO replaces hydrogen in the competition for the adsorption sites. However, as long as there are sufficient adsorption sites for CO available, the remaining hydrogen is not automatically desorbed, but is only pushed aside, and compressed and confined to a smaller surface area, whereby the overall adsorptive binding conditions become, of course, degraded. Novel low-temperature states in the H thermal desorption spectra clearly indicate that CO is able to displace hydrogen to sites with lower binding energy that are not characteristic of hydrogen. One should not think of such a binding site having a distinct new geometry, but rather, the H atoms in these sites suffer from very short mutual distances and experience very large repulsive energy contributions, resulting in the aforementioned low-temperature desorption states. It is quite striking, however, that upon removal of the most weakly held (i.e., most highly compressed) hydrogen the remaining species suddenly undergo *cooperative* effects, in other words, there appears a new tendency of the species (CO and H) to mix: new LEED patterns can be observed that are not visible with either of the adsorbed gases CO or H alone, thus corroborating the aforementioned *cooperative* effects; these are very likely provoked by the fact that CO and H particles are forced to adjacent adsorption sites by a high "surface pressure".

Figure 5.4 presents an example taken from our own work [8]. CO was post-adsorbed into a (1×1)-2H saturation phase on Ru(10$\bar{1}$0), and after ~20 L exposure the characteristic

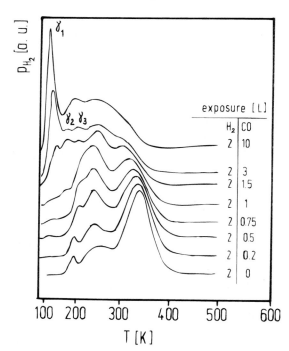

Fig. 5.4. Hydrogen thermal desorption spectra from a Ru(10$\bar{1}$0) surface always exposed first to 2 Langmuirs of hydrogen at 100 K, and then to increasing amounts of CO, as indicated in the panel. Besides the characteristic hydrogen states between 200 and 400 K, new γ-states appear, owing to a compression of adsorbed hydrogen by expanding CO, which are not observed with hydrogen alone. After Lauth et al. [8]

199

CO $(2\times1)p2mg$ structure with increased background intensity is formed. The various H desorption states denoted as $\gamma_1-\gamma_3$ can be thermally desorbed and the successive evacuation of these states causes the appearance of novel LEED phases [(4×1), (3×2), (3×1)]. Note that there is *no* desorption of CO taking place up to a temperature of 270 K, only the H coverage decreases via depopulation of the γ-states. Unfortunately, all these effects can only be studied under UHV conditions at very low pressures, and one is certainly extremely far away from those thermodynamical conditions that would favor an efficient chemical surface reaction between hydrogen and carbon monoxide on ruthenium (leading, perhaps, to appreciable yields of, for example, hydrocarbons or simple alcohols). Nevertheless, (and this was stressed in the introductory chapter), the aforementioned UHV single-crystal studies can often provide evidence of whether or not those reactions are principally possible and, if so, they may yield invaluable information as to how the reaction parameters are related to microscopic properties of the respective surfaces.

So far, we have tacitly assumed that the adsorption and surface reaction, respectively, occur on a *uniformly* composed and structured (single crystal) surface where all the particles will find, at least initially, *identical* adsorption sites. Although this matter is difficult enough, there are much more complicated systems used in practical (technical) catalysis. In the first instance, this comprises the surface morphology; secondly, chemical heterogeneity is concerned. No longer does one deal with homogeneous and equally composed flat surfaces, but instead heterogeneous, polycrystalline materials (often with laterally varying chemical composition) are almost entirely used in practical catalysis, and our considerations must be extended also to these materials. However, it is clear that due to their complexity, only qualitative remarks can be given, but it helps to remember the concepts and laws derived in the context of single-crystalline surfaces.

To begin with, it is certainly very helpful to consider a chemically homogeneous, but crystallographically heterogeneous surface, e.g., a polycrystalline metal or a uniformly composed polycrystalline chemical compound. Since the phase relations between adjacent surface atoms are lost, it is not possible to obtain well-ordered diffraction patterns, e.g., in a LEED experiment, and other means such as surface EXAFS must be invoked if one intends to determine the (local) surface structure (cf., Sect. 4.1.4). A more serious consequence in view of surface chemical reactions is that the surface disorder is also manifested in *disordered* adsorbate layers, and the formation of well-ordered surface complexes (as discussed before in the context of cooperative adsorption) is at least greatly impaired, if not suppressed. In addition, there may be a peculiar geometrical arrangement of the substrate surface atoms required for a reaction to run. These *structure-sensitive* (so-called *demanding*) reactions will, of course, be strongly affected by the surface geometry. An example is the catalytic trimerization of acetylene to benzene which runs well on hexagonal (111) Pd surfaces, but is not so efficient on squared (100) surfaces [14].

The previously mentioned *ensemble* effect (cf., Sect. 3.1.2) plays a decisive role here. On the other hand, *structure-insensitive* (so-called *facile*) reactions exhibit little response to surface structure, because the crucial reaction steps do not take place at characteristic surface ensembles, but rather at structure-independent *active centers*. These can, for example, be represented by certain foreign atoms distributed over the surface. Here we leave the crystallographical heterogeneity and definitely have to also admit *chemical* heterogeneity, and we must extend our considerations to the influence of foreign chemical compounds or atoms (so-called *additives*) on the reactivity, and study this influence as a function of the additives' surface concentration. For simplicity, we keep the geometrical surface structure fixed. It should then be possible to directly attribute any observed effects to kind and concentration of the respective additive (which is often also considered as chemical *modifier*). Numerous investigations have been devoted so far to the problem of chemically modifying surfaces in order to attain a peculiar reactivity, whereby noble metal additives were used, as well as alkali metal or halogen atoms. The latter modifiers combine a strong electrostatic action (alkali atoms are strongly electropositive, halogens are electronegative – even in the adsorbed state) with the ease of detectability in surface analysis. As will be shown later (cf., Sect. 5.4), coadsorption studies of the standard molecules such as carbon monoxide, nitric oxide, hydrogen, etc., plus alkali or halogen can provide a key understanding of the function of catalytic inhibitors and electronic promoters, respectively. This kind of coadsorption is naturally different from the one discussed in the beginning of this section, since here the coadsorbed additive represents a component of the surface and is not expected to directly participate in a bimolecular surface reaction.

Another case of significant practical interest concerns the formation of active "foreign" atoms or molecular species *in the course of* a chemical surface reaction. In this situation, an autocatalytic mechanism should evolve, which results in a considerable acceleration of the reaction rate. Of course, it may also be (and this is, by far, the more frequent situation) that these species generally *inhibit* the reaction or (less frequently) suppress only certain side-reactions, which would be, from the practical (selectivity) viewpoint, a favorable property. As an example, we recommend the study of Goodman et al. [15, 16] concerning the methanation reaction ($CO + 3\ H_2 = CH_4 + H_2O$) from synthesis gas on a Ni(100) surface, where it could be shown that the reaction-controlled precipitation of an active carbonaceous species, viz., *carbidic* carbon, increased the CH_4 yields considerably. In contrast, *graphitic* carbon (which was formed if the reaction was running too hot) did not have this beneficial effect, but even acted as an inhibitor and, finally, poisoned the respective activity of the nickel surface.

Another aspect that warrants some consideration here is the possible infuence of coadsorbed surface additives on the surface structure itself – there are many cases known, where coadsorbed species, for example, alkali metal atoms, induce or remove surface reconstructions with all the consequences that the related geometrical changes have on structure-sensitive reactions. We refrain here from presenting more data on ionic coadsorbates and refer the reader to Section 5.4, where this matter is again taken up.

5.2 Surface Model Reactions

The enormous importance of surface reactions may be judged from the fact that this subject is more and more extensively dealt with in modern textbooks of physical chemistry [17–19] and especially chemical kinetics. An excellent introduction is provided by Laidler [20], and some of the following essentials are based on his representation of surface kinetics. We would, however, like to emphasize that concise treatments of surface reaction mechanisms can also be found in many previous textbooks and monographs, e.g., by Bond [21], Ashmore [22], and Trapnell [23].

We recall that the process of adsorption (chemisorption) is the most important step prior to a surface reaction. It provides a certain concentration of reactive particles on the surface, which is governed, as was pointed out in Sect. 2.4, by the external gas pressure P and the surface temperature T. We have learned that the respective relation is called *adsorption isotherm*, and we recall the most important *Langmuir isotherm* for molecular and dissociative adsorption, which reads for the adsorbate coverage either

$\Theta = b_{(T)} \cdot P/(1 + b_{(T)} \cdot P)$ (in the case of molecular adsorption, cf., Eq. 2.52, Sect. 2.4, or $\Theta = \sqrt{b'_{(T)} \cdot P}/(1 + \sqrt{b'_{(T)} \cdot P})$ (dissociative adsorption, cf., Eq. 2.53, Sect. 2.4). In the case of *coadsorption*, especially *competitive coadsorption*, at least two different species (A and B in our case) compete for the available adsorption sites, and the simple Langmuir adsorption isotherm must be modified by introducing the fraction of sites covered by species A, Θ_A, and the sites occupied by species B, Θ_B. Accordingly, the respective bare fraction of the surface is given by $1 - \Theta_A - \Theta_B$; and for the situation that both gases are molecularly adsorbed, one can easily derive the simultaneous equations:

$$\frac{\Theta_A}{1 - \Theta_A - \Theta_B} = b_{A(T)} \cdot P_A , \qquad\qquad 5.10$$

with

$$b_{A(T)} = \frac{s_{0(A)} \exp\{(\Delta E_{des}^{(A)^*} - \Delta E_{ad}^{(A)^*})/kT\}}{\nu_A \cdot \sqrt{2\pi m_A kT}} ; \qquad\qquad 5.11$$

and

$$\frac{\Theta_B}{1 - \Theta_A - \Theta_B} = b_{B(T)} \cdot P_B , \qquad\qquad 5.12$$

with

$$b_{B(T)} = \frac{s_{0(B)} \exp\{(\Delta E_{des}^{(B)^*} - \Delta E_{ad}^{(B)^*})/kT\}}{\nu_B \cdot \sqrt{2\pi m_B kT}} . \qquad\qquad 5.13$$

One can simultaneously solve Eqs. 5.10 and 5.12 and obtain, for the surface fractions covered by A and B, respectively,

$$\Theta_A = \frac{b_{A(T)} P_A}{1 + b_{A(T)} P_A + b_{B(T)} P_B} , \qquad\qquad 5.14$$

and

$$\Theta_B = \frac{b_{B(T)} P_B}{1 + b_{A(T)} P_A + b_{B(T)} P_B} . \qquad\qquad 5.15$$

From these two equations it follows immediately that the fraction of the surface covered by one component (A) is reduced as soon as the amount of the other species (B) is increased. Since the intrinsic kinetic parameters (sticking probability s_0, frequency factor for desorption ν, activation energy for desorption ΔE_{des}^*, etc.) are implicit in these equations it is difficult to adjust the respective gas phase pressures P_A and P_B, respectively, such that stoichiometric amounts of A and B coexist on the surface. As both species compete for a limited number of adsorption sites, we have the prototype of a competitive adsorption. The situation may change completely if species A and B prefer *different* adsorption sites: in the simplest case both molecules adsorb without affecting each other and the simple Langmuir isotherm describes the adsorption of each species independently. A more complicated situation arises, however, if indeed both particles occupy different sites, but the adsorbate complexes interact with each other so as to modify the heat of adsorption. Here the simple Langmuir isotherm certainly no longer provides an adequate description of the adsorption process.

Regarding the problem of understanding the inhibiting function of a surface additive, formally, the coadsorption of an inhibiting agent I (which simply blocks adsorption sites irreversibly) can be treated in the same way as shown above (simply replace B by I). It is immediately seen that the surface concentration of a reactive species, say A, tends towards zero if the inhibiting species I wins the competition for the adsorption sites.

Quite similar equations to Eqs. 5.10–5.15 can be set up for atomic (dissociative) adsorption, whereby the most important change is that *one* initially adsorbing particle dissociates into *two* fragments, each of which demands an adsorption site. We end up with the expressions

$$\Theta_A = \frac{\sqrt{b'_A P_A}}{1 + \sqrt{b'_A P_A} + \sqrt{b'_B P_B}} \,, \qquad\qquad 5.16$$

and

$$\Theta_B = \frac{\sqrt{b'_B P_B}}{1 + \sqrt{b'_A P_A} + \sqrt{b'_B P_B}} \,. \qquad\qquad 5.17$$

Another important aspect which is only implicit in Eqs. 5.16 and 5.17 is that, in order for the dissociation to occur, two *adjacent* adsorption sites must be available, and owing to this additional requirement any dissociative adsorption reaction is at a disadvantage if it competes with a molecular (associative) adsorption. We shall see later that this is decisive for the catalytic CO oxidation reaction on transition metal surfaces (Sect. 5.3) in that it can lead to a preferential CO accumulation at the surface.

Let us now consider the possible *mechanisms* of surface reactions. However, before we actually start these considerations, we must remember what was pointed out in the introductory chapter, namely, that an ordinary surface reaction involves at least five discrete steps (trapping and sticking; adsorption (chemisorption); surface migration (diffusion); chemical reorganization of the reactants (chemical *reaction*); and desorption of the product(s)). In this sequence, adsorption and, particularly, desorption are the steps which usually involve the largest activation energy barriers and are therefore slowest, and it is justified to regard the overall reaction as a *single* step whose rate is governed by the step which exhibits the largest activation energy. This treatment was first suggested by Langmuir [24] and later adopted by Hinshelwood [25], and proceeds as follows: First, there is an expression for the surface concentration (coverage) of the educt molecules deduced, thereafter, the rate of product formation is calculated in terms of these coverages. For a more detailed consideration, it is necessary to distinguish a unimolecular from a bimolecular process. While there is no problem with a unimolecular reactive change (decomposition or isomerization, etc.) the bimolecular reaction deserves some attention, because it can occur in two different ways.

5.2.1 The Langmuir-Hinshelwood (LH) Reaction

We assume two reactants (educts) and call them A and B. In a first instance it does not matter whether A and B are single atoms or molecules, or whether the adsorption occurs associatively or dissociatively. It is only important that *both* these species enter the adsorbed state, i.e., form chemical bonds with the surface. This can be illustrated by the scheme

$$A_{(g)} + -S-S-S- = -\overset{\displaystyle A}{\overset{|}{S}}-S-S-$$

$$B_{(g)} + -S-S-S- = -S-\overset{\displaystyle B}{\overset{|}{S}}-S-$$

$$\Bigg\} = -S-\overset{\displaystyle A}{\overset{|}{S}}-\overset{\displaystyle B}{\overset{|}{S}}-S-S-S-$$

in which the symbol $-S-S-S-$ represents surface atoms with their free valencies. In this scheme, it is tacitly assumed that, after a smaller or larger number of diffusion steps, the A and B species reach *adjacent* surface sites.

Alternatively, one can describe the reaction by ordinary chemical equations (* denotes an empty surface site):

$$A_{(g)} + * = A_{(ad)} \quad \text{adsorption of species } A$$

and

$$B_{(g)} + * = B_{(ad)} \quad \text{adsorption of species } B \, .$$

The actual reaction towards the product Pr takes place from the adsorbed state:

$$A_{(ad)} + B_{(ad)} = Pr_{(ad)} \quad \text{surface reaction}$$

and

$$Pr_{(ad)} = Pr_{(g)} + 2 * \quad \text{product evolution by thermal desorption}$$

The *catalytic* mechanism of the overall process becomes immediately apparent from the fact that, at the end of the reaction, the initial surface conditions are completely restored (two empty adsorption sites are left behind). Thereby, it is tacitly assumed that the product species is so weakly adsorbed at the surface that it immediately desorbs into the gas phase after its formation; in other words, the product desorption is not rate-limiting. The beneficial role of the surface is best demonstrated by means of a reaction which involves dissociation of at least one reactant. The dissociation, which often requires appreciable activation energies, occurs automatically on many transition metal surfaces (cf., Sect. 3.2.1, Fig. 3.17), because the heat of adsorption overcompensates the dissociation energy of the gaseous molecule. The corresponding scheme is

$$A_{2(g)} + -S-S-S-S- = -S-\overset{\displaystyle A}{\overset{|}{S}}-\overset{\displaystyle A}{\overset{|}{S}}-S-$$

$$B_{2(g)} + -S-S-S-S- = -S-\overset{\displaystyle B}{\overset{|}{S}}-\overset{\displaystyle B}{\overset{|}{S}}-S-$$

$$\Bigg\} = -S-\overset{\displaystyle A}{\overset{|}{S}}-\overset{\displaystyle B}{\overset{|}{S}}-S-$$

In terms of chemical equations this reads:

$$A_{2(g)} + ** = 2A_{(ad)} \quad \text{and} \quad B_{2(g)} + ** = 2B_{(ad)} \, ;$$

chemical reaction at the surface with subsequent product desorption yields:

$$A_{(ad)} + B_{(ad)} = \{AB\}_{(ad)} = Pr_{(g)} + * * \, .$$

204

{AB} may be thought of as an activated complex that is formed at the surface (cf., Sect. 2.6) either as a hypothetical compound or as a short-living intermediate, and the entire reaction can often be treated by transition state theory. This leads to the actual *kinetics* of an LH reaction. We have seen that the rate of the crucial step, namely, the formation of a product molecule from coadsorbed species A and B, is proportional to the surface concentrations of both A and B. Accordingly, we have

$$\text{reaction rate} \quad \frac{d[Pr]}{dt} = k_r \Theta_A \Theta_B \, , \qquad\qquad 5.18$$

in which k_r is an ordinary rate constant of a bimolecular reaction which depends preferentially on temperature. $[Pr]$ stands for the (gas phase) concentration of the product molecules and may also be expressed in terms of its partial pressure P_{Pr}. Now, we make use of Eqs. 5.14 and 5.15 and write for an associative bimolecular LH reaction

$$\frac{d[Pr]}{dt} = \frac{k_r b_A b_B P_A P_B}{(1 + b_A P_A + b_B P_B)^2} \, . \qquad\qquad 5.19$$

Obviously, the overall reaction rate is largely governed by the partial pressures of the individual reactants P_A and P_B. It may be instructive to discuss the change of the reaction rate if the pressure of one component, for instance, P_B is held constant and the other component (P_B) is gradually increased. The situation is illustrated by means of Fig. 5.5. As predicted by Eq. 5.18, there is first a rise of the rate, which then passes through a maximum and, finally, at large surface concentrations of A decreases until the rate drops to zero. The reason is that one component displaces the other as its concentration increases (remember that we had explicitly assumed competitive adsorption). Right at the rate maximum there exist the maximum number of A–B pairs (stoichiometric conditions) whereby, of course, no island formation is taken into account (which would, however, frequently occur in this situation of immiscible components). From Eq. 5.18 the maximum condition can easily be established by choosing $\Theta_A = \Theta_B$ (equal surface concentrations).

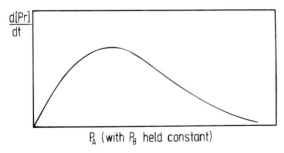

d[Pr]/dt

P_A (with P_B held constant)

Fig. 5.5. Variation of the rate d[Pr]/dt for a bimolecular surface reaction A + B $\rightarrow$ Pr, which proceeds via the Langmuir-Hinshelwood mechanism. For simplicity, the rate is plotted as a function of the (gas phase) concentration of one reactant only ([A]), with the corresponding concentration [B] being held fixed. After Laidler [20]

Quite a similar treatment can be executed for a reaction in which both reactant molecules adsorb dissociatively, and one ends up with the equation

$$\frac{d[Pr]}{dt} = \frac{k_r \sqrt{b'_A b'_B P_A P_B}}{(1 + \sqrt{b'_A P_A} + \sqrt{b'_B P_b})^2} \, . \qquad\qquad 5.20$$

The reaction rate constant k_r can be understood in the usual manner as being composed of a pre-exponential factor and an activation energy for *reaction*; it is, however, by no means trivial to accurately evaluate these energies from the usual Arrhenius plots, i.e., from the temperature dependence of the reaction rate. The reason is, of course, that also the b factors in the isotherm or rate expressions depend on T, and one actually measures an *apparent activation energy* that also contains activation energies for adsorption, diffusion, and desorption of the reactants and/or products. Only by carefully studying the individual aforementioned processes could it eventually be possible to disentangle the various energies and the corresponding pre-exponential factors.

It may be instructive to discuss some aspects of the equations derived above. One may, for example, look at a sparsely covered surface, where both reactants A and B exist in very low concentrations ($\Theta_{A,B} \ll 1$). Then, the denominator of Eqs. 5.19 or 5.20 approximately takes the value 1, resulting in the linear expressions for associative adsorption:

$$\frac{d[Pr]}{dt} \approx k_r b_A b_B P_A P_B ,$$

5.21

and

$$\frac{d[Pr]}{dt} \approx k_r \sqrt{b'_A b'_B} P_A^{1/2} P_B^{1/2} .$$

5.22

According to Eq. 5.21, we have for this limit of small coverages an overall second-order kinetics for associative adsorption, and first-order kinetics with respect to each reactant. Likewise, for dissociative adsorption (Eq. 5.22) a total reaction order of 1 is obtained, as is a square root order with respect to each reactant. It is thereby possible, by plotting the logarithm of the reaction rate as a function of the logarithm of the partial pressure of one reactant, to distinguish associative from dissociative adsorption of the reactants.

Other limiting cases can be discussed (strong differences in the adsorption energy of the reactants, for example) that render simplifications of the bimolecular expressions (Eqs. 5.19 and 5.20) possible, but for the sake of brevity, we expand no further on this matter here; rather, we turn to the second type of reaction mentioned in the beginning.

5.2.2 The Eley-Rideal (ER) Reaction

The principal ideas of this reaction mechanism can also be traced back to Langmuir; later, Rideal and Eley subjected the mechanism to another detailed consideration, and in honor of these surface chemists the reaction mechanism was given its name [26]. Sometimes, however, it is also called "Langmuir-Rideal" mechanism (for example, in [20]). Compared with the LH mechanism, where the adsorption of both reactants is the essential step, it is not necessary in an ER reaction that both partners coexist in the adsorbed state. It suffices if only one reactant is chemisorbed – the other one then reacts by impact from the gas phase. The situation can again be illustrated by a scheme which reads:

In terms of a chemical equation this process may be expressed as:

$$A_{(g)} + * \quad = \quad A_{(ad)} \qquad \text{adsorption of species } A$$

$$B_{(g)} + A_{(ad)} \quad = \quad \{A-B\}_{(ad)} \qquad \begin{array}{l} \text{reaction of species } B \text{ } directly \\ \text{from the gas phase} \end{array}$$

$$\{A-B\}_{(ad)} \quad = \quad Pr_{(g)} + * \qquad \text{desorption of product } Pr.$$

Accordingly, one has for a dissociative adsorption

$$A_{2(g)} + 2* \quad = \quad 2A_{(ad)} \qquad \text{(dissociative) adsorption}$$

$$2B_{(g)} + 2A_{(ad)} \quad = \quad 2\{A-B\}_{(ad)} \quad \text{reaction from the gas phase}$$

$$2\{A-B\}_{(ad)} \quad = \quad 2Pr_{(g)} + 2* \quad \text{product desorption.}$$

Of course, A and B may also be exchanged and there will be no difference in the net result. Again, as with the LH reaction, we consider the kinetics of the product formation. Note that now in a bimolecular reaction there is only one component adsorbed whose surface coverage Θ needs be considered. Accordingly, the reaction (product formation) rate can be stated as

$$\frac{d[Pr]}{dt} = k_r \Theta_A [B] , \tag{5.23}$$

whereby the brackets denote, as before, gas phase concentrations. Assuming ideal gas behavior, the actual concentrations c_i [mol/l] are interrelated with the partial pressures P_i, the absolute particle number N_i, Avogadro's constant N_L, the volume V, and the gas constant R via:

$$P_i = \left(\frac{N_i}{V}\right) \frac{RT}{N_L} = [c_i]\frac{RT}{N_L} , \tag{5.24}$$

and by redefining the reaction rate constant k_r as $k'_r = k_r(N_L/RT)$ we may write:

$$\frac{d[Pr]}{dt} = k'_r \Theta_A P_B . \tag{5.25}$$

The coverage Θ_A is, of course, governed by the adsorption isotherm and, hence, by the partial pressure of A in the gas phase. At this point, we must distinguish two cases, namely, case i) in which species B does not adsorb at all, leading always to $\Theta_B = 0$, and case ii) which admits coadsorption of species B, but rules out any LH reaction between adjacent A and B particles at the surface. Then, any action of adsorbed B comes about only by site blocking, to the disadvantage of species A. Whereas in the first case a rate equation can simply be formulated by replacing Θ_A in Eq. 5.25 by its adsorption isotherm to yield

$$\frac{d[Pr]}{dt} = k'_r \frac{b_A P_A}{1 + b_A P_A} P_B , \tag{5.26}$$

we obtain in the second case the expression

$$\frac{d[Pr]}{dt} = \frac{k'_r b_A P_A}{1 + b_A P_A + b_B P_B} P_B . \tag{5.27}$$

This equation 5.27 must be contrasted with Eq. 5.19, which described the LH type of reaction. Let us again assume that one component (for example, B) is held constant while the concentration of the other one (A) is linearly increased. The result on the rate is simply a non-linear increase which tends towards a saturation value, as shown in Fig. 5.6. Saturation is attained when the pressure of the adsorbing component A is so high that all available surface sites are actually occupied by A, and for a given concentration of B the reaction rate runs into limitation, because as soon as an A particle is reacted off, it will immediately be replaced by another A particle from the gas phase. In contrast to the LH mechanism, there is *no* maximum observable in the reaction rate! This phenomenon may be utilized to distinguish both reaction mechanisms from each other.

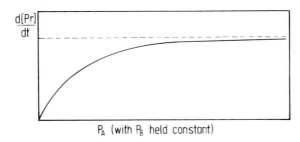

$\frac{d[Pr]}{dt}$

P_A (with P_B held constant)

Fig. 5.6. Variation of the rate d[Pr]/dt for a bimolecular surface reaction A + B $\rightarrow$ Pr, which proceeds via the Eley-Rideal mechanism. Again, as in Fig. 5.5, the concentration of one reactant [B] is kept constant, while the other ([A]) is varied. After Laidler [20]

In analogy to the discussion of the LH reaction, we may again consider the order of the kinetics of an ER reaction. According to Eq. 5.26, it will always be of first order with respect to P_B; the order with respect to P_A, however, will change from 1 to 0 as the reaction proceeds, and the term $b_A P_A$ in the denominator will become very large compared with 1.

Before we actually present some examples for surface reaction mechanisms, we would like to point to some difficulties concerning a clear-cut distinction of the two reaction mechanisms, i.e., the LH and the ER type. This is by no means as easy as one would think from looking at the above equations. Rather, a distinct mechanism may be largely obscured by several processes possibly occurring at the interface. We recall our remarks made in the kinetics section (cf., Sect. 3.3.2) about precursor states. It may very well be that for a certain reaction it is not required that both species actually enter the chemisorbed state – rather, just the population of a weakly held molecular state may be sufficient to induce a bimolecular reaction. The question arises as to whether this should be regarded as a real LH, or be better assigned as an ER mechanism. This ambiguity may also extent to those reactions in which trapped, but not fully accommodated molecules participate and exist in a vibrationally or rotationally excited state [27]. Again, this matter touches on reaction *dynamics* which shall, for obvious reasons, not be discussed here.

5.2.3 The Distinction between the LH and the ER Mechanism

Not only within our model considerations, but also in the context of practical heterogeneous reactions a frequent question arises as to an experimental means that could be able to positively distinguish between the possible reaction mechanisms. As pointed out before, one could, in principle, keep the concentration of one reactant constant and vary the concentration of the other in order to prove whether or not there appears a rate maximum. In practice, however, this is not a very convenient procedure, because it could in certain cases mean that one would have to cover an appreciable pressure range of a given reactant. In this situation, it is certainly worthwhile to briefly touch on a much more elegant experimental approach that was put forward some years ago by Jones et al. [28], Chang and Weinberg [29], Engel [30], and Lin and Somorjai [31] and that could substantially help to analyze the kinetics and mechanism of a given surface reaction. It basically consists of an amplitude modulation of a *molecular beam* of reactant molecules before impingement on the surface, combined with phase-sensitive detection utilizing, for example, lock-in techniques. These molecular beam (MB) studies are extremely advantageous when a determination of various relaxation times associated with the individual surface processes is attempted. In such an experiment a thermal molecular beam of reactant molecules is directed towards the surface, and the intensities of the incident and the off-scattered beams are measured and compared by means of a mass spectrometer. Probably the most important quantity here is the phase lag Φ between incoming and outgoing beams which directly reflects the relaxation or residence time of the particles at the surface. In turn, this residence time strongly depends on the surface temperature T. To illustrate the procedure, let us follow the mass (or particle) balance in a molecular beam experiment:

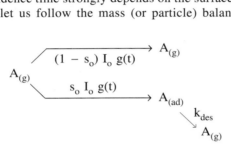

where I_0 is the unmodulated beam intensity, g is the gating function of the chopper (frequency ω), and k_{des} represents the well-known desorption rate constant.

Following Jones et al. [28], it can be shown that the normalized beam signal I/I_0 is given by

$$I/I_0 = \frac{k_{des} \cdot s_0}{k_{des} + i\omega} + (1 - s_0) .$$

5.28

The first term describes the fraction of particles that stick and then desorb, the second term arises from the fraction that is not trapped and directly reflected. Obviously, only the first term causes a phase lag Φ. The signal amplitude I approaches zero for this term only if $k_{des} \to 0$ (which corresponds to an infinitely long residence time of the particles at the surface). Note that k_{des} is the inverse of the residence time, i.e.,

$$\tau = \frac{1}{k_{des}} = \frac{1}{\nu} \exp\left(+\frac{\Delta E^*_{des}}{kT}\right) .$$

5.29

In practice, it suffices if τ only substantially exceeds the period of the chopping frequency ω. In other words, a long residence time is responsible for a pronounced phase lag and the demodulation of the beam signal. The second term of Eq. 5.28 accounts for elastic and direct-inelastic scattering (simply reflected particles) and does not produce any phase lag, owing to the extremely short residence time of the particle at the surface.

Engel [30, 32] studied the CO adsorption on Pd(111) and also the CO oxidation reaction at this surface by means of the modulated molecular beam technique described above. He used an MB-scattering chamber with an interior rotatable quadrupole mass filter and a mechanical chopper that allowed a given molecular beam of CO or O_2 molecules to be cut into small pressure pulses of frequency ω. A schematic drawing of the experimental set-up is presented in Fig. 5.7. He applied the principle of Jones et al. [28] (cf., Eq. 5.28), which rests on a measurement of the first Fourier-component of the scattered MB signal (frequency ω) as a function of the modulation frequency ω. An important experimental variable is thereby, as mentioned before, the surface temperature that sensitively governs any residence times of adsorbed or scattered particles. The procedure is to develop an appropriate kinetic model and to expand all time-dependent quantities into a Fourier series. If the model can be described in terms of *linear* algebraic equations (which is fortunately often the case) it suffices to consider only the ω-containing expressions obtained after insertion of the Fourier series. A subsequent comparison of the model's predictions with the actually observed ω and T dependences reveals whether or not the input model was correct. A simple example, viz., the associative desorption (rate constant k_{des}) can illustrate the method. The time-dependent change of the surface coverage $\Theta(t)$ in the modulated MB experiment is given by the differential equation:

$$\frac{d\Theta(t)}{dt} = s_0 I_0 g(t) - k_{des}\Theta(t) . \tag{5.30}$$

The time-dependent periodic functions $g(t)$ and $\Theta(t)$ can be expanded into a Fourier series, and one obtains (a_n, g_n being the Fourier coefficients):

$$\frac{d}{dt}\sum_{n=1}^{\infty} a_n e^{-in\omega t} = s_0 I_0 \sum_{n=1}^{\infty} g_n e^{-in\omega t} - k_{des}\sum_{n=1}^{\infty} e^{-in\omega t} . \tag{5.31}$$

Accordingly, Eq. 5.28 reduces, for the first Fourier component of frequency ω, to the expression:

$$-i\omega a_n e^{-i\omega t} = s_0 I_0 g_n e^{-i\omega t} - k_{des} a_n e^{-i\omega t} . \tag{5.32}$$

It is possible (after some algebraic manipulation) to express the first Fourier component J of the desorption rate as

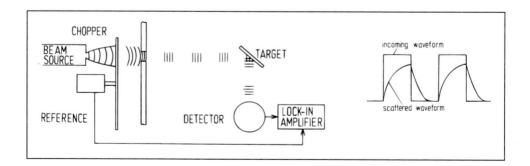

Fig. 5.7. Typical set-up for a chopped molecular beam experiment. The particle beam emitted by the source passes a chopper wheel which divides it into short particle pulses with frequency ω_0. These pulses interact with the surface (in the simplest case they are elastically reflected), and the particles emitted from the target are collected in the detector and analyzed with respect to their time dependence using lock-in technique. Inelastic interaction with the surface causes the initially rectangular shaped pulses to become distorted in a way shown in the right part of the figure. After Engel [32]

$$J = \frac{s_0 I_0 g_1}{\sqrt{1 + \omega^2/k_{des}^2}} \exp(-i\Phi) , \qquad\qquad 5.33$$

with the phase lag Φ being determined by $\tan \Phi = \omega/k_{des}$. Compared with the non-modulated experiment with $\omega = 0$, the modulated signal is phase-shifted by an angle Φ, and its intensity is reduced by the factor $1/\sqrt{1 + \omega^2/k_{des}^2}$. The advantage of this treatment concerning an evaluation of surface kinetic parameters becomes evident when we remember that the rate constant k_{des} and the particles' residence time τ are, via Eq. 5.29, inversely connected with each other, which also establishes a proportionality between $\tan \Phi$ and $1/v \exp(+ \Delta E_{des}^*/kT)$. Therefore, a plot of the logarithm of $\tan \Phi$ against $1/T$ yields a straight line with positive slope, from which one can immediately evaluate the activation energy for desorption, ΔE_{des}^*. Furthermore, since the phase lag Φ can be measured *absolutely*, an absolute determination of the pre-exponential factor v is also possible. Figure 5.8 shows an example for CO interaction with a Pt(111) surface, where τ values were determined by the modulated MB technique [43]. ΔE_{des}^* and the pre-exponential factor v came out as 138 kJ/mol and $10^{15}\,s^{-1}$, respectively.

While the first-order desorption is a relatively simple example for the application of this Fourier transform method, its full power becomes evident when we turn to actual

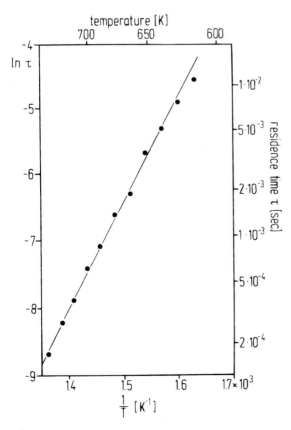

Fig. 5.8. Residence time τ for CO on platinum (111) as a function of surface temperature T, in an Arrhenius-like plot ($\ln \tau \triangleq \ln(\tan \Phi)$ vs $1/T$). Expectedly, high temperatures cause short residence times and vice versa. After Campbell et al. [47]

reaction kinetics and consider, particularly, the distinction between the LH- and the ER-types of reaction. Engel and Ertl [33] exploited the modulated MB method to study the CO oxidation reaction on Pd(111) surfaces. This reaction approximately obeys, as will be shown in the next section, the LH mechanism, i.e., molecularly adsorbed CO reacts with atomically adsorbed oxygen at the surface. This means that the (second-order) Eq. 5.18 can be applied. By adjusting a constant oxygen background pressure it was possible to make the reaction rate (i.e., the CO_2 formation rate) *independent* of the oxygen coverage, and there remained a "pseudo"-first-order type of reaction with Θ_{CO} as decisive parameter. CO, on the other hand, was admitted to the surface via the modulated molecular beam (frequency ω) in a manner as described before. Let us now set up the critical rate equations for the i) LH mechanism (note that Θ_O is constant in this type of experiment):

$$\frac{d[CO_2]}{dt} = k_{r(LH)} \cdot \Theta_{CO} \cdot \text{const} ,$$ 5.34

and ii) the ER mechanism

$$\frac{d[CO_2]}{dt} = k_{r(ER)} \cdot P_{CO} \cdot \text{const} .$$ 5.35

One can write for the modulated CO pressure (α and β are constants)

$$P_{CO} = P_{0,CO} + \alpha \exp(i\omega t)$$ 5.36a

and

$$\Theta_{CO} = \Theta_{0,CO} + \beta \exp(i\omega t) .$$ 5.36b

Inserting Eqs. 5.36a, b in Eqs. 5.34 and 5.35, respectively, and considering a phase lag

$$\tan \Phi = \frac{\Theta_{CO}}{k_{des(CO)} + k_{r(LH)}\text{const}} ,$$ 5.37

one obtains for the LH mechanism:

$$\frac{d[CO_2]}{dt} = k_{r(LH)}\text{const} \left(\frac{\Theta_{0,CO} + s_{0,CO}\alpha e^{i(\omega t - \Phi)}}{\sqrt{[k_{des(CO)} + k_{r(LH)}\text{const}]^2 + \omega^2}} \right) ,$$ 5.38

whereas the ER mechanism is described by

$$\frac{d[CO_2]}{dt} = k_{r(ER)}\text{const} \left(P_{0,CO} + \alpha e^{i\omega t} \right) .$$ 5.39

The above stated equations prove that the LH mechanism implies a *temperature-dependent* phase lag, while for the ER mechanism there should be no phase shift at all. Engel's results [30, 32] indeed show that there occurs an appreciable phase lag at surface temperatures where the adsorption-desorption equilibrium is adjusted for CO, but the oxygen coverage remains approximately constant due to the intentionally chosen pressure conditions. This is strong evidence that the catalytic CO oxidation reaction proceeds via the *Langmuir-Hinshelwood* mechanism. On this occasion, it should be added that so far there is hardly a single clear case known to the author where a catalytic surface reaction of the Eley-Rideal type has been proven. On the other hand, it is (as was demonstrated above) by no means trivial to evaluate a clear reaction mechanism, not even for a well-defined model reaction. This raises the legitimate question as to the practical applicability of the above considerations. The list of practical problems that enter the treatment of a realistic surface reaction such as ammonia synthesis, ammonia oxidation, SO_2 oxidation etc., is certainly extremely long. One should never forget that the Langmuir isotherms used so frequently in our discussion are, to a large extent, based on unrealistic *model assumptions* (for example, the neglect of coverage dependences of kinetic and energetic parameters). More seriously, practical surfaces never exhibit ideal flat single-crystal morphology, and there can be no hope that any reactant coverages could be predicted or adjusted by relying on an isotherm of the *Langmuir* type. It appears much more promising to determine isotherms experimentally and use them for the appropriate adjustments of pressure and temperature. Even then the matter is still very complicated, because sooner or later the catalyst morphology may very well change under reaction conditions as sintering or contamination phenomena come into effect. Here, the reaction temperature as well as the chemical conditions (oxidizing or reducing atmosphere) are certainly decisive. In this situation chemists have succeeded in carefully selecting special additives, the so-called *promoters* which can avoid sintering phenomena and often impose certain desired catalytic properties (enhancement of selectivity) on the catalyst material. This will be the subject of Section 5.4.

5.3 Examples for Surface Reactions; heterogeneous Catalysis

In the preceding section we have presented some basic, but fairly general kinetic considerations about surface reactions. Here, we are going to make some use of what we have worked out, and it is deemed useful to do this with the aid of some selected practical examples. In this section we will consider the CO oxidation reaction and, somewhat less extensively, the NH_3 synthesis reaction and the Fischer-Tropsch (in conjunction with the water-gas-shift) reaction. All these reactions are of great practical importance, either in the chemical industry for manufacturing basic chemicals, or for such special applications as automotive exhaust decontamination, etc. There are, of course, many other reactions for which this statement is true, but it is impossible within the framework of this book to deal with all of them.

5.3.1 The CO Oxidation Reaction

Several times (beginning in Sect. 1.2) we have referred to this reaction; here, we present some more data concerning activation energies, actual mechanisms, and peculiarities that make CO oxidation a fairly special reaction, such as the occurrence of kinetic oscillations. There exist numerous reports devoted to CO oxidation on metal surfaces, and we cite only a few of them here [32–46]. This reaction is readily catalyzed by transition metals, in particular platinum-group metals, and we have seen in the foregoing section that various single-crystal surfaces of palladium, platinum and iridium were investigated with respect to their activity in CO oxidation. Recently, also rhodium surfaces gained some interest, because Rh was found to be an essential constituent in the three-way automobile exhaust catalyst due to its ability to catalyze the reduction of nitrogen oxides NO_x.

It seems to be well-established that the catalytic CO oxidation can be regarded as a prototype for a Langmuir-Hinshelwood reaction (see the remarks made in Sect. 5.2.3). We have learned that this implies adsorption of both CO and O, whereby the spontaneous dissociation of the diatomic oxygen molecule is presumed. Accordingly, there exists a vast amount of literature concerning the adsorption of the individual reactants CO and O_2. This literature may serve as a firm data base regarding information about the adsorptive reaction steps. Not many reports, however, are available concerning the actual *reaction* between CO_{ad} and O_{ad}, especially the determination of the *reaction*-activation energy ΔE_r^*, and the way the CO_2 molecule forms. Again, coadsorption studies can provide some information here, and we refer to a work by Conrad et al. [38], where a Pd(111) single crystal surface was investigated. Without considering the details of this study we emphasize that there are *repulsive* interactions between neighboring adsorbed CO and O particles. Consequently, the individual adsorption energies of both species are *lowered*. This phenomenon also showed up in coadsorption studies performed with Ru(0001) by Thomas and Weinberg, where blue-shifts of the CO-stretching vibrations and red-shifts of the CO-metal vibrational frequencies were measured by means of HREELS [39]. The chemical reason for this behavior can be sought in the strong electronegative character of the oxygen atom which makes electron charge flow from its metallic environment right into the O-Me chemisorption bond. This electron transfer, however, causes a depletion of charge also in the environment of the adjacent CO molecule, whose chemisorptive binding requires a high electron concentration according to the back-bonding model of Blyholder [42] which we discussed briefly in Sect. 3.2.1. All in all, this restructuring of the electronic charge around a surface complex of a pair of particles which is, in principle, "ready" for reaction, can be made responsible for the existence of a reaction-activation barrier. Furthermore, the shape of the product (CO_2) molecule usually differs from the shape of the surface intermediate, i.e., the transition state or *activated complex* which is assumed to be bent in this case. The respective reorientation of atomic or molecular orbitals will also contribute to an activation energy barrier. Given a surface sparsely covered with CO and O, the activation energy for the step $CO_{ad} + O_{ad} = CO_{2(g)}$ could be determined to be around 100 kJ/mol for both a Pd(111) [33] and a Pt(111) [41] surface. Other related investigations concerned iridium [44,45] and rhodium [46] surfaces, where values of approximately 40 kJ/mol were reported. In all these cases, however, additional repulsive interaction energies come into play as the coverage of the reactants increases; they are responsible for a decrease of the respective adsorption energies. This, in turn, results in an additional decrease of the reaction-activation energy by 20–40 kJ/mol [47], along with a lowering of the pre-exponential factor, due to the compensation effect (cf., Sect. 3.3.3). This apparent coverage-dependence of the decisive kinetic and energetic parameters somewhat obscures the simple LH mechanism and is the reason why simple kinetic equations, such as Eq. 5.19, fail to describe the experimental situation accurately.

Nevertheless, it is possible from the available experimental data material to construct an approximate plot of the reaction energies involved in the CO_2 formation as a function of the reaction coordinate. We refer to the detailed and comprehensive report by Ertl [36], who communicated a corresponding energy diagram in a one- and two-dimensional representation. This is reproduced in Fig. 5.9, and provides, at the same time, an instructive example of the beneficial role of heterogeneous catalysis. In the left part of the figure we present a schematic one-dimensional potential energy diagram, whereby all energies are referred to the gaseous reactants ($CO + \frac{1}{2} O_2$) in molar concentrations. The adsorption of one mol of CO and O is accompanied by a net energy gain of 259 kJ. The actual reac-

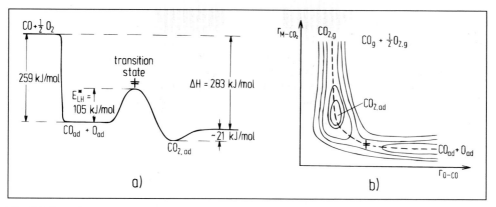

Fig. 5.9. Potential energy diagram of the CO oxidation proceeding (at low surface coverages) as a Langmuir-Hinshelwood reaction on a platinum (111) surface. In part a), a one-dimensional representation is shown with the involved energies being indicated; part b) illustrates the two-dimensional situation, whereby the CO_2–Pt distance r_{Me-CO_2} is plotted vs the mutual O–CO distance of the unreacted system r_{O-CO}. The reaction coordinate is marked by a broken line which corresponds to the solid line in Fig. 5.9a. After Ertl [36]

tion proceeds along an up-hill path and requires the aforementioned activation energy of ~ 100 kJ. The reaction product CO_2 is very weakly adsorbed (adsorption energy of ~ 20 kJ/mol) and usually immediately desorbs under reaction conditions. The gaseous CO_2 then exhibits the well-known thermodynamic reaction enthalpy of 283 kJ/mol as compared with the reactants.

The right part of Fig. 5.9 shows a corresponding two-dimensional representation of the potential-energy diagram, where the mutual distance r_{CO-O} between the reactants O_{ad} and CO_{ad} is plotted on the x-axis, and the adsorption bond length of the CO_2 product molecule on the y-axis. O_{ad} and CO_{ad} on the right side of the diagram reside in the deep potential energy minimum; as the reaction proceeds, both species must move up-hill along the dotted line to reach the transition state (denoted by the symbol ⁺), before CO_2 can actually be formed. This CO_2 is trapped in a shallow potential well from which it can easily escape and leave the surface. Its residence on the surface can, however, be very short. Time-of-flight measurements [48], measurements of the angular distribution of desorbing CO_2, as well as IR emission experiments [49, 50] with Pt surfaces revealed that the desorbing CO_2 molecules are translationally and even vibrationally excited and have not really equilibrated with the surface. With Pd(111) surfaces, however, there was strong evidence of a *cosine* distribution of the product molecules which must be understood as indicating complete translational accommodation and, hence, sufficiently long residence times of adsorbed CO_2 in the aforementioned shallow potential at the surface [33]. Very interesting data on the CO oxidation performed with Pd(110) and (111) surfaces were recently communicated by Matsushima [51]. The influence of the reaction site symmetry on the reaction mechanism was examined through analysis of the azimuthal distribution of the product (CO_2) molecule desorption, whereby angle-resolved thermal desorption spectroscopy experiments (cf., Sect. 4.4.1) were performed. Significant anisotropy in the spatial distribution of CO_2 was found for the Pd(110) surface, whereby the desorption perpendicular to the rows of Pd atoms (i.e., in [001] direction) was sharply collimated around the surface normal, according to a $\cos^n \vartheta$ function, with $n = 10 \pm 3$. Parallel to the troughs, however, a more broadened distribution was observed, with $n = 3 \pm 1$.

215

Interestingly, but in agreement with previous reports [30, 32], there was no anisotropy of the desorption flux found with a Pd(111) surface. Matsushima's conclusions were as follows: on Pd(110), the O atom is located in a long-bridge site (cf., Fig. 3.12d) *inside* the trough where it is much less mobile than coadsorbed CO molecules which are likely to diffuse towards the O atoms in order to react. In other words, the O atom is the actual reaction site. According to the local site geometry, the CO_2 being produced is initially immersed in the metal surface and is subjected to repulsive forces along the surface normal. The motion parallel to the surface is, in this stage, likely to be restricted across the troughs ([001]-direction), but much less *along* the troughs in [1$\bar{1}$0]-direction. These simple considerations of the local reaction-site geometry can fully account for the observed angular dependence and can also explain effects at higher coverages. The lack of anisotropy found with the CO_2 desorbing from Pd(111) is again easily understood in terms of the local adsorption geometry of the O atom, which resides in a threefold-coordinated site with its center of gravity being ~1 Å outside the metal plane. Consequently, the CO_2 produced is *not* immersed in the surface and vibrates in either direction with similar probability.

So far, we have concentrated on the interactions and elementary processes that occur within and around a reaction complex consisting of an individual pair of adsorbed CO and O. In a typical steady-state reaction experiment, however, the sample which is mounted in a pumped reaction vessel (= reactor) is subjected to a flow of CO and O_2 gas (as provided by stationary pressures of carbon monoxide and oxygen), and the turnover is measured mass-spectrometrically by following the partial pressure of carbon dioxide as a function of sample temperature. In Fig. 5.10, we have compiled several curves of this kind, viz., the rate of CO_2 formation (P_{CO_2} = d[CO_2]/dt) as a function of the sample temperature T, whereby we refer to various experiments performed with Pd, Pt, and Ir surfaces [52–54]. A common feature of all these curves seems to be a very low reaction rate at temperatures below ca. 450 K, a sharp rise as T exceeds the range of 470–500 K, and the subsequent formation of a relatively broad reaction maximum, because the CO_2 partial pressure declines again for elevated temperatures ($T > 550$ K). Remembering the microscopic processes discussed above it is not difficult to understand the overall behavior. For $T < 450$ K, both reactants CO and O readily adsorb and compete for adsorption sites, whereby the associative CO adsorption is superior to the dissociative O_2 adsorption (cf., our remarks made in Sect. 5.1). Thus, after a while the surface will preferentially be covered by CO molecules, resulting in a very low reaction rate. Around 450 K, however, the situation changes as the adsorption-desorption equilibrium of CO shifts towards desorption (due to the lower activation energy for desorption, cf., Eqs. 2.45 and 2.46), whereas this equilibrium is, for oxygen, still well on the adsorption side, because ΔE^*_{des} of O is appreciably larger. Due to the enhanced CO desorption, a larger fraction of bare sites is formed, and the probability for *adjacent* bare sites increases markedly. The existence of these adjacent sites is, however, the prerequisite for dissociative oxygen adsorption, and accordingly, the rate of O adsorption increases steeply. Thus, more and more *pairs of neighboring adsorbed CO and O species* are formed, especially since the rate of surface diffusion for CO is also facilitated at the higher temperatures. Hence, the Langmuir-Hinshelwood reaction can readily occur, resulting in high turnovers, as evident from the sharp rise of the CO_2 production. The decrease of the reaction rate at still higher temperatures occurs, because more and more oxygen atoms populate the available surface sites, but less and less CO molecules, and the number of CO–O pairs increasingly deviates from stoichiometry. As the temperature rises further, also the oxygen desorption becomes dominant, making the reaction rate decline further. In this T-range, other factors

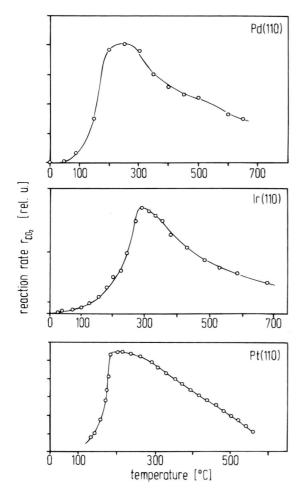

Fig. 5.10. Temperature dependence of the steady-state reaction (CO_2 formation) rate for the catalytic oxidation of CO over three different (110) transition metal surfaces. Top curve: Pd(110) [52], middle curve: Ir(110) [54], bottom curve: Pt(110) [53]. Note the steep rise of the rate at or shortly above 100°C which is caused by the onset of CO desorption. See text for further details

gain importance, such as oxygen bulk diffusion and oxide formation processes accompanied by restructuring (e.g., facetting) phenomena which may eventually even destroy the catalytic activity of the entire metal surface.

In the experiment described above we varied the sample temperature and kept the partial pressure of oxygen P_{O_2} and carbon monoxide P_{CO} constant. It is now interesting to hold T constant and vary the gas phase composition of the reaction mixture, i.e., vary P_{CO} at fixed P_{O_2}. We have performed a series of corresponding experiments with a Pt(210) surface [55, 58] and summarize the results in Fig. 5.11, where we plot the rate of CO_2 formation at fixed oxygen partial pressure as a function of the CO partial pressure (which was increased from approximately 10^{-7} mbar up to 10^{-5} mbar, while the oxygen pressure remained constant $\sim 2.7 \times 10^{-6}$ mbar). The temperature of the sample was chosen as 500 K, which ensured that, with the pressures given, appreciable turnovers were accessible. Together with the partial pressures, also surface coverages Θ could be monitored: Θ_{CO} was measured by laser-induced thermal desorption, and Θ_O could be deduced from work function measurements. Further details of this study can be taken from the original publi-

217

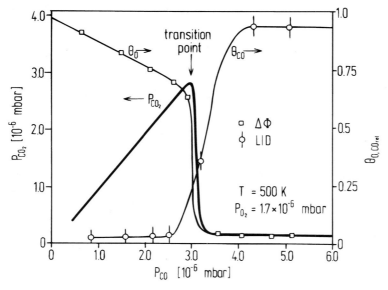

Fig. 5.11. Rate of reaction ($\cong P_{CO_2}$) during CO oxidation on a Pt(210) surface, as a function of CO partial pressure at constant oxygen pressure ($P_{O_2} = 1.7 \times 10^{-6}$ mbar) and temperature ($T = 500$ K). Relative coverages of CO and oxygen were determined using laser-induced thermal desorption and work function measurements during the reaction. In the high-reactive region near the transition point nearly every CO molecule impinging on the surface adsorbs and reacts. After Ehsasi et al. [58]

cation [55]. As evident from Fig. 5.11, the reaction rate, i.e., the partial pressure of CO_2 as a function of P_{CO} with all other parameters held fixed, exhibits a pronounced triangular shape. Expectedly, there is almost no or only very little product formed for CO pressures below 10^{-6} mbar. However, as P_{CO} starts growing there is an almost linear rise of P_{CO_2} until a maximum is reached, followed by a steplike decrease of the reactivity into a regime that is characterized by only small CO_2 yields. In a narrow CO pressure range, there is an almost *bistable* behavior: a region of maximum reactivity borders on an area that is practically non-reactive. The explanation of these phenomena is straightforward. At small CO partial pressures, the surface is predominantly covered with oxygen atoms. As more and more CO molecules collide from the gas phase, CO adsorption increasingly takes place (interestingly, an O-covered surface does *not* inhibit CO coadsorption). The number of adjacent CO-O complexes increases, each such pair immediately reacts off to CO_2, and a situation will be created, where the maximum possible number of reactive pairs is formed. This corresponds to the reactivity maximum of Fig. 5.11. However, with P_{CO} being increased further, more and more adsorption sites will become covered with CO, and the number of reactive pairs will decrease in accordance with the drop in surface reactivity. It is important to note here that oxygen is *unable* to find adsorption sites within a dense CO adlayer. Finally, this is the reason why practically the whole surface is covered with CO (CO *poisoning*), and an LH reaction can no longer proceed. In turn, this behavior may also be taken as a hint to the operation of the LH mechanism. It is perhaps worth to note that this triangular reaction behavior could recently be modeled by computer simulation based on a cellular automata technique [56]. The authors Ziff, Gulari, and Barshad ("ZGB model") allowed the LH mechanism to operate and assumed a surface reaction between CO and O on a lattice with square symmetry. They could show, by

218

introducing different probabilities for CO adsorption (*one* site is required) and oxygen adsorption (*two adjacent* sites are required) that there exists a transition in the reaction for a critical ratio of the gas phase concentration of reactants (in very satisfactory agreement with our experimental results). In a continuation of the ZGB study, several refinements were introduced (such as thermal desorption, surface diffusion and participation of precursor states) and led to an even better agreement between experiment and theory [55,57].

The existence of the aforementioned bistable behavior can be connected with an extremely interesting *non-linear* phenomenon, namely, kinetic oscillations of the CO_2 production rate, which can occur in spite of constant reactant pressures. One can easily rationalize that the reaction rate of the CO oxidation will periodically oscillate between maximum and minimum values, if one could establish a servo-mechanism that provides a self-sustained shift of the CO partial pressure back and forth in the small pressure interval of the transition. On several Pd and Pt surfaces, there exist such driving mechanisms, which consist, for example, of hystereses in structural phase transformations, along with structure-dependent sticking probabilities of one or more reactants. In other words, in the *adsorption* path a certain phase requires a critical coverage Θ_{ad} to form. In the *desorption* path, however, a different (and usually lower) critical coverage Θ_{des} only makes this phase disappear again. It would be far beyond the scope of this book to expand on these interesting phenomena. Nevertheless, we would like to provide the reader with some brief examples. One can imagine that it is fairly difficult to adjust the reaction parameters such that the system undergoes the kinetic oscillations, because the corresponding parameter space is rather narrow. In our laboratory, Ehsasi et al. succeeded in generating very regular kinetic oscillations in the CO oxidation over a Pt(210) and Pd(110) surface [58,59] by carefully adjusting the sample temperature and the CO partial pressure at fixed oxygen concentration. An example for the onset of the oscillations is presented in Fig. 5.12; if the reactants are sufficiently purified these oscillations can last for more than 24 h. Ertl and collaborators have been working on these non-linear phenomena for about 10 years and found oscillatory behavior with a variety of Pt and Pd surfaces (Pt(100), Pt(110), Pd(110)). A particularly intriguing example was the Pt(100) surface, where the oscillations could convincingly be correlated with surface-structure hysteresis effects [60].

In the following, we present only a very crude and schematic description of the atomistic processes leading to self-sustained kinetic oscillations with the system Pt(100)/CO + O, based primarily on reports by Ertl [61] and Imbihl et al. [60,62]. In the clean state, the Pt(100) surface exhibits a (5×20) LEED pattern which is caused by a surface reconstruction in which the topmost Pt surface atoms are displaced so as to form a hexagonal-close packed layer on top of the Pt bulk with cubic (100) symmetry. The relatively complicated (5×20) pattern is due to a superposition of LEED electrons scattered at the respective periodic gratings of the (1×1) and the hexagonal ("hex") overlayer. This surface is now exposed to a mixture of CO and O_2, and the rate of CO_2 formation is monitored as a function of time. On the "hex"-reconstructed surface, the sticking coefficient s_0 for CO is appreciable, while oxygen does not stick very effectively ($s_0 < 0.1$). Consequently, CO preferentially adsorbs until, at a critical coverage $\Theta_{CO} = 0.5$, a $c(2\times2)$ CO phase is formed, accompanied by an adsorbate-induced removal of the "hex"-phase. In this process, the Pt atoms of the topmost layer become displaced so as to gain the same periodicity as in the bulk, and a (1×1)-phase forms. The crucial point now is that the sticking probability of oxygen on this phase is considerably higher, and reactive CO–O complexes are increasingly produced which react off immediately. Hence, CO is depleted to

219

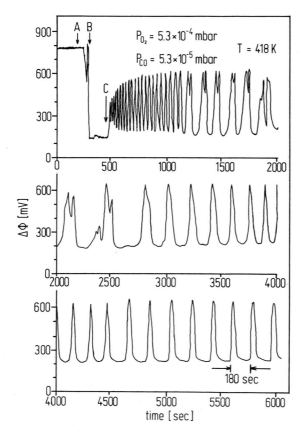

Fig. 5.12. Example for the onset of kinetic oscillations during catalytic CO oxidation on Pt(210). The oscillations were monitored through the work function changes. The overall pressure and temperature conditions are indicated in the figure. Oscillations were started from the CO-covered surface (point A) by introducing oxygen to the system. The amplitude of oscillation grows gradually (point B) and becomes limited at both ends while the initially high frequency drops until the system reaches the region of regular periodic behavior with periods of ~200s. After Ehsasi [57]

such an extent that, at the lower bound of the critical CO coverage, the reconstructed (5×20) Pt surface phase is restored. Note that this *lower* bound is below the aforementioned upper limit of the critical CO coverage which induced the phase transformation to the unreconstructed state, resulting in a pronounced hysteresis. We remember that on the (5×20) phase the oxygen sticking coefficient s_0 was considerably lower, and CO again has the chance to win the competition for the adsorption sites. It can and will adsorb, until the structural phase transformation (5×20) → (1×1) occurs a second time. Hence, a cycle is completed, and one can easily understand the accompanying changes in reactivity. Clearly, rate maxima occurring shortly after lifting of the (5×20) structure, and minima associated with the oxygen-poisoning alternate in a periodic fashion, thus giving rise to the kinetic oscillations. In Sect. 4.4, we briefly mentioned that the work function change can be utilized to follow surface-concentrations of adsorbates also during kinetic oscillations. In Fig. 5.13, we present an example of such a measurement, taken from Ertl's work on Pt(100) oscillations [61]. In this figure the reactivity is monitored by $\Delta\Phi_{CO}$ (which could be shown to parallel the CO_2 formation), along with the intensities of the LEED spots of the Pt(5×20) and (1×1) surface phase. Clearly, oscillations also occur in these LEED intensities, thus confirming our statements made above that a *structural hysteresis* is primarily responsible for the kinetic oscillations. We should add that the macroscopic observability of the oscillations demands rigid *phase relations* of neighboring oscillating

220

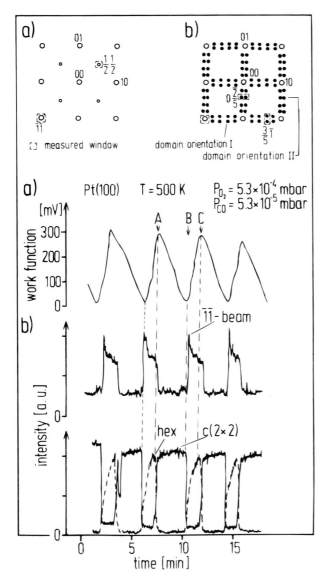

Fig. 5.13. Kinetic oscillations during CO oxidation on a Pt(100) surface.
Top part: LEED patterns observed a) for Pt(100) covered with 0.5 monolayers of CO (c(2×2)-structure, Pt surface unreconstructed); b) for clean Pt(100) ((5×20) "hex"-reconstructed phase). The important beams are indicated; also shown are the electronic frames placed around the decisive beams by the video method. The intensities of the corresponding beams are electronically integrated and monitored as a function of time.
Bottom part: a) Time dependence of the work function change [mV] during CO oxidation at 500 K. P_{CO} = 5.3×10^{-5} mbar, and P_{O_2} = 5.3×10^{-4} mbar; b) Variation of the LEED intensities measured by the video method of the $(\bar{1},\bar{1})$ beam and of the hexagonal and c(2×2) CO-induced beams. The phase relations are such that a maximum reactivity ($\Delta\Phi$ maximum, point A) coincides with a maximum of the hexagonal (reconstructed) phase, whereas minimum reactivity (point B) corresponds to a steep increase of the (1×1) intensity of the unreconstructed surface. After Imbihl et al. [60]

221

patches of the surface, i.e., there must exist a communication between different surface areas corresponding to a spatial self-organization. This communication can also be anisotropic thus leading to reaction fronts or chemical reaction waves at the surface. As we have seen in Sect. 4.5 there are experimental means to image even the respective surface structures.

Although this is an extremely interesting topic that has strong relations to a field called *synergetics*, we refrain from further considerations here and refer the reader to the relevant literature [63–67].

In the final part of our discussion of CO oxidation a few words are necessary about *practical aspects* of this reaction. So far, we concentrated on thermodynamic conditions that are relatively far away from actual catalysis, and a question pertinent to these model studies is whether the determined mechanisms also hold if much higher reactant pressures and highly disperse, polycrystalline catalysts (active surface areas of typically $100–200\,\mathrm{m^2/g}$) are used. In the following, we refer to work on the CO oxidation (and NO–CO reduction-oxidation) performed in the group of Fisher in which Rh single-crystal data and results obtained with alumina-supported rhodium catalysts were compared with each other [68]. In this study, two aspects of CO oxidation were elaborated, namely, i) the structure sensitivity, i.e., the dependence of the reaction rates on the morphology of the catalyst surface, and ii) the influence of the higher pressures under realistic conditions. Among others, apparent activation energies and turnover rates were measured and compared. The turnover number is defined as the number of CO_2 product molecules per surface metal atom and unit time. The reaction experiments were carried out in a combined UHV-cell–high-pressure reactor, and the products could be determined using gas chromatography with flame-ionization detection.

As regards structure sensitivity, kinetic parameters were determined under somewhat more realistic pressure conditions ($P_{O_2} = P_{CO} = 10^{-2}\,\mathrm{atm}$) for both Rh(111) surfaces and Al_2O_3-supported Rh catalysts. In Fig. 5.14a, site-normalized turnover numbers are plotted in an Arrhenius diagram vs reciprocal temperature. The plot yields an apparent activation energy of $120–125\,\mathrm{kJ/mol}$, practically regardless of the catalyst surface morphology. This result confirms that CO oxidation is a structure-insensitive reaction. In the second part of the experiment, the role of the CO partial pressure was examined. Again, the reaction was performed in a parallel fashion with Rh(111) single crystals and alumina-supported Rh catalysts, at fixed oxygen partial pressure ($10^{-2}\,\mathrm{atm}$) and constant catalyst temperature (500 K). The results are shown in Fig. 5.14b, where the site-normalized turnover number is plotted against CO partial pressure. Obviously, the comparison reveals extraordinary similarities in kinetic data measured over the two rhodium catalyst surfaces. As can be seen from the plot, the reaction rate decreases practically linearly with increasing CO partial pressure, in accordance with kinetic equations derived by the authors on the basis of a Langmuir-Hinshelwood mechanism. These considerations assume steady-state conditions and the limiting case $\Theta_{CO} \approx 1$, and lead to the expression

$$\text{Rate} \approx \frac{2k_{O_2,\text{ads}}\,k_{CO,\text{des}}(T)}{k_{CO,\text{ads}}} \cdot \frac{P_{O_2}}{P_{CO}} , \qquad 5.40$$

where the k_i represent constants describing the adsorption and desorption kinetics, respectively. They contain, in the usual manner, the sticking probabilities, kinetic factors, and pre-exponentials, etc., as explained in Sect. 2.6. Thus, under pressure conditions of practical interest, the reaction rate is approximately proportional to the oxygen partial pressure, but *inversely proportional* to the CO pressure. This again underlines the inhibiting function of high CO reactant concentrations.

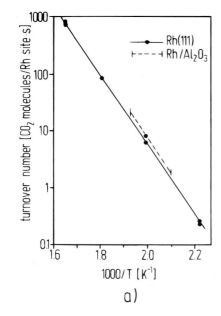
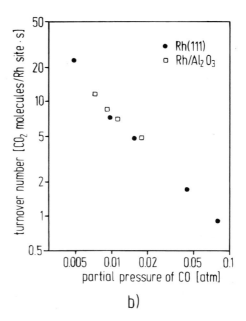

a)

b)

Fig. 5.14. Comparison of the specific rates of the catalytic CO oxidation measured over a Rh(111) single crystal surface and an alumina-supported rhodium catalyst at $P_{CO} = P_{O_2} = 10^{-2}$ atm. The left panel shows Arrhenius plots for both cases, whereby the striking coincidence of the two curves indicates a structure-independent reaction mechanism. Right panel: Measured rates of the CO oxidation reaction (site-normalized turnover-number) for Rh(111) and Rh/Al$_2$O$_3$ as a function of CO partial pressure. The oxygen pressure (10^{-2} atm) and the temperature (500 K) were kept constant during the experiments. After Oh et al. [68]

One may further ask about the influence of contaminants, such as carbon or sulfur, which are abundant impurities in technical materials. Here, we refer to exemplary work by Ertl and Koch, who studied contaminated polycrystalline palladium wires exposed to large pressures of oxygen and carbon monoxide [52]. The somewhat surprising result was that there occurred a self-cleaning process under reaction conditions, i.e., the substantial amount of sulfur and carbon at the beginning were almost completely removed at the end of the reaction. Apparently, the high oxygen pressures provide extensive oxidation of C and S to the respective dioxides CO$_2$ and SO$_2$. Processes of this kind may very well also occur under technical conditions.

Concerning the practical importance of CO oxidation, we simply mention that this reaction is essential in all equilibria involving oxidation of carbon in general, for example, in blast furnace processes. As mentioned before, CO oxidation has become *the* catalytic reaction in the context of automotive exhaust decontamination, because the removal of CO from the exhaust gas is one of the major tasks when designing the respective catalysts. It is only mentioned here that, besides CO decontamination, the automobile exhaust catalyst must also oxidize hydrocarbons and reduce NO$_x$ compounds. Technically, this is achieved by means of the so-called three-way catalyst. Many catalyst compositions are in use; a common feature is that they contain mixtures of several noble metals (rhodium, platinum, and palladium) supported on alumina. In some cases, ceramic monoliths coated with a thin washcoat of alumina are also used; criteria here are high activity (high surface area), attrition resistance (cf., Sect. 3.1), structure stability under typical

223

exhaust conditions, and favorable pore structure. Problems currently under investigation are the reduction of the noble metal concentration and improvement of deterioration behavior against the typical catalyst poisons sulfur, phosphorus, and lead. However, we cannot delve into this interesting matter here and must refer to the literature; an excellent survey of chemical and catalytical aspects of automobile exhaust decontamination is provided in the article by Taylor [69].

5.3.2 The Ammonia Synthesis Reaction

The gas-phase equilibrium reaction

$$\tfrac{1}{2}N_2 + \tfrac{3}{2}H_2 \rightleftharpoons NH_3 \qquad\qquad 5.41$$

is well-known to chemists as the basis of the famous Haber-Bosch process, which represents *the* example for a heterogeneously catalyzed reaction. All our remarks made before about kinetics and therrnodynamics of a heterogeneous reaction apply again. A reaction according to Eq. 5.41, from left to right, leads to a reduction of the number of moles (Δn = 1). Furthermore, the reaction is slightly exothermic, ΔH_r = 46 kJ/mol. Hence, high reactant pressures and low temperatures should be beneficial for NH_3 formation. This is evident from Fig. 5.15 showing the ammonia yield as a function of temperature and pressure for the gas-phase reaction. However, the existence of a reaction activation barrier prohibits too-low reaction temperatures, and one must find a compromise between the desired yield and sufficiently fast reaction rate. In *catalytic* NH_3 synthesis this rate is accelerated by the catalyst, and the reaction can run at lower temperatures which, in turn, influences the yield favorably. Again, as with many other catalytic reactions, *model experiments* were performed using single crystals and low pressures. There is, however, a serious principal shortcoming of these studies: thermodynamics require high pressures to obtain noticeable yields; in an UHV experiment, therefore, only low or extremely low turnovers are expected, which can lead to problems with the detection of ammonia. There exist a vast number of review articles on NH_3 synthesis which concern the industrial-technological aspects, as well as principal physical-chemical problems. Our list of literature relevant to that subject is necessarily incomplete [70–77]. Especially worth mentioning are model studies using single crystals which were, to a great extent, performed by Ertl and his group. In the following we will repeatedly refer to his work which is explained in the original publications in much greater detail [78–89].

Practical catalysts used in NH_3 synthesis are *multifunctional* catalysts; they consist mainly (in the unreduced form) of magnetite Fe_3O_4 doped with small percentages of Al_2O_3, potassium oxide K_2O, CaO, SiO_2, and MgO [90]. Alumina and potassium act as structural and electronic promoters, respectively, and we will devote a short section later (cf., Sect. 5.4) to the role of these additives. There have been studies in the group of Emmett [91,92] showing that, despite their nominal amounts, the promoters can cover large fractions of the actual catalyst surface and must, therefore, not be neglected. This introduces some complications for eventual model studies which can no longer be performed with clean single crystal surfaces *alone*, but must explicitly also take into account these chemical additives. Nevertheless, one may apply a similar strategy to gain access to the fundamental reaction steps, as we did with CO oxidation, i.e., to study in some detail the individual adsorption, dissociation, reaction, and desorption steps involved, and to attain as complete information as possible, in particular about reaction *intermediates* and

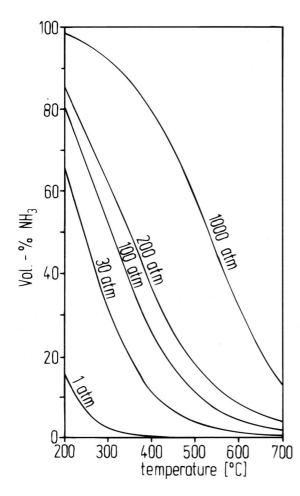

Fig. 5.15. Pressure and temperature dependence of the ammonia yield (vol. %) under thermodynamic equilibrium conditions for the gas phase reaction $N_2 + 3H_2 \rightleftharpoons 2NH_3$. Note that the yield is close to 100% at 200°C and 1000 atm pressure

activation energies. This can help to find out the rate-determining reaction steps. We direct the reader's attention again to the principle of microscopic reversibility which allows conclusions about the synthesis path to be made simply on the basis of decomposition studies of ammonia.

From the foregoing, one is inclined to suggest adsorption/desorption experiments with nitrogen and hydrogen on clean and oxidized iron single-crystal surfaces, as well as decomposition studies of ammonia on these surfaces. The complication as concerns the promoter action can perhaps be accounted for by coadsorbing various amounts of potassium and studying its influence on the reaction steps. Actually, many of these experiments have been and still are performed with clean and potassium-doped iron [78–87]. For the sake of brevity, we cannot explain all of these; instead, we illustrate some essential points.

In the beginning, we simply stated that there is ample evidence for a Langmuir-Hinshelwood type of reaction [71]. The following sequence of steps for the reaction of Eq. 5.41 was proposed:

$$N_{2(g)} + 2* \underset{k_{-1}}{\overset{k_1}{\rightleftharpoons}} 2N_{(ad)} \qquad\qquad 5.42$$

$$H_{2(g)} + 2* \underset{k_{-2}}{\overset{k_2}{\rightleftharpoons}} 2H_{(ad)} \qquad\qquad 5.43$$

$$N_{(ad)} + H_{(ad)} \underset{k_{-3}}{\overset{k_3}{\rightleftharpoons}} NH_{(ad)} \qquad\qquad 5.44$$

$$NH_{(ad)} + H_{(ad)} \underset{k_{-4}}{\overset{k_4}{\rightleftharpoons}} NH_{2(ad)} \qquad\qquad 5.45$$

$$NH_{2(ad)} + H_{(ad)} \underset{k_{-5}}{\overset{k_5}{\rightleftharpoons}} NH_{3(ad)} \qquad\qquad 5.46$$

$$NH_{3(ad)} \underset{k_{-6}}{\overset{k_6}{\rightleftharpoons}} NH_{3(g)} + n* \;. \qquad\qquad 5.47$$

While this mechanism, namely, the stepwise hydrogenation of nitrogen, had been proposed previously by Emmett [71], only recently has there been spectroscopic evidence reported for an imino-intermediate NH_{ad}, based on SIMS investigations [93]. From various nitrogen-adsorption studies performed in Ertl's group it soon became clear that step 5.42, the dissociative nitrogen adsorption, was rate-limiting for the overall reaction sequence. Interestingly, extremely small sticking coefficients of the order of 10^{-6} were found for the clean, densely packed, low-index Fe single crystal surfaces [78, 79, 84]. Actually, it became clear from low-temperature adsorption studies [84, 89] that the dissociative nitrogen adsorption occurred via a molecular precursor intermediate $N_{2(ad)}$ so that Eq. 5.42 must actually be split up as follows:

$$N_{2(g)} + * = N_{2(ad)} \qquad\qquad 5.42a$$

and

$$N_{2(ad)} + * = 2N_{(ad)} \qquad\qquad 5.42b$$

The molecular nitrogen intermediate is only very weakly adsorbed; hence, its surface concentration will always be quite small, particularly at higher temperatures. It is evident from Eq. 5.42a that $\Theta_{N_{2(ad)}}$ will be directly governed by the nitrogen partial pressure in the reactant mixture. On the other hand, as predicted by Eq. 5.42b, a high concentration of this molecular species is crucial for the build-up of appreciable amounts of atomic nitrogen, in order to continue the reaction sequence of Eqs. 5.43–5.47. Here, the adsorption energy ΔE_{ad} of N_2 is a crucial parameter in that a high ΔE_{ad} increases the surface concentration of N_2 for a given set of temperature and pressure values. The adsorption of hydrogen, however, occurs readily, with appreciable sticking coefficients; if a molecular H_2 precursor is involved cannot easily be decided experimentally because of its very low binding energy; at any rate, it does not noticeably affect the kinetics.

When considering the overall reaction kinetics of the ammonia synthesis one must distinguish two situations: first, only the reactants H_2 and N_2 are present, i.e., the reaction is far from equilibrium, and, second, the reaction has approached equilibrium conditions with the consequence that also large amounts of NH_3 product molecules exist in the reaction mixture and compete for adsorption sites on the catalyst surface. Far from equilibrium, we may, to a first approximation, neglect all backward reactions in the scheme described by Eqs. 5.42–5.47, and obtain for the decisive surface concentration of atomic nitrogen, applying the steady-state approximation:

$$[N_{(ad)}] \approx 2k_1[N_{2(g)}]/(k_3[H_{(ad)}])^{-1} = \frac{k'}{\Theta_H} P_{N_2} , \qquad 5.48$$

where k_1 and k_3 represent adsorption and reaction rate constants for N_2 adsorption and imino (NH) formation, respectively (cf., Eqs. 5.42 and 5.44). For the conditions given, the rate of ammonia formation can be described by the Temkin theory [74, 90], whereby this rate equals the rate of N atom formation on the surface, i.e.,

$$\frac{d[NH_3]}{dt} = -\frac{d[N_{(ad)}]}{dt} = k'' P_{N_2} ; \qquad 5.49$$

In this equation, the rate constant k'' can be replaced by the equilibrium constant for the N_2 adsorption K times the rate constant k_1 for dissociative N adsorption. Evidence for a rate law of Eq. 5.49, namely, a linear dependence on the gas-phase nitrogen pressure, was repeatedly found under conditions of low NH_3 formation efficiency.

As one gets closer to equilibrium, the steady-state concentration of N_{ad} will increase, due to partial product decomposition, as the back reaction steps gain importance, and more complicated kinetic rate laws apply. A recent consideration of these problems was offered by Nørskov and Stoltze [94], based on the assumption that each of the reaction steps indicated before are in equilibrium (cf., Eqs. 5.43–5.47) except the nitrogen dissociation (Eq. 5.42). Then, the law of mass action provides an expression for the output concentration of ammonia in terms of the input concentrations of the reactants. The various coverages of interest can be obtained from the equilibrium constants of the individual sub-reactions which, in turn, can be calculated using statistical mechanics (i.e., based on the partition functions of the gas phase and adsorbate species). The latter are accessible from measured binding energies, and vibrational and rotational excitation energies. Nørskov and Stoltze have compared the resulting "theoretical" ammonia production with experimental data obtained in a test reactor fed with commercial catalyst under a broad range of conditions. In order to mimic the plug-flow test reactor, it was necessary to divide the reactor into small segments, whereby the gas composition in each segment was given by the conversion in all previous segments. As far as the catalyst is concerned, the authors claim that the only important parameter in the calculation is the area of metallic iron (which could independently measured by titration). The result is shown in Fig. 5.16, from which one can immediately deduce a striking 1:1 relationship between calculated and measured ammonia output, over the surprisingly large pressure range from 1 to 300 atmospheres. This provides good confidence that the model describes the essential ingredients of the catalytic ammonia synthesis reaction.

So far, we have not commented on the activation energies for the individual reaction steps. By careful experiments using the single-crystal approach, Ertl and coworkers succeeded in determining most of these energies, and similar to CO oxidation a potential energy reaction diagram was constructed [76] which we present in Fig. 5.17. The energy-zero point is chosen, according to Eq. 5.41, to 0.5 moles of molecular nitrogen and 1.5 moles molecular hydrogen. By passing a small activation barrier of ~21 kJ dissociative adsorption takes place which lowers the energy of the system by 259 kJ. The stepwise catalytic hydrogenation of the nitrogen and imino-species occurs uphill (whereby the individual activation energies are yet unknown except the hydrogenation of $NH_{2(ad)}$), until adsorbed ammonia is formed with the aforementioned reaction enthalpy of 46 kJ/mol.

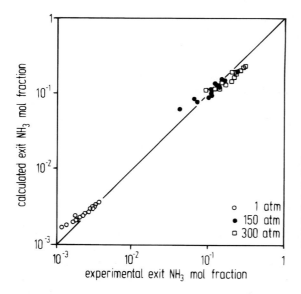

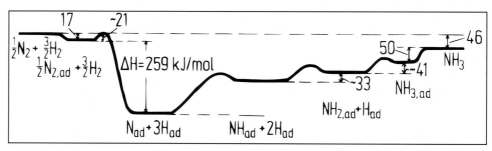

Fig. 5.16. Comparison of the calculated and the experimental NH_3 production over an industrial ammonia catalyst. The data extend over a range from 1 to 300 atm reactant pressures, and from 375° to 500°C temperature, with a 1 : 3 stoichiometry of $N_2 : H_2$. After Nørskov and Stoltze [94]

○   1 atm
●   150 atm
□   300 atm

Fig. 5.17. Schematic potential energy diagram as proposed by Ertl [76] for the catalytic NH_3 synthesis on iron surfaces at low coverages. All energy values are given in kJ/mol.

A final comment should be devoted to practical aspects of NH_3 synthesis, especially concerning the nature of the catalysts. Between Mittasch's early considerations [70] and Nielsen's work [90] there are certainly numerous other reports or specification of patents, all of which deal with catalyst optimization. A major problem in all these treatments has been (and still is today) the correct determination of the *actual surface composition* of the catalyst, and it was not until the introduction of surface-sensitive analytical tools (cf., Chapter 4) that this important issue could be more or less successfully tackled. There are various examples in the literature where the surface composition of a catalyst was studied by means of scanning Auger or Auger-microprobe techniques [95], however, always under ex-situ conditions, because it is impossible to characterize a catalyst surface analytically under working conditions. Nevertheless, one could characterize the catalyst prior to and after the reaction, in the oxidized state and in the reduced state, and the Auger scans revealed interesting lateral inhomogeneities and element distributions which were largely correlated with grain boundaries, surface defects, etc. The data further indicate that Al_2O_3 and CaO tend to form separate phases, whereas potassium and oxygen complexes more or less uniformly cover the iron surface. XPS measurements by Ertl and Thiele [96] showed that, despite the presence of K and O on the surface, the iron remained in the metallic state, and hence justified the use of clean Fe single crystals as model catalysts.

Another noteworthy point concerns the role of the additives. We said before that they act as *promoters* that either help stabilize the catalyst morphology (structural promoters) in that they prevent sintering effects, or they facilitate certain reaction steps in a particular manner and thus render to the catalyst only the peculiar activity. In these cases usually chemical bonding strengths are affected, due to redistribution of surface electronic charges caused by the additive: these are then called *electronic* promoters. Potassium is a typical example, and we will briefly expand on its action in Sect. 5.4.

5.3.3 The Catalytic CO-Hydrogenation

We will end our brief description of catalytic reactions with an excursion to an extremely important chemical process that is widely used in industry to produce all sorts of organic compounds (fuels, alcohols, etc.), namely, the catalytic hydrogenation of carbon monoxide. There is a long history of this reaction, beginning in 1902 with the investigations of Sabatier and Senderens [97], who succeeded in producing methane from a CO and hydrogen mixture over a Ni catalyst. For this contribution and his work on catalytic hydrogenation, Sabatier was awarded the Nobel Prize in chemistry in 1912 [98]. One year later, the German Badische Anilin- und Soda-Fabrik (BASF) started to fabricate longer-chain hydrocarbons and oxygenated hydrocarbons with the aid of transition-metal Co-Os catalysts promoted with alkali metals. In 1923, BASF performed the first successful exclusive methanol synthesis and 3 years later, Fischer and Tropsch described the formation of longer-chain hydrocarbons from $CO + H_2$ at $200°$ C and atmospheric pressures, using combined iron-cobalt catalysts doped with K and Cu as promoters [99]. To clarify the terminology, the CO hydrogenation process carried out under these conditions is called *Fischer-Tropsch synthesis*.

Since then especially the synthesis of light hydrocarbons, of gasoline, or of methanol as convenient carriers of chemical energy has largely dominated the chemical industry, although in times of low oil prices after World War II the synthetic fabrication of chemicals from $CO + H_2$ was not intensively pursued, unlike the situation during the war when more than 100,000 barrels a day of synthetic fuel were produced in Germany. The two oil crises in the 1970s somewhat revived interest in fabricating hydrocarbons and other basic chemicals from coal, which is an abundant mineral resource in many countries. (We recall what is certainly well-known to any chemist: that hot coal can be gasified by exposing it to water steam, whereupon the so-called *water gas* or *synthesis gas* is formed, which is essentially a mixture of hydrogen and carbon monoxide). In the 1980s, however, the price for crude oil remained at a relatively moderate level, so efforts to intensify hydrocarbon synthesis waned or were at least curtailed: now it is expected that the presently still threatening political situation in the Persian Gulf region and escalating oil prices will certainly prompt increased activity in hydrocarbon synthesis. How many different chemical compounds can be catalytically synthesized from a mixture of CO and H_2 can actually be seen from a diagram that was published by Pichler and Hector [100], reproduced as Fig. 5.18. Because of its obvious importance, a vast amount of literature dealing with gasification of coal and, particularly, with catalytic conversion of synthesis gas has been accumulated over the past 50 years. It is not possible to cite all the respec-

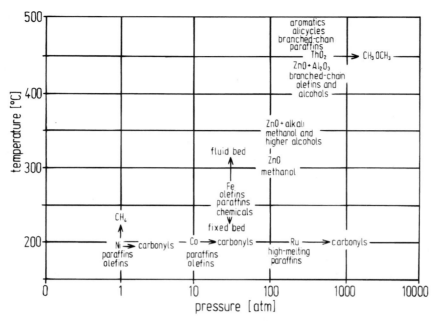

Fig. 5.18. Range of temperatures, pressures, and catalyst materials for reactions of synthesis gas $CO + H_2$. After Pichler and Hector [100]

tive publications, nor can we even hope to present an appropriate selection of them. We must satisfy the interested reader by quoting just a very few such papers, which may serve as encouragement for a further literature search.

Dry [101] comprehensively reviewed the state of the art of Fischer-Tropsch synthesis up to 1980, with emphasis on catalyst characterization and chemical-technological aspects of the CO hydrogenation. Among others, he described activities of the Republic of South Africa to build hydrocarbon and oxygenated hydrocarbon plants based on Fischer-Tropsch *technology*. A related paper on technological perspectives of synthesis gas conversion was written by King et al. [102]. Here, we do not attempt to comment too much on this technological subject, instead we intend to emphasize possible mechanisms of selected reactions as they were developed primarily on the basis of single crystal investigations. In this context it is noteworthy to point to an interesting similarity between ammonia synthesis and the Fischer-Tropsch process which concerns the nature of the catalyst: in both cases Fe-containing multi-functional materials with potassium (oxide), alumina, chromia, etc., as electronic and structural promoters are in use, and it is felt that a separate discussion of the promoting, inhibiting, or poisoning effects of these catalyst additives is worthwhile. This will be presented in Sect. 5.4.

But firstly, some further general remarks about hydrogenation of CO should be offered. There are two characteristic properties of the Fischer-Tropsch reaction, namely, i) the often unavoidable production of a fairly wide range of hydrocarbons (lack of selectivity), and ii) the high degree of heat accompanying the conversion. By means of suitable choices of the catalyst materials and by appropriate design of the reactors (heat-exchange features, etc.) one can control the reaction and direct it to the desired channel(s), as will be shown later. From the chemical viewpoint, we note that there are *two*

230

principal reaction routes of CO hydrogenation: Either the C-O bond is preserved, which results in products containing oxygen, or the C-O bond dissociates leading to water formation and hydrocarbon products. These two routes can be described schematically by the two simple net reaction equations

$$2H_2 + CO = CH_3OH \quad \text{methanol synthesis} \qquad \qquad 5.50$$

$$(\Delta H = -91 \, kJ)$$

or

$$3H_2 + CO = CH_4 + H_2O \quad \text{(methanation)} . \qquad \qquad 5.51$$

$$(\Delta H = -206 \, kJ)$$

Also competing may be the (unwanted) oxidation, accompanied by carbon ("coke") precipitation

$$2H_2 + 4CO = CO_2 + 2H_2O + 3C , \qquad \qquad 5.52$$

as well as many other side reactions leading to carbon (graphite) precipitation or formation of many higher molecular-mass hydrocarbons, among which are olefins and paraffins. Really, the task is to "tailor" a plant fabrication line such that the desired products appear at the reactor exit. In the temperature range usually employed in Fischer-Tropsch synthesis, it is well-known that the actual attainable selectivity is far from the one expected from thermodynamic calculations. Tillmetz [103] calculated the composition of a 1:1 ratio CO/H_2 mixture at 100 kPa pressure and predicted large amounts of methane, CO_2, and graphitic carbon to form under these conditions. However, the experimental observation was low CH_4, negligible C, and a major fraction of higher hydrocarbon species. One must, therefore, conclude (and this makes all further considerations complicated) that the reaction is, under the usual conditions, definitely not in equilibrium. For the sake of convenience, it is common to number the carbon atoms of the hydrocarbons or alcohols formed in Fischer-Tropsch synthesis ($n = 1$ means CH_4, $n = 2$ means C_2H_6 or C_2H_4, etc.).

In the following, we will mainly be concerned with C_1 processes (which are already sufficiently complicated) and will separately discuss the methanol formation (Eq. 5.50) and the methanation reaction (Eq. 5.51). We are interested, as pointed out above, in possible reaction mechanisms, whereby we mostly rely on single-crystal studies performed in combined UHV-atmospheric pressure apparatus equipped with coupled surface analytical instrumentation, and mass spectrometers or gas chromatography of the kind described in Chapter 1.

Somorjai devoted several articles to the catalytic "C_1"-chemistry, and in the following, we will repeatedly refer to his work [105–108]. Other authors have also examined CO hydrogenation from a similar viewpoint, by using either mono-crystalline or polycrystalline samples, together with combined UHV/high-pressure cells and surface-sensitive diagnostic tools, and we selectively cite communications by Bonzel's group [109–114], Goodman and coworkers [15, 16, 114–117], Wedler and Körner [118], Palmer and Vroom [119], and Hirschwald and coworkers [120].

We also recall the many studies that are and were concerned with coadsorption studies of hydrogen and carbon monoxide on single-crystal samples, without considering actual reactions and reaction yields, and refer again to our remarks made in the beginning of this chapter (Section 5.1). Investigations of coadsorption, nevertheless, can have great impact on an understanding of the reaction mechanisms pertinent to CO hydrogenation, and we therefore present, for the sake of completeness, some references that cover this field [8, 10–13, 121–127].

It is unnecessary to quote all the many studies that explored CO adsorption or hydrogen adsorption on transition metals *alone*; they are, of course, also quite helpful in the context of CO hydrogenation, because they provide a firm data base from which heats of adsorption can be judged (which, in turn, may govern surface concentrations under reaction conditions). This aspect was, for example, emphasized in the article by Bonzel and Krebs [113].

To begin with, it may be very informative (although by no means surprising) that there exist huge differences in activity for CO hydrogenation among the various metals. Vannice [128, 129] has systematically compared these activities for a variety of metals supported on silica, and arrived at a so-called vulcano-curve (this terminology was introduced by Balandin [130]), which we reproduce in Fig. 5.19. Plotted therein is the turnover number of methane (number of CH_4 molecules produced per second and surface site) against the heat of adsorption of CO on the various metals. As can be seen, there is an activity variation over five (!) orders of magnitude. Cobalt (which we have learned is a good Fischer-Tropsch catalyst) is in the top position, while Cu and Pd exhibit little activity. On the group V and VI metal surfaces, for instance, Nb, Re, or Mo and W, still higher binding energies result, which – according to Blyholder's model [42] – favor *dissociative* CO adsorption. Such dissociation can, during a catalytic reaction, easily lead to the rapid build-up of inactive carbonaceous overlayers and delete the catalyst activity. One could therefore argue that a medium binding energy of CO is beneficial, probably because too

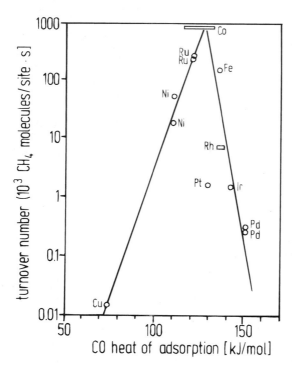

Fig. 5.19. Typical "vulcano"-curve, in which the turnover-number for methane formation is plotted against the CO heat of adsorption for Cu and some Group VIII transition metals. The curve exhibits a maximum for Co, Ru, and Fe. After Vannice [129]

232

high heats of CO adsorption cause complete CO and/or C poisoning (we remind the reader of the similar situation in CO oxidation), whereas the opposite is true, namely, vanishing CO coverages, for too low adsorption energies. This then entails low activities, for example, in methanation reactions.

Similar considerations must be invoked also for hydrogen chemisorption, where again Cu exhibits fairly low values around 50–75 kJ/mol, while on the typical group VIII metals Fe, Co, Ni, Ru, Rh, Pd, relatively similar heats of adsorption of the order of 100 kJ/mol are observed [131]. On Mo and W, as in the CO adsorption case, even larger heats of adsorption around 150 kJ/mol are the rule, and we take the opportunity to point to this interesting parallelism in the trend of CO and H_2 chemisorption energies as a function of the position of the aforementioned metals in the Periodic Table.

Next, one must inquire into the *coadsorption behavior* of CO and H and there particularly examine whether CO and H_2 compete for the adsorption sites and form separate islands, or if they exhibit more attractive interactions that give rise to {CO-H} surface complexes. From our examples presented in Section 5.1, we obtained the impression that both situations can occur with almost equal probability, depending on the chemical nature of the metal, but depending also on the crystal face orientation and hence on the local surface geometry. Here, the crystallographically open fcc (110) surfaces bear a tendency of *cooperative* coadsorption, while the dense faces, especially the very smooth (111) planes, instead sustain formation of separate islands.

As regards the *chemical nature* of the catalytic material, one must certainly scrutinize its electronic structure in the first place, because the apparent existence of closed or practically closed electron shell configurations could play a decisive role. (For a detailed discussion of the relation between catalytic activity and the electronic orbital configuration of a given material we refer to the excellent representation in Bond's book [132]). Closely related to the electronic catalyst structure is the question of whether or not the adsorption of CO and H_2 occurs *molecularly* or *dissociatively*, which depends crucially on the electronic properties of the substrate metal (copper, for example, does not spontaneously dissociate molecular hydrogen, while CO dissociation on the other hand, is not difficult to achieve on various Fe surfaces, on Ni or on Co, especially at elevated temperatures). On many $4d$ and $5d$ electron metals, CO dissociation is the rule, for example, on Mo or W, being a consequence of the large CO-metal binding energies stated above. A somewhat more detailed inspection of the CO dissociation behavior on metal surfaces in the context of CO hydrogenation was made by Bell [133], who found that the tendency to dissociation increases when moving from right to left in the Periodic Table. Although the structure of the transition state leading to dissociated CO is not known, it is reasonable to assume simultaneous interactions of both the carbon and the oxygen ends of the molecule with two or even more sites of the catalyst. Since the initial bonding of chemisorbed CO to the surface occurs via the carbon atom, it is very likely that a bent Me-C-O configuration finally provides the coupling between the oxygen end and a neighboring surface atom of the catalyst, possibly by vibrational deformation.

Related to this is also the influence of chemical *additives* that can either be formed in the course of the reaction itself, that is to say, directly from the reactants (carbidic carbon represents a striking example here [15, 16]), or that may be deliberately added as electronic promoters, such as potassium. As will be demonstrated in Section 5.4, it is relatively easy to establish a direct relation between coadsorption of an electropositive element (K, Cs) and an enhanced tendency to CO dissociation.

Let us now more closely inspect the well-known methanation reaction (cf., Eq. 5.51). Methane is almost selectively formed over Ni catalysts, while with other metals longer-chain hydrocarbons are additionally produced under the same conditions. The activation energy for methane formation from CO + H_2 in the temperature range 500–750 K is approximately 90–100 kJ/ mol, quite similar for most group VIII metals (Fe, Co, Ni, Ru, Rh), thereby likely indicating a very parallel (and structure-insensitive) mechanism on these surfaces. The more active catalyst materials provide TONs of up to 10 CH_4 molecules/(site and second). Based on a variety of investigations which we will discuss below, this common mechanism includes dissociation of adsorbed CO into carbon and oxygen; either directly, according to the equation

$$CO_{(ad)} = O_{(ad)} + C_{(ad)} , \qquad 5.53$$

or via CO disproportionation, the well-known Boudouard reaction:

$$2 CO_{(ad)} = C_{(ad)} + CO_{2(g)} , \qquad 5.54$$

followed by subsequent direct hydrogenation of the active carbon surface atoms to methane. This process was extensively examined for Ni and Ni-based catalysts in the group of Goodman [15, 16], and on Fe and oxidized Fe by Bonzel and coworkers [109–112]. In both these investigations the CO hydrogenation could be studied at comparatively large pressures, ranging well into the mbar regime, and site-normalized turnover numbers for methane formation were derived as a function of temperature. Clearly, the presence of a reactive carbidic carbon species was found as the reactive principle which carried the CH_4 production. Other groups reached the same conclusions for different systems; among others, Rabo et al. [134] succeeded in directly identifying and quantifying the surface carbon by means of titration techniques. Once the active surface carbon is formed, its subsequent hydrogenation to methane occurs as a relatively rapid step. Accordingly, the overall rate of methanation is largely governed by the carbidic carbon deposition which, in turn, depends on the rate of CO dissociation.

We may now formulate the reaction steps of the methanation mechanism as they were comprised by Bell [133], whereby we leave out the necessary preceding CO and H_2 adsorption steps:

$$CO_{(ad)} + * \rightleftharpoons C_{(ad)} + O_{(ad)} , \qquad 5.55$$

$$C_{(ad)} + H_{(ad)} \rightleftharpoons CH_{(ad)} + * , \qquad 5.56$$

$$CH_{(ad)} + H_{(ad)} \rightleftharpoons CH_{2(ad)} + * , \qquad 5.57$$

$$CH_{2(ad)} + H_{(ad)} \rightleftharpoons CH_{3(ad)} + * , \qquad 5.58$$

$$CH_{3(ad)} + H_{(ad)} \rightleftharpoons CH_{4(g)} + * . \qquad 5.59$$

Apparently, the methanation consists of a series of subsequent hydrogenation steps of chemisorbed methyne, methylene or methyl fragments via the LH mechanism. Only the saturated hydrocarbon CH_4 interacts so weakly with the surface that it immediately desorbs after its formation. There are also other possible reaction channels. The hydrogen may react with coadsorbed oxygen (formed by step 5.55) or, CH fragments at various stages of hydrogenation may react with each other, thus forming hydrocarbons with longer chains, for example, via

$$CH_{3(ad)} + CH_{2(ad)} \longrightarrow CH_3-CH_{2(ad)} , \qquad 5.60$$

followed by hydrogenation

$$C_2H_{5ad} + H_{ad} \longrightarrow C_2H_{6g} + n * . \qquad 5.61$$

234

Although successive hydrogenation of active carbon is the dominant mechanism, there are reports whereafter methane or higher aliphatic hydrocarbons can be formed at a surface in a different way. We refer to the work by Poutsma et al. [135], who observed CH_4 over palladium surfaces that definitely adsorb CO *molecularly*. Similar findings were reported by Vannice [128, 129] with platinum, palladium, and iridium surfaces which also do not dissociate CO. It is therefore feasible that there can occur a *direct* insertion of hydrogen into adsorbed CO, followed by the elimination of water under the reaction (temperature, pressure) conditions. If this subsequent separation step does not occur, we obviously arrive at *oxygenated* species, methanol in the first instance, but also aldehydes, or ketones and related compounds. It is important to note that these oxygen-containing products are preferentially obtained with all those metal surfaces that tend to adsorb carbon monoxide molecularly, also at higher temperatures. Fe, Co, and Ni very likely do not belong to this category, but Ru, Rh, and Pd do. Below, we will therefore briefly deal with the methanol synthesis reaction, again from the viewpoint of possible reaction mechanisms, and not so much under the aspect of industrial catalysis.

The catalytic hydrogenation of CO to methanol dates back to the early 1920s, when CH_3OH was, without significant side reactions, obtained for the first time in pure form from synthesis gas at BASF. This catalytic methanol synthesis reaction is widely exploited in chemical industriy; at temperatures between 300° and 400° C and pressures around 200 atm methanol forms in large quantities over ZnO/Cr_2O_3 (zinc-chromite) catalysts. It is interesting to note that raising the temperature by only 40° C and adding alkali metal promoters to the catalyst changes the product pattern to higher alcohols, preferentially, iso-butanole (these being valuable basic chemicals with a wide range of applications). Today, methanol synthesis is performed over mixed $ZnO/CuO/Cr_2O_3$ catalysts at lower pressures (< 100 atm) and temperatures (220–250° C). The catalyst morphology was investigated by Mehta et al. [136], who reported significant changes of this property under reaction conditions. It appears as if Cu^+ ions are formed in the ZnO matrix and then act as the essentially effective species. Chromium, on the other hand, is beneficial, because it enhances the solubility of Cu^+ in zinc oxide. It is no wonder that there exist numerous studies of CO + H_2 interaction with Cu, CuO, and particularly ZnO surfaces. We refer to the comprehensive report by Hirschwald et al. [137], and to investigations by Herman et al. [138], Poutsma et al. [135], as well as to review articles by Klier [139] and Kung [140]. Despite all these studies there is, not yet, complete clarity about the mechanism of methanol synthesis over copperoxide/zincoxide/chromiumoxide catalysts. Apart from the aforementioned direct insertion of H_2 into CO and subsequent hydrogenation the interest today concentrates on a possible hydrogenation of CO_2 which may actually be an unavoidable constituent of the reaction mixture, owing to the operation of the watergas shift reaction

$$CO + H_2O \rightleftharpoons CO_2 + H_2 . \qquad 5.62$$

This mechanism was first proposed by Rozovskij [141], who interpreted the experimental observation (whereafter low CO_2 concentrations added deliberately to the reaction mixture markedly increased the rate of CH_3OH formation) as direct CO_2 hydrogenation, according to

$$CO_2 + 3H_2 \rightleftharpoons CH_3OH + H_2O . \qquad 5.63$$

In this situation reaction studies with well-defined surfaces under realistic pressure and temperature conditions are needed, as well as coadsorption studies of CO, CO_2, and H_2 on ZnO or CuO surfaces, in order to identify possible reaction intermediates (which could consist of formaldehyde, methoxide, or formate species). Furthermore, exploring the adsorption and decomposition behavior of methanol on those surfaces is deemed very useful in this context, and there exists, indeed, a wealth of related work that is impossible to cite here completely. Lüth et al. [142] studied H + CO coadsorption on ZnO by means of UPS and found that precovered H was necessary to also make CO adsorb. Ueno [143] examined the coadsorption of CO_2 and H_2 and could detect formate (HCOO) production using infrared spectroscopy. The importance of CO_2 in the synthesis of methanol from $CO + H_2$ was also emphasized by Hirsch and Hirschwald [144, 145] in their UPS/XPS study in a combined UHV/high-pressure reactor.

Compared to this work on oxide surfaces there exists an even better data base in the field of methanol synthesis over *metal* surfaces. A comprehensive article on this subject was written by Vannice [129], in which many single crystal studies were also reviewed. $CO + H_2$ reaction over supported metal catalysts (alumina, silica supports) is also frequently investigated, among others, by Kölbel and coworkers [146] using iron-based catalysts and infrared spectroscopy. They could isolate a surface complex with the stoichiometry H_2CO, and also carboxyl, carbonate, and carbonyl surface species. On Pd-supported catalysts, Poutsma et al. [135] reported on an exclusive methanol formation as long as the reaction conditions (P, T) were thermodynamically favorable; similar results were obtained with Pt and Ir. In contrast, Ni-based catalysts revealed, under the same thermodynamic conditions, primarily hydrocarbons (methane, in particular), thus corroborating the aforementioned ideas, whereafter the metal's ability or tendency to dissociate carbon monoxide is decisive for the route of reaction. In this respect, Ni and Pd are chemically very different. Quite in line with this are also adsorption and decomposition studies of methanol on Ni and Pd single crystal surfaces performed by Rubloff and Demuth [147], Ibach and Demuth [148], and Christmann and Demuth [149]. On Ni(111), only methoxide (CH_3O) was formed during annealing of a chemisorbed methanol layer, as confirmed by a UPS and HREELS analysis, while on Pd at least two other surface intermediates with a similar stoichiometry, but different molecular structure could be identified. Despite the vast literature that has been accumulated up to now, a breakthrough in understanding the elementary processes of methanol and oxygenated hydrocarbon formation on metal surfaces has not yet been achieved, and an appropriate explanation is still lacking. Nevertheless, several possible mechanisms have been proposed [133]. There is a well-founded likelihood for a mechanism involving the successive hydrogenation of the -C-O- skeleton [133], which we reproduce here (*Me, Me'* different metal surface atoms):

$$
\left.
\begin{array}{l}
\mathrm{Me} = \mathrm{C} = \mathrm{O} + \mathrm{Me'} = \mathrm{Me} \qquad \begin{matrix} \diagup \mathrm{C} = \mathrm{O} \diagdown \\ \mathrm{Me'} \end{matrix} \\[3em]
\mathrm{Me} = \mathrm{C} = \mathrm{O} \cdots \mathrm{Me'} + \mathrm{H_{(ad)}} = \mathrm{Me} - \overset{\overset{\textstyle H}{|}}{\mathrm{C}} = \mathrm{O} \cdots \mathrm{Me'} \\[3em]
\mathrm{Me} - \overset{\overset{\textstyle H}{|}}{\mathrm{C}} = \mathrm{O} \cdots \mathrm{Me'} + \mathrm{H_{(ad)}} = \mathrm{Me} - \overset{\overset{\textstyle H}{|}}{\underset{\underset{\textstyle H}{|}}{\mathrm{C}}} - \mathrm{O} - \mathrm{Me'} \\[3em]
\mathrm{Me} - \mathrm{CH_2} - \mathrm{O} - \mathrm{Me'} + \mathrm{H_{(ad)}} = \mathrm{CH_3} - \mathrm{O} - \mathrm{Me'} + \mathrm{Me} \\[2em]
\mathrm{CH_3} - \mathrm{O} - \mathrm{Me'} + \mathrm{H_{(ad)}} = \mathrm{CH_3OH_{(g)}} + \mathrm{Me'}
\end{array}
\right\} \qquad 5.64
$$

In a similar way, the formation of aldehydes or higher alcohols also occurs by stepwise hydrogenation of methoxy-intermediates. Further considerations of kinetic laws and rate expressions can be found in Bell's article [133]. As before, the reaction sequence of Eq. 5.64 is essentially based on an overall Langmuir-Hinshelwood mechanism; under no circumstance was there any evidence for an Eley-Rideal mechanism. Also, similar to the methanation reaction, one cannot a priori assume that the reactant mixture is in thermodynamic equilibrium. Furthermore, cross-reactions between various oxygenated intermediates must be taken into account; they can lead to a wealth of different oxygenated products at the reactor exit.

In summary, the CO hydrogenation appears by far as the most complicated case among our few examples of heterogeneously catalyzed reactions discussed hitherto, despite the simplicity of the input molecules CO and H_2. In chemical practice, however, there are still much more complex heterogeneous reactions; we recall all the various reforming and platforming processes (hydrocarbon conversion) used to refine crude oil or aliphatic hydrocarbons. These processes are often carried out over platinum or platinum metal catalysts and have great practical importance. A thriving branch of the chemical industry is the production of chemicals, especially organic chemicals, that are often used as input materials for polymerization reactions. In Table 5.1 (taken from Somorjai's book [105]) we comprise some practically relevant heterogeneous processes.

Table 5.1. Some heterogeneously catalyzed chemical reactions and appropriate catalysts [105]

Reaction	Catalyst Material
CO, C_xH_y oxidation in car exhaust	Pt, Pd on alumina
NO_x reduction in car exhaust	Rh on alumina
Cracking of crude oil	Zeolites
Hydrotreating of crude oil	Co-Mo, Ni-Mo, W-Mo
Reforming of crude oil	Pt, Pt-Re, and other bimetallics on Al_2O_3
Hydrocracking	Metals on zeolites Al_2O_3
Hydrogenation of oils	Ni
Steam reforming	Ni on support
Water gas shift reaction	Fe-Cr, CuO, ZnO, Al_2O_3
Methanation	Ni on support
Ammonia synthesis	Fe + promoters
Ethene oxidation	Ag on support
Ammonia oxidation	Pt, Rh, Pd
SO_2 oxidation (sulfuric acid)	Vanadium oxide(s)
Acrylonitrile from propene	Bi, Mo-oxides
Vinyl chloride from ethene	Cu-chloride
Polyethene	Cr, Cr_2O_3 on SiO_2

At the end of this section we will draw the reader's attention to the additional application of surface-chemical principles in *physical* technologies that is to say, in semiconductor fabrication – a steadily growing field. We simply mention that chemical reactions play a significant role there, among others in *chemical vapor deposition* (CVD) – to be more precise – metalorganic chemical vapor deposition (MOCVD), where compound semiconductors are produced by co-precipitation of, e.g., galliumtrimethyl ($Ga(CH_3)_3$) and arsine AsH_3 on various target surfaces which stimulate simultaneous decomposition and help to produce well-ordered and ultra-clean semiconductor materials [150]. Other processes where surfaces and surface reactions predominate are chemical *etching* of semiconductor materials, for example, by using fluorinated agents or atomic hydrogen; during the recent past a profound interest has arisen with regard to the preparation of high-temperature superconducting materials by means of molecular beam epitaxy (MBE) – here another facet of surface physical chemistry becomes apparent, namely, exploiting surfaces simply as host substrates in order to obtain a desired crystallographic orientation of a certain material. Of course, this field can no longer be regarded as *catalysis* in a strict sense; nevertheless, it is the *surface* that helps achieving special material properties. In spite of this wide spectrum of utilizing surfaces many processes are not yet well understood, not even apparently "simple" processes such as unimolecular decomposition or isomerization reactions. It is an open secret that much more work still needs to be done, especially on model single-crystal systems, but also on real supported catalyst materials, on metal and on oxide surfaces. This holds especially for understanding of the promoting action of alkali-metal and alkaline earth-metal additives. Again, there is an overwhelmingly large number of related studies being performed that concentrate on the elucidation of the processes behind promotion, and we will close our considerations with some remarks on the role of promoters and catalyst poisons.

5.4 Promotion and Poisoning in Heterogeneous Catalysis

On various occasions we saw that practical catalysis is seldom carried out over homogeneous elemental or mono-compound materials, but real catalysts usually consist of a mixture of several chemical compounds or elements, supported on a carrier material that provides, in the first instance, spreading of the catalyst which results in high-surface areas. High-surface areas, or large degrees of dispersion, which persist also under reaction conditions, are certainly an important prerequisite for catalytic activity; additionally, the strong metal-support interaction (SMSI) can play a role. In this final section, we want to expand somewhat on these so-called synergetic effects that are often the basis for a specific catalytic action. (Synergism [151] means here that specific parameters supplement each other in a way so as to create new chemical or physical properties). It is almost a platitude when we emphasize that mere catalytic *activity* is a necessary, but not sufficient condition for successful chemistry; truly valuable catalyst materials provide activity *and selectivity*, that means, they catalyze only a certain reaction path out of many possible routes. General reports about surface science and catalysis in which examples are presented for generally and specifically catalyzed reactions and in which the role of chemical additives is considered, are numerous; here we list only a few of them [113, 152–158]. In the following, we will concentrate mainly on two aspects of chemical additives, i) promoting action evoked either by electronic or morphological alterations of the catalyst material, and ii) inhibiting or poisoning action which can either *selectively* or *generally* suppress chemical activity. In recent years there have been tremendous research efforts in this field, but we can only touch on most of the relevant work. However, it is our intention to provide the reader with a somewhat more fundamental understanding in terms of our former successful microscopic view, that is to say, of adsorption and coadsorption of reactant particles at surfaces.

Turning to catalytic *promoters* first, we want to briefly recall where we encountered promoters so far: in ammonia synthesis, we mentioned the addition of potassium, calcium, aluminum; in the Fischer-Tropsch reaction potassium also played a significant role as catalyst additive, and in methanol synthesis again alkali metals, and also chromium, calcium, etc., were used to direct the reaction to the desired route. Many more examples could be cited wherein chemical syntheses are specifically influenced even by small amounts of these chemical modifiers. It is, however, striking that alkali metals, potassium in particular, are most frequently named in this context. The first real valuable information about the underlying physical processes arose in the late 1970s from model single-crystal studies performed in various laboratories, among others by Bonzel [110–114, 159–165], Somorjai [106, 108, 166, 167], and Ertl [82–88] (but many other researchers could be listed). It soon became evident that adsorbed alkali metal atoms change the electronic charge distribution in their vicinity dramatically; we recall the fact that (long known from phototube multipliers or photocells) alkali metals make the work function of metals decrease dramatically, often by more than two electron volts. Alkali metal adsorption onto metal single-crystal surfaces has since been frequently and fairly systematically explored; for more details, we recommend the review article by Bonzel [168] on that subject. As far as the *local* interaction of K (Na, Cs) atoms with metal surfaces is concerned, there were density-functional calculations performed by Lang et al. [169] which confirmed a strong charge transfer from the alkali metal atom to the underlying substrate metal, thus resulting in a fairly strong positively polarized alkali species embedded in a negatively polarized environment. Band structure calculations of the surface electronic charge density distribution for the systems Ni(100)/K + CO and Ni(100)/S + CO (see below) were performed in the group of Freeman [170] and supported, at least for the potassium case, the aforementioned back-donation model, while the effect of the sulfur indicated that electrostatics alone cannot completely explain the observed phenomena, but that an influence on the local density of electron states (LDOS) must also be taken into account. However, there is, at least for metals and their conduction electrons, a fairly rapid and effective screening of charge gradients possible and expected, and there is ongoing discussion about the actual range of operation of these screening effects, i.e., whether they have a local or a long-range character [165]. The idea is to pinpoint this range of operation of these charge fluctuations by post-adsorbing molecules (such as CO, H_2, or N_2) whose chemisorption properties are well-known, onto various surface coverages of alkali metal atoms, and to subsequently monitor any alterations of binding states, vibratonal frequencies, and surface potentials of the probe molecules as a function of the alkali metal pre-adsorption.

In the context of promotion effects in heterogeneous catalysis it was Ertl who drew attention to the possible role of potassium additives in ammonia synthesis; he undertook a number of systematic studies of nitrogen adsorption onto iron single-crystal surfaces that were chemically modified by various K precoverages. Details of these studies can be found in the literature [82, 83, 85, 87]; here we concentrate on the interesting potential energy model offered by Ertl et al. [82] that explains the modified nitrogen surface binding and kinetic properties. Essential here is the existence of the molecular nitrogen precursor that was already discussed in the context of the NH_3 synthesis model reaction (cf., Section 5.3.2). It appeared that only a high concentration of this molecular precursor provided a sufficiently rapid uptake of atomic nitrogen on the Fe surfaces (which, in turn, carried the subsequent stepwise hydrogenation reactions). Most noticeable was the increase of the apparent nitrogen sticking probability by almost two orders of magnitude when going from the bare to the K-precovered Fe(100) surface. This increase is illustrated in Fig. 5.20, taken from [82]. We refer the reader to Sect. 3.2.1, where we discussed activated and non-activated adsorption and pointed out that extremely low initial sticking coefficients (meaning very slow rates of adsorption) very likely indicate activated adsorp-

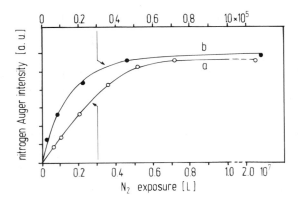

Fig. 5.20. Variation of the nitrogen Auger signal intensity with N_2 exposure for a clean (a) and a potassium-promoted (b) Fe(100) surface. Note the difference in the horizontal scale. Apparently, the velocity of the uptake of atomic nitrogen is about two orders of magnitude larger for the K-promoted surface. After Ertl et al. [82]

tion. Accordingly, the dissociative nitrogen adsorption on Fe(100) is thermally activated by approximately 12–13 kJ/mol. The one-dimensional potential energy situation is schematically shown in Fig. 5.21, curve a; it resembles, for obvious reasons, Fig. 3.20. The activation barrier of height E_{ad}^* can be lowered in two different ways: either the chemisorptive bonding weakens somewhat (which makes the chemisorption potential shallower and shifts the minimum outwards toward longer adatom-surface distances), or, with the chemisorption energy curve remaining unchanged, the (rather more physisorptive) bind-

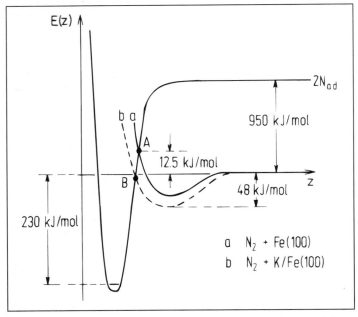

Fig. 5.21. One-dimensional potential energy diagrams for nitrogen N_2 interacting with a clean (full line) and a potassium-promoted (broken line) Fe(100) surface. On the clean surface (a) molecular N_2 exhibits a relatively low adsorption energy, resulting in an intersection point A between atomic and molecular potential energy curves well above the energy zero line. Clearly the N_2 dissociation is activated by ≈12.5 kJ/mol. With potassium preadsorbed, molecular nitrogen is more strongly bound, and the intersection point is shifted close to the energy zero line, point B, leading to almost vanishing activation energy for dissociative adsorption. After Ertl et al. [82]

240

ing energy of the molecular precursor becomes reinforced, resulting in a deepening of the respective potential energy curve, with a concomitant shift of the minimum towards shorter surface bond distances. The measurements by Ertl and coworkers did indeed reveal evidence for this latter phenomenon; they could convincingly show that molecular nitrogen exhibited a stronger heat of adsorption when potassium was preadsorbed. This is documented in Fig. 5.21, curve b. The beneficial effect of the potassium on the adsorption energy of N_2 remains to be explained. Here, the isoelectronic structure of CO and N_2, i.e., the similarity of their molecular orbital arrangement, not only suggests an alike upright adsorption geometry, but also allows a direct transfer of the back bonding chemisorption mechanism developed by Blyholder [42]. The accumulation of negative charge around an adsorbed K atom provides particularly effective back bonding and, hence, a strengthening of the Fe-(N_2) bond, because of the decreased work function of Fe (= the highest occupied d electron level) in relation to the vacuum level E_{vac} which the coupling N_2 molecule and its molecular orbitals are pinned to.

Quite a related view exists in terms of the promoting action of potassium in the Fischer-Tropsch (methanation) reaction. Here, the influence of K on the chemisorption behavior of *carbon monoxide* must be considered. The procedure to tackle this problem is analogous to what we discussed in the preceding paragraph, namely, examining the Blyholder model for an enhanced charge density around a potassium atom. The net result is expected to resemble that of N_2 adsorption in all respects, i.e., a reinforced back bonding for CO on those surface sites adjacent to a K atom. This reinforced occupation of CO's antibonding $2\pi^*$ orbitals by metal d electrons can occur to such an extent that the internal molecular C–O bond becomes dissociated, and there have been numerous studies published, in which a linear dependence between K precoverage and CO dissociation was reported [165, 171–174]. The resulting potential energy diagram for CO adsorption and dissociation, respectively, was first proposed by Brodén et al. [161], who studied the influence of alkali metals systematically, particularly on Fe (depicted in Fig. 5.22). In the following, we refer to this combined UPS, XPS, and thermal desorption work which deals especially with CO adsorption on potassium-promoted Fe(110) surfaces. Consequently, it was found that CO adsorbs molecularly at room temperature with a somewhat larger binding energy than on clean Fe(110), whereas the sticking coefficient was lower on the K-covered surface. Nevertheless, the authors observed an increased CO saturation coverage with potassium. Upon heating the molecularly adsorbed CO underwent partial dissociation, whereby the probability for dissociation increased with K surface concentration. The potential energy diagram of Fig. 5.22 can sufficiently explain this behavior, for quite a similar reason as discussed above for N_2 dissociation. It should be added that Brodén et al. could rule out the formation of (chemically possible) K-O-C≡C-O-K complexes. If there is, however, no *specific* CO-K interaction, the predominant role of potassium should only consist in its electropositive character, and it should also be possible to replace K by any other alkali metal. Indeed, potassium is not the only electropositive metal that can be used as a promoter for the CO hydrogenation. The promoting effect in Fischer-Tropsch synthesis increases in the order Li, Na, K, and Cs [101], and one immediately realizes that the number of electrons and, hence, the polarizability is responsible for this sequence. Then the one and only promoting effect rests on a charge transfer from K to the (Fe) substrate: CO acts as an electron acceptor and the increased charge on the surrounding Fe atoms strengthens the Fe-C and simultaneously weakens the C–O bond, making it susceptible to hydrogenation, completely in line with the back-bonding mechanism discussed above. This weakening effect of coadsorbed K becomes nicely evident from vibrational loss studies, where the strength of the C-O bond is directly monitored by

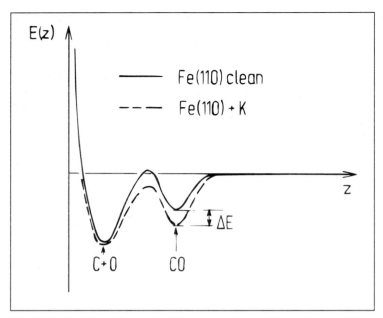

Fig. 5.22. One-dimensional potential energy diagram illustrating the influence of potassium on the strength of the CO chemisorption bond. The left potential well corresponds to dissociated CO, while the right well indicates molecularly held CO. The solid line denotes a clean unpromoted Fe(110) surface, the broken line the same surface covered with potassium. The stronger heat of CO adsorption (difference ΔE) in this case causes a reduction of the activation barrier for CO dissociation. After Brodén et al. [161]

its stretching vibrational frequency. On clean platinum(111), for example, v_{CO} appears at a markedly higher wavenumber (doublet around 2100 and 1860 cm^{-1}, cf., Fig. 4.35) than on a surface precovered with potassium ($v_{CO} = 1380$ cm^{-1}), as HREELS investigations by Wesner et al. [165] demonstrate.

Some remarks should be made concerning the *practical* aspects of catalytic promotion. It is mandatory that the promoter retains its activity under reations conditions (which are, however, hard to realistically model in an UHV/single-crystal experiment). On industrial ammonia synthesis catalysts, Ertl and collaborators studied the chemical state of potassium additives by means of AES and found that, with the unreduced catalysts, potassium was present in oxidized form [95]. It is expected that potassium oxides of varying stoichiometry are also stable under working conditions of the catalyst, whereby a composite K + O layer exhibits a much larger binding energy to the iron of the catalyst than K alone, one further reason for the remarkable thermal stability of the potassium promoter. We must not forget the CO hydrogenation and would like to selectively refer to a work by Rieck and Bell [175] in which the interaction of H$_2$ and CO with silica-supported Pd catalysts promoted with Li, Na, K, Rb, and Cs was studied by means of temperature-programmed thermal desorption in a microreactor under flow conditions. Reduction of the promoted catalysts removed a significant quantity of oxygen from the alkali; nevertheless, there was relatively little effect of the promoter on the amounts of H$_2$/CO that could he adsorbed on the metal. Rather, the distributions of CO adsorption states were greatly influenced, but not so with the hydrogen adstates. (This is, by the way, in line with observations of Ertl's group concerning ammonia synthesis, where the H adstates were also much less affected than was the nitrogen [83]). Rieck and Bell further report that an increase of the catalyst reduction temperature led to an enlarged CO dissociation.

All in all, we learned that this kind of promoting function is closely tied to modifications of the *electronic structure* of the catalyst metal, while the actual surface geometry is not so much affected. One, therefore, refers to this type of promoting material as *electronic* promoters, in contrast to those additives that simply stabilize surface morphology, those being known as *structural* promoters. From a physical-chemical viewpoint, the corresponding compounds are perhaps somewhat less interesting. Relevant questions in the context of structure-stabilization are the formation of intermetallic or inorganic compounds with an enhanced lattice energy, for example, spinels $(Fe,Mg)(Al,Fe)_2O_4$ or related materials. ESCA investigations on NH_3 catalysts seem to support this conclusion, because a comparison of the $O1s$ level position in the reduced and oxidized state revealed energy shifts that were consistent with a transformation of iron spinel $FeAl_2O_4$ in the oxidized catalyst to aluminum oxide Al_2O_3 in the reduced state [96]. Although this matter has great practical importance for the endurance of technical catalysts, we do not have space here to expand further on it; rather, we turn to the (closely related) subject of catalyst inhibition and poisoning, for which, again, a vast number of reports exist.

Exhaustive review articles by Hegedus and McCabe [158], and Kiskinova [176] are recommended for further reading. In order to explain the inhibiting or poisoning function, it is straightforward to tentatively employ the same microscopic chemisorption models that were able to elucidate the promoting effect. The simplest consideration here would be to first replace the electropositive alkali metals by adsorbates that are electro*negative*: oxygen, sulfur, or halogens (chlorine, bromine, etc.). Indeed, it is known that, particularly, sulfur is an effective poisoning impurity in catalysis that must be completely removed from the input reactant stream of the reactor. Accordingly, there have been many single-crystal and supported metal catalyst investigations carried out that focused on the role of sulfur in Fischer-Tropsch synthesis. An example is provided by Kelemen et al. [177], who studied the binding energy of CO on clean and sulfur-covered platinum (111) and (100) surfaces. On both surfaces sulfur was found to decrease the initial adsorption-energy for CO, and to introduce a much stronger coverage dependence than was observed with the clean surfaces. These findings are absolutely parallel to results reported by Bonzel and Ku [159], and Erley and Wagner [178], where enlarged repulsive forces between the ad-particles in the presence of sulfur were made responsible for the rapid decrease of the CO binding energy. In terms of Blyholder's model, the electronegative S atom withdraws charge from its neighboring Pt atoms and, hence, deteriorates the CO binding conditions on the corresponding sites. Apparently, a single S atom will affect at least four, if not more, adjacent metal atoms, which also explains the drastic reduction of the adsorption energy of CO as the sulfur coverage increases. (Related investigations were carried out by Rhodin and Brucker [179] with CO adsorption on clean and sulfur-covered Fe(100) surfaces, and similar conclusions were drawn). Johnson and Madix [180] studied the sulfur-induced changes of selectivity that occurred over Ni(100) in the methanol decomposition. The sulfur decreased the amount of adsorbed methanol which reacted, and a pronounced interaction of S with high-temperature hydrogen produced by CH_3OH decomposition arose. On a Ni(100)-$c(2\times2)$S structure methoxy species and formaldehyde reaction products were observed. The selective blocking of H adsorption sites was made responsible for the poisoning effect of sulfur. In regard to this site blocking theoretical calculations were also carried out in which the role of adatom geometry in the strength and range of catalyst poisoning was scrutinized [181]. While poisoning C atoms were found to only change the density of states on nearest-neighbor sites, sulfur atoms affected also the next-nearest neighbors. For more real catalyst systems, we recommend the extensive review article by Madon and Shaw [182], which shows the complexities of sulfur poisoning of the Fischer-Tropsch synthesis reaction. Shelef et al. [183] have also discussed poisoning effects of S, Pb, and P impurities in catalysis, especially considering automobile exhaust catalyst technology.

Not only sulfur, but also oxygen can play the role of an electronegative agent on a surface, although it is seldom regarded as a particular catalyst poison. Nevertheless the chemical functioning of coadsorbed O, for example, in CO_2 coadsorption, is completely along this line. Preadsorbed O can, for example, entail molecular CO_2 adsorption (which would otherwise suffer dissociation) by a partial charge transfer to the C atom, thus leading to a kind of surface carbonate complex [184].

There are certainly many other aspects and peculiarities of oxygen coadsorption which would be worth mentioning here, however, available space does not allow further considerations and makes us proceed to another topic that is well-known in catalyst-poisoning, namely, the formation of inactive carbonaceous layers, the so-called *coke formation*. This process frequently occurs in catalyzed hydrocarbon reactions and is often responsible for the degradation of catalyst activity as was briefly mentioned in our remarks on CO hydrogenation, cf., Eq. 5.52. In this case, a specific alteration of the surface-electron-charge distribution is not responsible for the deactivation. Rather, there occurs a kind of neutralization of the chemisorptive forces of the catalyst surface atoms by irreversible adsorption of carbon. The coke thereby consists of largely *inert* graphitic carbon layers which simply block any chemical contact between the reactants and the catalyst surface. This kind of carbon is definitely much different from the *carbidic* species that we encountered in the context of CO hydrogenation and that was even *responsible* for the peculiar catalytic activity. It is self-evident that it is a prominent concern of catalytic chemists to avoid coking (which can often be governed by reactant pressures, catalyst temperature, and reactor flow conditions). Precursors for coke formation are deeply dehydrogenated species that remain and accumulate on the surface during hydrocarbon reactions. Frequently among these are CH_x surface radicals, open-chain surface polyenes, or unsaturated C_5 cyclics. Expectedly, the formation of these poisoning species can be slowed down substantially in the presence of hydrogen, i.e., by increasing the hydrogen content in the reaction mixture [185].

We move on to a phenomenon that is closely linked to poisoning, namely, *selective* poisoning or *inhibition*, and we refer the reader especially to our brief explanation of the terms: *bimetallic catalysts, ligand effect*, and *ensemble effect* in Sect. 3.1. Apart from the already cited work by Sachtler [154] and Ponec [157] on alloy catalysts, we draw the reader's attention again to Sinfelt's publications, in which it was convincingly shown that catalyst selectivity could be sensitively influenced by additives such as copper [155, 156, 186]. A striking example here is provided by a study of hydrogenolysis of ethane to methane, according to the equation

$$C_2H_6 + H_2 = 2CH_4 \qquad\qquad 5.65$$

and the dehydrogenation of cyclohexane to benzene [187]:

$$C_6H_{12} = C_6H_6 + 3H_2 \qquad\qquad 5.66$$

The catalyst material used by Sinfelt et al. consisted of a copper-nickel alloy, whose composition, i.e., the Cu content, was systematically varied from 0% to 100%. The interesting result of these investigations was a pronounced difference in activity for the two reactions as the amount of copper was increased. As can be seen from Fig. 5.23, there is a

244

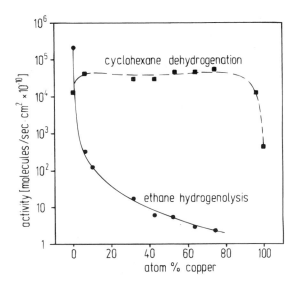

Fig. 5.23. Variation in the activity of Cu-Ni supported catalysts for two typical hydrocarbon reactions, viz., cyclohexane dehydrogenation, and ethane hydrogenolysis, as a function of the copper content of the catalyst material. Interestingly, the activity of the hydrogenolysis declines sharply (over five orders of magnitude) with increasing Cu content, while the dehydrogenation activity is hardly affected up to 80 % Cu. After Sinfelt et al. [187]

decrease of hydrogenolysis activity of the alloy over more than four orders of magnitude when the copper content reaches 50%, while the dehydrogenation activity is hardly affected; only when the composition approaches pure copper does the dehydrogenation activity also decline. The addition of Cu to the generally very active nickel has obviously improved the selectivity of the catalyst remarkably. The hydrogenolysis of ethane is believed to occur via the Langmuir-Hinshelwood mechanism, which means that both C_2H_6 and H must be present in the adsorbed state. The adsorption of ethane thereby involves the rupture of the C-H bonds, resulting in a hydrogen-deficient dicarbon surface species. In a next step, also the C-C bond is cleaved, and there remain monocarbon fragments on the surface which may then be successively hydrogenated by coadsorbed hydrogen to yield methane [188]. Particularly the cleavage of the C-C bond requires the hydrogen-deficient intermediate to be linked to two adjacent metal atoms, in other words, only a specially structured ensemble provides the favorable reaction geometry. At this point the inhibiting function of the copper comes into play: the alloy consists of a catalytically very active (Ni) and a relatively inactive metal (Cu), and the lateral distribution of active and inert atoms in the surface can dramatically influence the overall adsorption behavior. The reader will certainly realize that we again encounter the ensemble-effect that we have already mentioned in Sect. 3.1. Fairly basic statistical considerations about ensemble size effects in chemisorption and catalysis on binary alloys were made by Dowden [189], and later by Burton and Hyman [190], and have since then always played an important role in heterogeneous catalysis on multi-component systems. Applied to our Cu-Ru system above, the hydrogenolysis reaction may also be hindered if the necessary coadsorption of hydrogen is suppressed. We pointed out earlier (cf., Sect. 3.1.) that dissociative adsorption of diatomics may require ensembles of a relatively large number of atoms. Since H_2 does not spontaneously dissociate on copper, it is additionally possible that the uptake of hydrogen is impaired, because the necessary Ru ensembles are destroyed by the statistically mixed-in copper atoms. There have been studies in our own laboratory dealing with Cu/Ru model catalysts that exactly show this effect, i.e., a strong ensemble effect with hydrogen chemisorption [191, 192]. This result was also completely in line with investigations of Yu et al. [193], who determined the ensemble size required for hydrogen chemisorption to be around 4. In brief, copper plays the part of the inhibiting agent in these hydrocarbon and/or hydrogen reaction systems, and there are a great deal of other bimetallic or alloy systems in which a noble metal is mixed in, in order to furnish a particular selectivity. We have studied the adsorption of ethene C_2H_4 between 100 and 500 K on ruthenium(0001) single-crystal surfaces covered with various amounts of gold, ranging from submonolayer to multilayer coverages [194, 195]. On the active transition metal Ru, ethene is known to be di-σ-bonded, that is to say, the sp^2 hybridization of the C atoms changes to sp^3 hybridization, with the C–C bond being at least weakened, if not ruptured. As the temperature is increased to beyond 350 K, only hydrogen desorption is found; the C_2H_4 molecule is completely destroyed. The other extreme situation is provided by a thick Au multilayer film – exposure to ethene even at 100 K hardly leads to any adsorption. However, if a *monoatomic homogeneous* Au layer is deposited on top of the Ru surface, ethene becomes (weakly) adsorbed in a π-bonded configuration; the molecule does not rehybridize as it did before. This proves that noble metal atoms can be chemically activated by neighboring transition metal atoms, often for the benefit of catalytic selectivity. A convincing example here was delivered again from Somorjai's group: Sachtler et al. [196-198] investigated bimetallic Au-Pt(111) and Au-Pt(100) single-crystal surfaces with respect to their structure and chemical reactivity, and they used the cyclohexane [198] and cyclohexene [197] dehydrogenation as test reactions. They systematically varied the Au/Pt surface atom ratio and found a striking enhancement of the catalytic activity of the bimetallic systems for certain Au surface coverages. Particularly interesting and parallel to our own observations was the behavior of the Au-on-Pt(100) system, where a steep activity maximum occurred when just a single Au monolayer was deposited. This is illustrated in Fig. 5.24. Several factors were made responsible for this peculiar behavior. The creation of a special hollow-site structure of

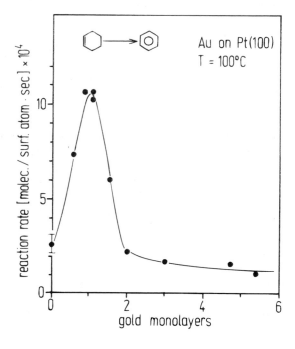

Fig. 5.24. Rate of dehydrogenation of cyclohexene to benzene on ordered Au(100) oriented films deposited in vacuo onto a Pt(100) crystal. The number of deposited Au monolayers is plotted on the abscissa. Most striking is the activity maximum after deposition of a single Au monolayer. After Sachtler et al. [197]

the Au-Pt surface was considered, but also electronic effects caused by a partial charge transfer between Au and Pt are believed to influence the hydrocarbon bonding.

In continuation of Sinfelt's pioneering work on bimetallics (we recall here also his *structural* investigations [199–204] using EXAFS, cf., Sect. 4.1.4), a variety of groups studied these systems either in UHV or at higher pressures, whereby preferentially Ru (in some cases also Rh) catalysts in combination with Cu [205-220], Ag [221–223] Au [194, 224, 225], Fe [226, 227] were investigated with various adsorbates, for example, CO, H_2, O_2, N_2O, CH_3OH. Other bimetallic systems moved to the center of interest, due to their practical application in hydrocarbon conversion, in particular Pt-Re[228] and other Pt-based bimetallic materials [229]. In most of these studies, ensemble effects (but also changes of the valence states of the components by mutual electronic charge transfer) were invoked to explain the peculiar activity and selectivity effects. Although citation of the corresponding literature on bimetallic and alloy catalysis would require several pages of this book, there remains a need for further model studies in order to gain a better understanding of the underlying synergetic effects that are deemed essential for catalytic activity and selectivity.

We have seen in this final chapter, and emphasize again, that the research interest in heterogeneous model catalysis increasingly includes *ternary* model systems, whereby preadsorbed electropositive or electronegative atoms are frequently studied, in addition to noble metal or transition metal additives. So far, preferentially simple diatomic molecules have been used as probes for adsorption site changes, or for modifications of the surface electronic structure. Hydrogen is a particularly suited test molecule, the more so as it is a reactant in almost any practically important catalytic reaction. Accordingly, researchers have devoted much of their attention to hydrogen effects in catalysis, and we recommend a monograph [230], which covers practically the complete field of "hydrogen catalysis" and which can supply the reader with much more information on this subject, including ensemble-size effects, spill-over, strong metal-support interaction, etc.. We also recall our presentation of general literature pertinent to catalysis and surface chemistry in Chapter 1.

In the context of selective inhibition and ensemble size effects we must briefly touch again the phenomenon of *spill-over* which describes the transfer of surface species (e.g., H atoms) from active to inactive sites. Consider a binary alloy consisting of an active and an inert material (Cu-Ni, Au-Ru etc.). Hydrogen molecules can dissociate and chemisorb only on the Ni(Ru) atoms or ensembles, but once the H atoms are formed, they may well diffuse to Cu(Au) sites, where they are trapped and adsorbed, however with markedly lower binding energy. This reduced energy is believed to render these sites a particular reactivity in catalytic reactions, and there is also evidence that hydrogen spillover is responsible for the aforementioned strong-metal-support interaction. The interested reader will find more information on spillover in a recently published monograph on that subject [231].

Relating the end of this chapter with Chapter 1, our message is that it is in many cases possible and advantageous to bridge the gap between well-defined model reactions and real technical conditions and applications. Yet, there is still a great number of practically very important heterogeneous reactions that are too complicated to allow simple modeling. In this situation, there remain many challenging problems, and surface chemists and physicists still have to invest a large amount of work towards understanding the mystery of "heterogeneous catalysis", but they can and certainly will benefit from the steadily improving surface analytical instrumentation and procedures, some of which were described in Chapter 4.

References

1. Ostwald W (1897) Grundriß der allgemeinen Chemie. 2nd edn. Verlag von Wilhelm Engelmann, Leipzig
2. Ertl G (1969) In: Drauglis E, Gretz RD. Jaffee RI (eds) Molecular Processes on Solid Surfaces. McGraw-Hill, New York, p. 147
3. May JW (1972) Heterogeneous Catalysis at Line Defects I. Active Sites in Periodic Adlayers. Proc Roy Soc (London) A331:185–193
4. May JW (1972) Heterogeneous Catalysis at Line Defects II. Simple Reactions. Proc Roy Soc (London) A331:195–202
5. Ertl G, Koch J (1972) Adsorption Studies with a Palladium (111) Surface. In: Ricca F (ed) Adsorption – Desorption Phenomena. Academic Press, London-New York, p. 345–357
6. Conrad H, Ertl G, Latta EE (1974) Coadsorption of Hydrogen and Carbon Monoxide on a Pd(110) Surface. J Catal 35:363–368
7. Lauth G (1989) Die Wechselwirkung von Kohlenmonoxid und Wasserstoff mit einer Ruthenium (10$\bar{1}$0) Oberfläche, PhD Thesis, Freie Universität Berlin, p. 140f
8. Christmann K, Lauth G, Dokenwald R (1991) Coadsorption of H$_2$ and CO on a Ru(10$\bar{1}$0) Surface (to be published)
9. Mate CM, Somorjai GA (1985) Carbon Monoxide-induced Ordering of Benzene on Pt(111) and Rh(111) Crystal Surfaces. Surface Sci 160:542–560
10. Jakob P, Menzel D (1990) Coadsorption of Benzene and Oxygen on Ru(001). Surface Sci 235:197–208
11. Koel BE, Peebles DE, White JM (1981) The Interaction of Coadsorbed Hydrogen and Carbon Monoxide on Ni (100). Surface Sci 107:L367–L373
11a. Mitchell GE, Gland JL, White JM (1983) Vibrational Spectra of Coadsorbed CO and H on Ni(100) and Ni(111). Surface Sci 131:167–178
12. Peebles HC, Peebles DE, White JM (1983) Temperature-induced Structural Changes for Coadsorbed H and CO on Ni(100). Surface Sci 125:L87–L92
13. Goodman DW, Yates Jr. JT, Madey TE (1980) Interaction of Hydrogen, Carbon Monoxide and Methanol with Ni(100). Surface Sci 93:L135–L142
14. Sesselmann W, Woratschek B, Ertl G, Küppers J, Haberland H (1983) Low Temperature Formation of Benzene from Acetylene on a Pd(111) Surface. Surface Sci 130:245–258
15. Goodman DW, Kelley RD, Madey TE, Yates Jr. JT (1980) Kinetics of the Hydrogenation of CO over a Single Crystal Nickel Catalyst. J Catal 63:226–234
16. Kelley RD, Goodman DW (1982) Catalytic Methanation over Single Crystal Nickel and Ruthenium; Reaction Kinetics on Different Crystal Planes and the Correlation of Surface Carbide Concentration with Reaction Rate. Surface Sci 123:L743–L749
17. Wedler G (1982) Lehrbuch der Physikalischen Chemie. Verlag Chemie, Weinheim, p. 813–821
18. Atkins PW (1982) Physical Chemistry, 2nd edn. Oxford University Press, London, Ch. 29
19. Moore WJ, Hummel DO (1983) Physikalische Chemie, 3rd edn. de Gruyter, Berlin, p. 594 ff
20. Laidler KJ (1987) Chemical Kinetics, 3rd edn. Harper and Row, New York, Ch. 7
21. Bond GC (1962) Catalysis by Metals. Academic Press, London – New York
22. Ashmore PG (1963) Catalysis and Inhibition of Chemical Reactions. Butterworth, London Ch.7
23. Trapnell BMW (1955) Chemisorption. Butterworth, London
24. Langmuir I (1921) Chemical Reactions on Surfaces. Trans Faraday Soc 17:607–620
25. Hinshelwood CN (1940) The Kinetics of Chemical Change. Clarendon Press, Oxford
26. Eley DD (1948) The Catalytic Activation of Hydrogen. Adv Catal Rel Subj 1:157–199
27. Harris J, Kasemo B (1981) On Precursor Mechanisms for Surface Reactions. Surface Sci 105:L281–L287
28. Jones RH, Olander DR, Siekhaus WJ, Schwarz JA (1972) Investigation of Gas-solid Reactions by Modulated Molecular Beam Mass Spectrometry. J Vac Sci Technol 9:1429–1441
29. Chang HC, Weinberg WH (1977) An Analysis of Modulated Molecular Beam Mass Spectrometry Applied to Coupled Diffusion and Chemical Reaction. Surface Sci 65:153–163

30. Engel T (1978) A Molecular Beam Investigation of He, CO, O_2 Scattering from Pd(111). J Chem Phys 69:373–385
31. Lin TH, Somorjai GA (1981) Modulated Molecular Beam Scattering of CO and NO from Pt(111) and the Stepped Pt(557) Crystal Surfaces. Surface Sci 107:573–585
32. Engel T (1978) Molekularstrahluntersuchungen der Adsorption und Reaktion von H_2, O_2, und CO an einer Pd(111) Oberfläche. Habilitationsschrift, Universität München
33. Engel T, Ertl G (1978) A Molecular Beam Investigation of the Catalytic Oxidation of CO on Pd(111). J Chem Phys 69:1265–1281
34. Engel T, Ertl G (1979) Elementary Steps in the Catalytic Oxidation of Carbon Monoxide on Platinum Metals. Adv Catal Rel Subj 28:1–78
35. Engel T, Ertl G (1982) Oxidation of Carbon Monoxide. In: King DA, Woodruff DP (eds). The Chemical Physics of Solid Surfaces and Heterogeneous Catalysis, vol 4. Elsevier, Amsterdam pp 73–93
36. Ertl G (1983) Kinetics of Chemical Processes on Well-defined Surfaces. In: Anderson JR, Boudart M (eds) Catalysis, Science and Technology. Springer, Berlin Heidelberg New York pp 209–282
37. Imbihl R (1984) Nichtgleichgewichts-Phasenübergänge bei der katalytischen Oxidation von CO an Pt(100). PhD thesis, Universität München
38. Conrad H, Ertl G, Küppers J (1978) Interactions between Oxygen and Carbon Monoxide on a Pd(111) Surface. Surface Sci 76:323–342
39. Thomas GE, Weinberg WH (1979) High-resolution Electron Energy Loss Spectroscopy of Chemisorbed Carbon Monoxide and Oxygen on the Ruthenium (001) Surface. J Chem Phys 70:954–961
40. Matsushima T (1979) Kinetic Studies on the CO Oxidation over Platinum by means of Carbon 13 Tracer. Surface Sci 79:63–75
41. Bonzel HP, Ku R (1972) Mechanisms of the Catalytic Carbon Monoxide Oxidation on Pt(110). Surface Sci 33:91–106
42. Blyholder G (1964) Molecular Orbital View of Chemisorbed Carbon Monoxide. J Phys Chem 68:2772–2778
43. Campbell CT, Ertl G, Küppers H, Segner J (1980) A Molecular Beam Study of the Catalytic Oxidation of CO on a Pt(111) Surface. J Chem Phys 73:5862–5873
44. Taylor JL, Ibbotson DE, Weinberg WH (1980) The Oxidation of Carbon Monoxide over the (110) Surface of Iridium. J Catal 62:1–12
45. Zhdan PA, Boreskov GK, Egelhoff WF, Weinberg WH (1976) The Application of XPS to the Determination of the Kinetics of the CO Oxidation Reaction over the Ir(111) Surface. Surface Sci 61:377–390
46. Campbell CT, Shih SK, White JM (1979) The Langmuir-Hinshelwood Reaction between Oxygen and CO on Rh. Appl Surface Sci 2:382–396
47. Campbell CT, Ertl G, Kuipers H, Segner J (1981) A Molecular Beam Investigation of the Interactions of CO with a Pt(111) Surface. Surface Sci 107:207–219
48. Becker CA, Cowin JP, Wharton L, Auerbach DJ (1977) CO_2 Product Velocity Distributions for CO Oxidation on Platinum. J Chem Phys 67:3394–3395
49. Mantel DA, Ryali SB, Halpern BL, Haller GL, Fenn JB (1981) The Exciting Oxidation of CO on Pt. Chem Phys Lett 81:185–187
50. Bernasek SL, Leone SR (1981) Direct Detection of Vibrational Excitation in the CO_2 Product of the Oxidation of CO on a Platinum Surface. Chem Phys Lett 84:401–404
51. Matsushima T (1990) Crystal Azimuth Dependence of the Desorption Flux of Carbon Dioxide Produced on Palladium (110) and (111) Surfaces. Vacuum 41:275–277
52. Ertl G, Koch J (1972) The Catalytic Oxidation of CO on Pd Surfaces. In: Hightower JW (ed) Proceedings Vth International Congress on Catalysis, Miami Beach, vol 2 pp 67–969
53. Bonzel HP, Ku R (1972) Carbon Monoxide Oxidation on a Pt(110) Single Crystal Surface. J Vac Sci Technol 9:663–667
54. Christmann K, Ertl G (1973) Interactions of CO and O_2 with Ir(110) Surfaces. Z Naturf 28a:1144–1148

55. Ehsasi M, Matloch M, Frank O, Block JH, Christmann K, Rys FS, Hirschwald W (1989) Steady and Non-steady Rates of Reaction in a Heterogeneously Catalyzed Reaction: Oxidation of CO on Platinum, Experiment and Simulations. J Chem Phys 91:4949–4960

56. Ziff RM, Gulari E, Barshad Y (1986) Kinetic Phase Transitions in an Irreversible Surface-Reaction Model. Phys Rev Lett 56:2553–2556

57. Ehsasi M (1989) Kinetic Instabilities and Oscillations in a Surface Reaction: Oxidation of CO over Platinum Group Metals. PhD thesis, Freie Universität Berlin

58. Ehsasi M, Rezaie-Serej S, Block JH, Christmann K (1990) Reaction Rate Oscillations of CO Oxidation on Pt(210). J Chem Phys 92:7596–7609

59. Ehsasi M, Seidel C, Ruppender H, Drachsel W, Block JH, Christmann K (1989) Kinetic Oscillations in the Rate of CO Oxidation on Pd(110). Surface Sci 210:L198–L208

60. Imbihl R, Cox MP, Ertl G (1985) Kinetic Oscillations in the Catalytic CO Oxidation on Pt(100): Experiments. J Chem Phys 84:3519–3534

61. Ertl G (1986) Reactive Transformation of Surface Structure. Ber Bunsenges Phys Chem 90:284–291

62. Imbihl R, Cox MP, Ertl G, Müller H, Brenig W (1985) Kinetic Oscillations in the Catalytic CO Oxidation on Pt(100): Theory. J Chem Phys 83:1578–1587

63. Sheintuch M (1985) Nonlinear Kinetics in Catalytic Oxidation Reactions: Periodic and Aperiodic Behavior and Structure Sensitivity. J Catal 96:326–346

64. Razon LF, Schmitz RA (1986) Intrinsically Unstable Behavior during the Oxidation of Carbon Monoxide on Platinum. Catal Rev Sci Eng 28:89–164

65. Schüth F, Wicke E (1989) IR Spectroscopic Investigation during Oscillations of the CO/NO and the CO/O$_2$ Reaction on Pt and Pd Catalysts. I. Platinum. Ber Bunsenges Phys Chem 93:191–201

66. Field JR, Burger M ((eds) (1985) Oscillations and Travelling Waves in Chemical Systems. Wiley, New York

67. Nicolis G, Prigogine I (1977) Self-Organization in Non-equilibrium Systems. Wiley, New York

68. Oh SH, Fisher GB, Carpenter JE, Goodman DW (1986) Comparative Kinetic Studies of CO-O$_2$ and CO-NO Reactions over Single Crystal and Supported Rhodium Catalysts. J Catal 100:360–376

69. Taylor Kathleen C (1984) Automobile Catalytic Converters. In: Anderson JR, Boudart M (eds) Catalysis – Science and Technology, vol 5. Springer, Berlin Heidelberg New York, p. 120–171

70. Mittasch A (1950) Early Studies of Multicomponent Catalysts. Adv Catal Rel Subj 2:81–104

71. Emmett PH (1975) Fifty Years of Progress in the Study of the Catalytic Synthesis of Ammonia. In: Drauglis E. Jaffee RI (eds) The Physical Basis for Heterogeneous Catalysis. Plenum, New York pp 3–34

72. Vancini CA (1971) Synthesis of Ammonia. McMillan, London

73. Grunze M (1982) Synthesis and Decomposition of Ammonia. In: King DA, Woodruff DP (eds) The Chemical Physics of Solid Surfaces and Heterogeneous Catalysis, vol 4. Elsevier, Amsterdam pp 143–194

74. Ozaki A, Aika, K (1981) Catalytic Activation of Dinitrogen. In: Anderson JR, Boudart M (eds) Catalysis – Science and Technology, vol 1. Springer, Berlin Heidelberg New York pp 87–158

75. Boudart M (1981) Kinetics and Mechanism of Ammonia Synthesis. Catal Rev Sci Eng 23:1–15

76. Ertl G (1980) Studies on the Mechanism of Ammonia Synthesis – The P.H. Emmett Award Address. Catal Rev Sci Eng 21:201–230

77. Ertl G (1990) Elementary Steps in Ammonia Synthesis: The Surface Science Approach. In: Catalytic Ammonia Synthesis – Fundamentals and Practice. Plenum, New York Ch. 5

78. Bozso F, Ertl G, Grunze M, Weiss M (1977) Interaction of Nitrogen with Iron Surfaces. I. Fe(100) and Fe(111). J Catal 49:18–41

79. Bozso F, Ertl G, Weiss M (1977) Interaction of Nitrogen with Iron Surfaces. II. Fe(110). J Catal 50:519–529

80. Grunze M, Bozso F, Ertl G, Weiss M (1978) Interaction of Ammonia with Fe(111) and Fe(100) Surfaces. Appl Surface Sci 1:241–265

81. Weiss M, Ertl G, Nitschke F (1979) Adsorption and Decomposition of Ammonia on Fe(110). Appl Surface Sci 2:614–635

82. Ertl G, Weiss M, Lee SB (1979) The Role of Potassium in the Catalytic Synthesis of Ammonia. Chem Phys Lett 60:391–394

83. Ertl G, Lee SB, Weiss M (1981) The Influence of Potassium on the Adsorption of Hydrogen on Iron. Surface Sci 111:L711–L715

84. Ertl G, Lee SB, Weiss M (1982) Kinetics of Nitrogen Adsorption on Fe(111) Surface Sci 114:516–526

85. Lee SB, Weiss M, Ertl G (1981) Adsorption of Potassium on Iron. Surface Sci 108:357–367

86. Ertl G, Huber M (1980) Mechanism and Kinetics of Ammonia Decomposition on Iron. J Catal 61:537–539

87. Ertl G, Lee SB, Weiss M (1982) Adsorption of Nitrogen on Potassium-promoted Fe(111) and (100) Surfaces. Surface Sci 114:527–545

88. Ertl G, Huber M, Lee SB, Paal Z, Weiss M (1981) Interactions of Nitrogen and Hydrogen on Iron Surfaces. Appl Surface Sci 8:373–386

89. Grunze M, Golze M, Hirschwald W, Freund HJ, Pulm H, Seip U, Tsai MC, Ertl G, Küppers J (1984) π-bonded N_2 on Fe(111): The Precursor for Dissociation. Phys Rev Lett 53:850–853

90. Nielsen A (1968) An Investigation of Promoted Iron Catalysts for the Synthesis of Ammonia. J Gjellerups Forlag, Copenhagen

91. Brunauer S, Emmett PH (1940) Chemisorption of Gases on Iron Synthetic Ammonia Catalysts. J Am Chem Soc 62:1732–1746

92. Solbakken V, Solbakken A, Emmett PH (1969) The Exchange of $H_2^{18}O$ with the Oxygen of Promoters on the Surface of Iron Catalysts. J Catal 15:90–98

93. Drechsler M, Hoinkes H, Kaarmann H, Wilsch H, Ertl G, Weiss M (1979) Interaction of NH_3 With Fe(110): Identification of Surface Species by Means of Secondary Ion Mass Spectroscopy (SIMS). Appl Surface Sci 3:217–228

94. Nørskov JK, Stoltze P (1987) Theoretical Aspects of Surface Reactions. Surface Sci 189/190:91–105

95. Ertl G, Prigge D, Schloegl R, Weiss M (1983) Surface Characterization of Ammonia Synthesis Catalysts. J Catal 79:359–377

96. Ertl G, Thiele N (1979) XPS Studies with Ammonia Synthesis Catalysts. Appl Surface Sci 3:99–112

97. Sabatier P, Senderens JB (1902). CR Acad Sci Paris 134:514

98. Sabatier P (1947) Die Hydrierung durch Katalyse. (Rede, gehalten am 11.12.1912 in Stockholm bei der Empfangnahme des Nobelpreises für Chemie) In: (Schuller H, ed) Mittasch A, Lebensprobleme und Katalyse, Reihe: Forschung und Humanität. J Ebner Verlag Ulm/Donau

99. Fischer F, Tropsch H (1926) Über die direkte Synthese von Erdöl-Kohlenwasserstoffen bei gewöhnlichem Druck (1. und 2. Mitteilung). Chem Ber 59:830–836

100. Pichler H, Hector A. In: Kirk-Othmer Encyclopedia of Chemical Technology, vol 4. Wiley, New York, p 446

101. Dry ME (1981) The Fischer-Tropsch Synthesis. In: Anderson JR, Boudart M (eds) Catalysis – Science and Technology, vol 1. Springer, Berlin Heidelberg New York, pp 159–251;

101a. Dry ME, Hoogendoorn JC (1981) Technology of the Fischer-Tropsch Process. Catal Rev Sci Eng 23:264–278

102. King DL, Cusamano JA, Garten RL (1981) A Technological Perspective for Catalytic Processes Based on Synthesis Gas. Catal Rev Sci Eng 23:233–263

103. Tillmetz KD (1976) Über thermodynamische Simultangleichgewichte bei der Fischer-Tropsch-Synthese. Chem Ing Techn 48:1065

104. Somorjai GA (1981) The Catalytic Hydrogenation of Carbon Monoxide. The Formation of C_1 Hydrocarbons. Catal Rev Sci Eng 23:189–202

105. Somorjai GA (1981) Catalytic Hydrogenation of Carbon Monoxide. In: Chemistry in Two Dimensions: Surfaces, Cornell University Press, Ithaca NY pp 516–544

106. Castner DG, Blackadar R, Somorjai GA (1980) CO Hydrogenation over Clean and Oxidized Rhodium Foil and Single Crystal Catalysts. Correlations of Catalyst Activity, Selectivity, and Surface Composition. J Catal 66:257–266

251

107. Sexton BA, Somorjai GA (1977) The Hydrogenation of CO and CO_2 Over Polycrystalline Rhodium: Correlation of Surface Composition, Kinetics, and Product Distributions. J Catal 46:167–189

108. Dwyer DJ, Somorjai GA (1979) The Role of Readsorption in Determining the Product Distribution during CO Hydrogenation Over Fe Single Crystals. J Catal 56:249–257

109. Krebs HJ, Bonzel HP, Gafner G (1979) A Model Study of the Hydrogenation of CO Over Polycrystalline Iron. Surface Sci 88:269–283

110. Bonzel HP, Krebs HJ (1980) On the Chemical Nature of the Carbonaceous Deposits on Iron after CO Hydrogenation. Surface Sci 91:499–513

111. Krebs HJ, Bonzel HP (1980) Hydrogenation Kinetics of Carbonaceous Layers on Polycrystalline Iron. Surface Sci 99:570–580

112. Krebs HJ, Bonzel HP, Schwarting W, Gafner G (1981) Microreactor and Electron Spectroscopy Studies of Fischer-Tropsch Synthesis on Magnetite. J Catal 72:199–209

113. Bonzel HP, Krebs HJ (1982) Surface Science Approach to Heterogeneous Catalysis: CO Hydrogenation on Transition Metals. Surface Sci 117:639–658

114. Bonzel HP, Broden G, Krebs HJ (1983) X-ray Photoemission Spectroscopy of Potassium Promoted Fe and Pt Surfaces after H-Reduction and CO/H_2 Reaction. Appl Surface Sci 16:373–394

115. Goodman DW, Kelley RD, Madey TE, White JM (1980) Measurement of Carbide Buildup and Remocal Kinetics on Ni(100). J Catal 64:479–481

116. Madey TE, Goodman DW, Kelley RD (1979) Surface Science and Catalysis: The Catalytic Methanation Reaction. J Vac Sci Technol 16:433–434

117. Peden CHF, Goodman DW (1985) Hydrocarbon Synthesis and Rearrangement over Clean and Chemically Modified Surfaces. In: Deviney ML, Gland JL (eds) Catalyst Characterization Science, ACS Symposium Series 288 The American Chemical Society, Washington DC, p. 185–198

118. Wedler G, Körner H (1981) Decomposition of Carbon Monoxide and its Reaction with Hydrogen on Iron Films under Static Conditions at Pressures between 2 and 25 mbar at Temperatures between 373 and 573 K. Ber Bunsenges Phys Chem 85:283–288

119. Palmer RL, Vroom DA (1977) Mass-spectrometric Measurements of Enhanced Methanation Activity Over Cobalt and Nickel Foils. J Catal 50:244–251

120. Hirsch W, Hofmann D, Hirschwald W (1984) Interaction of CO, C_2H_4, CH_2O and HCOOH with Single Crystal Faces of ZnO and ZnO(Cu) Catalyst Surfaces Studied by Surface Spectroscopies. Proceedings 8th International Congress on Catalysis, Berlin, vol IV pp 251–262

121. Kawasaki K, Shibata M, Miki H, Kioka T (1979) Coadsorption of Carbon Monoxide and Hydrogen on Polycrystalline Rhodium. Surface Sci 81:370–378

122. Baldwin Jr. VH, Hudson JB (1971) Coadsorption of Hydrogen and Carbon Monoxide on (111) Platinum. J Vac Sci Technol 8:49–52

123. Kim Y, Peebles HC, White JM (1982) Adsorption of D_2, CO and the Interaction of Co-adsorbed D, and CO on Rh(100). Surface Sci 114:363–380

124. Berton JC, Imelik B (1979) Coadsorption of Carbon Monoxide and Hydrogen on a Ni(111) Surface: Influence of the "Surface Carbide". Surface Sci 80:586–592

125. Ibbotson DE, Wittrig TS, Weinberg WH (1980) The Coadsorption of Hydrogen and Carbon Monoxide on the (110) Surface of Iridium. Surface Sci 97:297–308

126. Peebles DE, Schreifels JA, White JM (1982) The Interaction of Coadsorbed Hydrogen and Carbon Monoxide on Ru(001) Surface Sci 116:117–134

127. Canning NDS, Chesters MA (1986) The Co-adsorption of H_2 and CO on Ni(110) Surface Sci 175:L811–L816

128. Vannice MA (1975) The Catalytic Synthesis of Hydrocarbons from H_2/CO Mixtures over the Group VIII Metals. I. The Specific Activities and Product Distributions of Supported Metals. J Catal 37:449–461; II.The Kinetics of the Methanation Reaction over Supported Metals. J Catal 37:462–473

129. Vannice MA (1976) The Catalytic Synthesis of Hydrocarbons from Carbon Monoxide and Hydrogen. Catal Rev Sci Eng 14:153–191

130. Balandin AA (1969) Modern State of the Multiplet Theory of Heterogeneous Catalysis. Adv Catal Rel Sub. 19:1–210

131. Christmann K (1988) Interaction of Hydrogen with Solid Surfaces. Surface Sci Repts 9:1–163

132. Bond GC (1962) Catalysis by Metals. Academic Press, London, New York

133. Bell AT (1981) Catalytic Synthesis of Hydrocarbons over Group VIII Metals. A Discussion of the Reaction Mechanism. Catal Rev Sci Eng 23:203–232

134. Rabo JA, Risch AP, Poutsma ML (1978) Reactions of Carbon Monoxide and Hydrogen on Co, Ni, Ru, and Pd Metals. J Catal 53:295–311

135. Poutsma ML, Elek LF, Ibarbia P, Risch H, Rabo JA (1978) Selective Formation of Methanol from Synthesis Gas Over Palladium Catalysts. J Catal 52:157–168

136. Mehta S, Simmons GW, Klier K, Herman RG (1979) Catalytic Synthesis of Methanol from CO/H_2. II. Electron Microscopy (TEM, STEM, Microdiffraction, and Energy-dispersive Analysis) of the Cu/ZnO and $Cu/ZnO/Cr_2O_3$ Catalysts. J Catal 57:339–360

137. Hirschwald W (1981) Catalysis on Zinc Oxide. In: Kaldis E (ed) Current Topics in Materials Science, vol 7. North Holland, Amsterdam, pp 448–467

138. Herman RG, Klier K, Simmons GW, Finn BP, Bulko SB (1978) Catalytic Synthesis of Methanol from CO/H_2. I. Phase Composition, Electronic Properties, and Activities of the $Cu/ZnO/Cr_2O_3$ Catalysts. J Catal 56:407–429

139. Klier K (1982) Methanol Synthesis. Adv Catal Rel Subj 31:243–313

140. Kung HH (1980) Methanol Synthesis. Catal Rev Sci Eng 22:235–259

141. Rozovskij A (1980) New Data on the Mechanism of Catalytic Reactions with the Participation of Carbon Oxides, Kinetics and Catalysis 21:78–87

142. Lüth H, Rubloff GW, Grobman WD (1976) Ultraviolet Photoemission Studies of Formic Acid Decomposition on ZnO Nonpolar Surfaces. Solid State Commun 18:1427–1430

143. Ueno A, Onishi T, Tamaru K (1970) Dynamic Technique to Elucidate the Reaction Intermediate in Surface Catalysis – Water Gas Shift Reaction. J Chem Soc Faraday Transact 66:756–763

144. Hirsch W (1987) Photoelektronenspektroskopische Untersuchung der Adsorption kleiner Moleküle an Zinkoxid– und Zinkoxid/Kupfer-Oberflächen, PhD thesis, Freie Universität Berlin

145. Hirschwald W, Hirsch W (1991) in preparation

146. Kölbel H, Hanus D (1974) Zum Reaktionsmechanismus der Fischer-Tropsch-Synthese. Chem Ing Techn 46:1042–1043

147. Rubloff GW, Demuth JE (1977) Ultraviolet Photoemission and Flash-desorption Studies of the Chemisorption and Decomposition of Methanol on Ni(111) J Vac Sci Technol 14:419–423

148. Demuth JE, Ibach, H (1979) Observation of a Methoxy Species on Ni(111) by High-resolution Electron Energy Loss Spectroscopy. Chem Phys Lett 60:395–399

149. Christmann K, Demuth JE (1982) The Adsorption and Reaction of Methanol on Pd(100). I. Chemisorption and Condensation. J Chem Phys 76:6308–6317; II. Thermal Desorption and Decomposition. J Chem Phys 76:6318–6327

150. Dupuis RD (1984) Metalorganic Chemical Vapor Deposition of III-V Semiconductors, Science 226:623–635

151. Haken H (1982) Synergetik. Springer, Berlin Heidelberg New York

152. Fisher TE (1974) Catalysis and Surfaces. J Vac Sci Technol 11:252–260

153. Bonzel HP (1977) The Role of Surface Science Experiments in Understanding Heterogeneous Catalysis. Surface Sci 68:236–258

154. Sachtler WMH (1976) Chemisorption Complexes on Alloy Surfaces. Catal Rev Sci Eng 14:193–210

155. Sinfelt JH (1977) Heterogeneous Catalysis: Some Recent Developments, Science 195:641–646

156. Sinfelt JH (1977) Catalysis by Alloys and Bimetallic Clusters. Acc Chem Res 10:15–20

157. Ponec V (1979) Surface Composition and Catalysis on Alloys. Surface Sci 80:352–366

158. Hegedus LL, McCabe RW (1981) Catalyst Poisoning. Catal Rev Sci Eng 23:377–476

159. Bonzel HP, Ku R (1973) Adsorbate Interactions on a Pt(110) Surface. I. Sulfur and Carbon Monoxide. J Chem Phys 58:4617–4624

160. Broden G, Gafner G, Bonzel HP (1977) A UPS and LEED/Auger Study of Adsorbates on Fe(110). Appl Phys 13:333–342
161. Broden G, Gafner G, Bonzel HP (1979) CO Adsorption on Potassium-promoted Fe(110). Surface Sci 84:295–314
162. Broden G, Bonzel HP (1979) Potassium Adsorption on Fe(110). Surface Sci 84:106–120
163. Kiskinova MP, Pirug G, Bonzel HP (1983) Coadsorption of Potassium and CO on Pt(111). Surface Sci 133:321–343
164. Pirug G, Winkler A, Bonzel HP (1985) Multilayer Growth of Potassium on a Pt(111) Surface. Surface Sci 163:153–171
165. Wesner DA, Pirug G, Coenen FP, Bonzel HP (1986) A Structural and Vibrational Study of CO Adsorbed on Potassium-covered Pt(111). Surface Sci 178:608–617
166. Garfunkel EL, Crowell JE, Somorjai GA (1982) The Strong Influence of Potassium on the Adsorption of CO on Platinum Surfaces. A Thermal Desorption Spectroscopy and High-resolution Electron Energy Loss Spectroscopy Study. J Phys Chem 86:310–313
166. Garfunkel EL, Somorjai GA (1982) Potassium and Potassium Oxide Monolayers on the Platinum(111) and Stepped (755) Crystal Surfaces: A LEED, AES, and TDS Study. Surface Sci 115:441–454
167. Crowell JE, Tysoe WT, Somorjai GA (1985) Potassium Coadsorption Induced Dissociation of CO on the Rh(111) Crystal Surface: An Isotope Mixing Study. J Phys Chem 89:1598–1601
168. Bonzel HP (1987) Alkali-metal-affected Adsorption of Molecules on Metal Surfaces. Surface Sci Repts 8:43–125
169. Lang ND, Holloway S, Nørskov JK (1985) Electrostatic Adsorbate-Adsorbate Interactions: The Poisoning and Promotion of the Molecular Adsorption Reaction. Surface Sci 150:24–38
170. Wimmer E, Fu CL, Freeman AJ (1985) Catalytic Promotion and Poisoning: All-electron Local-density-functional Theory of CO on Ni(001) Surfaces Coadsorbed with K or S. Phys Rev Lett 55:2618–2621
171. Benziger J, Madix RJ (1980) The Effects of Carbon, Oxygen, Sulfur, and Potassium Adlayers on CO and H_2 Adsorption on Fe(100). Surface Sci 94:119–153
172. Hoffmann FM, Hrbek J, DePaola RA (1984) The Observation of Direct Attractive Interactions between Potassium and Carbon Monoxide Coadsorbed on Ru(001). Chem Phys Lett 106:83–86
173. DePaola RA, Hrbek J, Hoffmann FM (1985) Potassium-promoted C-O Bond Weakening on Ru(001). I. Through-metal Interaction at Low Potassium Precoverage. J Chem Phys 82:2484–2498
174. Weimer JJ, Umbach E, Menzel D (1985) The Properties of K and Coadsorbed CO + K on Ru(001). I. Adsorption, Desorption, and Structure. Surface Sci 155:132–152; II.Electronic Structure. Surface Sci 159:83–107
175. Rieck JS, Bell AT (1986) Studies of the Interactions of H_2 and CO with Pd/SiO$_2$ Promoted with Li, Na, K, Rb, and Cs. J Catal 100:305–321
176. Kiskinova MP (1988) Electronegative Additives and Poisoning in Catalysis. Surface Sci Repts 8:359–402
177. Kelemen SR, Fischer TE, Schwarz JA (1979) The Binding Energy of CO on Clean and Sulfur-covered Platinum Surfaces. Surface Sci 81:440–450
178. Erley W, Wagner H (1978) Sulfur Poisoning of Carbon Monoxide Adsorption on Ni(111). J Catal 53:287–294
179. Rhodin TN, Brucker CF (1977) Effect of Surface Deactivation on Molecular Chemisorpton: CO on α-Fe(100). Solid State Commun 23:275–279
180. Johnson S, Madix RJ (1981) Sulfur Induced Selectivity Changes for Methanol Decomposition on Ni(100) Surface Sci 103:361–396
181. MacLaren JM, Pendry JB. Joyner RW (1986) The Role of Adatom Geometry in the Strength and Range of Catalyst Poisoning. Surface Sci 165:L80–L84
182. Madon RJ, Shaw H (1977) Effect of Sulfur on the Fischer-Tropsch Synthesis. Catal Rev Sci Eng 15:69–106

254

183. Shelef M, Otto K, Otto NC (1978) Poisoning of Automotive Catalysts. Adv Catal Rel Subj 27:311–365

184. Bartos B, Freund HJ, Kuhlenbeck H, Neumann M, Lindner H, Müller K (1987) Adsorption and Reaction of CO, and CO_2/O Coadsorption on Ni(110): Angle-resolved Photoemission (ARUPS) and Electron-energy Loss Studies. Surface Sci 179:59–89

185. Paal Z (1988) Hydrogen Effects in Skeletal Reactions of Hydrocarbons over Metal Catalysts. In: Z Paal, PG Menon (eds) Hydrogen Effects in Catalysis. Dekker, New York pp 449–497

186. Sinfelt JH (1983) Bimetallic Catalysts. Wiley, New York

187. Sinfelt JH, Carter JL, Yates DJC (1972) Catalytic Hydrogenolysis and Dehydrogenation over Copper-nickel Alloys. J Catal 24:283–296

188. Sinfelt JH (1974) Catalysis by Metals: The P.H. Emmett Award Address. Catal Rev Sci Eng 9:147–168

189. Dowden DA (1972) Electronic Structure and Ensembles in Chemisorption and Catalysis by Binary Alloys. In: Hightower JW (ed) Proceedings Vth International Congress on Catalysis, Palm Beach (USA) North Holland Amsterdam pp 41–621

190. Burton JJ, Hyman E (1975) Surface Segregation in Alloys: Agreement between a Quantitative Model and Experimental Data for Ethane Hydrogenolysis Over Copper-Nickel-Alloys. J Catal 37:114–119

191. Christmann K, Ertl G, Shimizu H (1980) Model Studies on Bimetallic Cu/Ru Catalysts. I. Cu on Ru(0001). J Catal 61:397–411

192. Shimizu H, Christmann K, Ertl G (1980) Model Studies on Bimetallic Cu/Ru Catalysts. II. Adsorption of Hydrogen. J Catal 61:412–429

193. Yu KY, Ling DT, Spicer WE (1976) Thermal Desorption Studies of CO and H_2 from the Cu-Ni Alloy. J Catal 44:373–384

194. Harendt Chr, Christmann K, Hirschwald W, Vickerman JC (1986) Model Bimetallic Catalysts: The Preparation and Characterization of Au/Ru(0001) Surfaces and the Adsorption of Carbon Monoxide. Surface Sci 165:413–433

195. Sakakini B, Swift AJ, Vickerman JC, Harendt Chr, Christmann K (1987) A Comparison of Cu and Au on the Surface Reactivity of Ru(0001). J Chem Soc Faraday Trans 83:1975–2000

196. Sachtler JWA, Biberian JP, Somorjai GA (1981) The Reactivity of Ordered Metal Layers on Single Crystal Surfaces of Other Metals: Au on Pt(100) and Pt on Au(100). Surface Sci 110:43–55

197. Sachtler JWA, van Hove MA, Biberian JP, Somorjai GA (1980) Enhanced Reactivity of Ordered Monolayers of Gold on Pt(100) and Platinum on Au(100). Single-crystal Surfaces. Phys Rev Lett 45:1601–1603

198. Sachtler JWA, Somorjai GA (1984) Cyclohexane Dehydrogenation Catalyzed by Bimetall Au-Pt(111) Single-crystal Surfaces. J Catal 89:35–43

199. Lytle FW, Via GH, Sinfelt JH (1977) New Application of Extended X-ray Absorption Fine Structure (EXAFS) as a Surface Probe-nature of Oxygen Interaction with a Ruthenium Catalyst. J Chem Phys 67:3831–3832

199a. Sinfelt JH, Via GH, Lytle FW (1978) Extended X-ray Absorption Fine Structure (EXAFS) of Supported Platinum Catalysts. J Chem Phys 68:2009–2010

200. Lytle FW, Wei PSP, Greegor RB, Via GH, Sinfelt JH (1979) Effect of Chemical Environment on Magnitude of X-ray Absorption Resonance at L_{III} Edges. Studies on Metallic Elements, Compounds, and Catalysts. J Chem Phys 70:4849–4855

201. Via GH, Sinfelt JH, Lytle FW (1979) Extended X-ray Absorption Fine Structure (EXAFS) of Dispersed Metal Catalysts. J Chem Phys 71:690–699

202. Sinfelt JH, Via GH, Lytle FW (1980) Structure of Bimetallic Clusters. Extended X-ray Absorption Fine Structure (EXAFS) Studies of Ru-Cu Clusters. J Chem Phys 72:4832–4844

203. Sinfelt JH, Via GH, Lytle FW, Greegor RB (1981) Structure of Bimetallic Clusters. Extended X-ray Absorption Fine Structure (EXAFS) Studies of Os-Cu Clusters. J Chem Phys 75:5527–5537

204. Sinfelt JH, Via GH, Lytle FW (1982) Structure of Bimetallic Clusters. Extended X-ray Absorption Fine Structure (EXAFS) Studies on Pt-Ir Clusters. Chem Phys 76:2779–2789

205. Bond GC, Turnham BD (1976) The Kinetics and Mechanism of Carbon Monoxide Hydrogenation over Silica-Supported Ruthenium-Copper Catalysts. J Catal 45:128–136

206. Helms CR, Sinfelt JH (1978) Electron Spectroscopy (ESCA) Studies of Ru-Cu Catalysts. Surface Sci 72:229–242

207. Richter L, Bader SD, Brodsky MB (1981) Thermal Desorption and UPS Study of CO Adsorbed on Cu-covered Ru(0001). J Vac Sci Technol 18:578–580

209. Vickerman JC, Christmann K, Ertl G (1981) Model Studies on Bimetallic Cu/Ru Catalysts. III. Adsorption of Carbon Monoxide. J Catal 71:175–191

210. Shi SK, Lee HI, White JM (1981) Chemisorption of O_2 and N_2O on Cu/Ru(001) Surface Sci 102:56–74

211. Vickerman JC, Christmann K (1982) Model Studies on Bimetallic Cu/Ru Catalysts. IV. Adsorption of D_2 and Co-adsorption of CO and D_2. Surface Sci 120:1–18

212. Boerner D, Moffat JB (1982) The Interaction of Hydrogen with Copper-Ruthenium Surfaces: Application of a Pairwise-Additive Model. Surface Sci 122:L606–L612

213. Vickerman. JC, Christmann K, Ertl G, Heimann P, Himpsel FJ, Eastman DE (1983) Geometric Structure and Electronic States of Copper Films on a Ruthenium(0001) Surface. Surface Sci 134:367–388

214. Brown A, Vickerman JC (1984) The Characterisation of Model Cu/Ru(001) Bimetallic Catalysts by Static SIMS with XPS and TPD. Surface Sci 140:261–274

215. Lai SY, Vickerman JC (1984) Carbon Monoxide Hydrogenation over Silica-Supported Ruthenium-Copper Bimetallic Catalysts. J Catal 90:337–350

216. Yates JT, Peden CHF, Goodman DW (1985) Copper Site Blocking of Hydrogen Chemisorption on Ruthenium. J Catal 94:576–580

217. Paul J, Hoffmann FM (1986) TDS and EELS Observations for CO, O_2, and CH_3OH Bound to Ru(0001)/Cu. Surface Sci 172:151–173

218. Goodman DW, Peden CHF (1985) Hydrogen Spillover from Ruthenium to Copper in Cu/Ru Catalysts: A Potential Source of Error in Active Meta Titration. J Catal 95:321–324

219. Houston JE, Peden CHF, Feibelman PJ, Hamann DR (1986) Observation of a True Interface State in Strained-Layer Cu Adsorption on Ru(0001). Phys Rev Lett 56:375–377

220. Harendt C, Sakakini B, van den Berg JA, Vickerman JO (1986) A Combined HREELS and SSIMS Study of the Adsorption of CO on Ru/Cu and Ru/Au Bimetallic Catalysts. J Electr Spectr Rel Phen 39:35–44

221. Wandelt K, Markert K, Dole P. Jablonski A, Niemantsverdriet JW (1987) Microscopic Properties of Two-dimensional Silver and Gold Metal– and Alloy-Films on Ru(001). Surface Sci 189/190:114–119

222. Lenz P, Christmann K (1991) The Interaction of Silver with a Ru(10$\bar{1}$0) Surface (to be published)

223. Daniel WM, Kim Y, Peebles HC, White JM (1981) Adsorption of Ag, O_2, and N_2O on Ag/Rh(100). Surface Sci 111:189–204

224. Bassi IW, Garbassi F, Vlaic G, Marzi A, Tauszik GR, Cocco G, Galvagno S, Parravano G (1980) Bimetallic Ruthenium-Gold-on-Magnesia Catalysts: Chemicophysical Properties and Catalytic Activity. J Catal 64:405–416

225. Galvagno S, Schwank J, Parravano G, Garbassi F, Marzi A, Tauszik GR (1981) Bimetallic Ru-Au Catalysts: Effect of the Support. J Catal 69:283–291

226. Ott GL, Fleisch T, Delgass WN (1979) Fischer-Tropsch Synthesis Over Freshly Reduced Iron-Ruthenium Catalysts. J Catal 60:394–403

227. Bhasin MM, Bartley WJ, Ellgen PC, Wilson TP (1978) Synthesis Gas Conversion Over Supported Rhodium and Rhodium-Iron Catalysts. J Catal 54:120–128

228. Sachtler WMH (1984) Selectivity and Rate of Activity Decline of Bimetallic Catalysts. J Molec Catal 25:1–12

229. Guczi L (1984) Mechanism and Reactions on Multimetallic Catalysts. J Molec Catal 25:13–29

230. Paal Z, Menon PG (eds) (1988) Hydrogen Effects in Catalysis. Dekker, New York

231. Pajonk GM, Teichner SJ, Germain JE (eds) (1983) Spillover of Adsorbed Species, Studies in Surface Science and Catalysis, vol 17. Elsevier, Amsterdam

256

6 General Conclusions

In the five preceding chapters, we have tried to provide the reader with an unbiased representation of surface physical chemistry. These have spanned (macroscopic) surface thermodynamics, (microscopic) surface physics, a selection of spectroscopic tools to investigate surfaces and, finally, the application of surface physical chemistry to a relevant practical subject, namely, heterogeneous catalysis. Our intention was to inform generally trained chemists or physicists about certain facets of the exciting field of surface science. Enthusiasm is a symptom of the people working in this discipline, and this is best apparent in the many short review papers written for the community of natural scientists. We selectively cite an article by Somorjai, one of the most distiguished surface chemists of today, who has over the past 30 years inspired the field with many new ideas and experiments [1]. It is our hope that we have aroused interest in surface science among the readers of this book. Of course, we are aware that we have presented only a selection of material, due to the limited space, and we have not included such interesting topics or aspects as, for example, semiconductor surfaces and their physical and chemical properties. Electrochemists will certainly be disappointed that their field was not discussed, in spite of the many relations that exist between surface science and electrochemistry. And there may be those researchers who will find the presentation or selection of diagnostic surface sensitive tools not comprehensive, because just their special technique has not been mentioned or properly described here.

Before we present some statements and perspectives regarding the future role of surface physics and chemistry, we offer some philosophical remarks concerning surface science in particular, but also applying to other fields of natural science – nuclear physics, biochemistry, biotechnology, information technology, low-temperature physics, etc. How quickly the interest and research activity in surface science is growing (deliberately avoiding the term "progressing") can perhaps best be estimated by following the increase in related publications. Consequently, a review of papers appearing each year under the topic: Surface Science can be very informative. In a recent publication Jahrreiss [2] has reflected on the future development of surface physics and has tallied "Surface Science" papers. We reproduce his revealing findings in Fig. 6.1. Apparently, there has been a steady growth in the number of scientific papers which becomes noticeable around 1960, and which were correctly ascribed to the technological progress made in ultra-high vacuum technology around that time [2]. Even at a first glance, the curve of Fig. 6.1 closely resembles an exponential function, and as Jahrreiss has pointed out, it appears as if there is, since 1975, a doubling of the annual number of publications every 8 years. Whether or not the curve of Fig. 6.1 is really of exponential character cannot be said with certainty, one must wait for the next 10 to 20 years. It could very well be that the rate will increase further (as is known to be true for most technical disciplines, initiated mainly by the impetus of information technology in a wider sense).

What are the consequences of this development? The most important question is whether this increase in numbers really parallels or indicates a progress in knowledge which one could suppose at a first glance. To be honest, this author is unable to give an affirmative answer – to the contrary, one must fear that this accumulation of (mostly extremely specialized) facts might overwhelm the individual scientist and prevent him

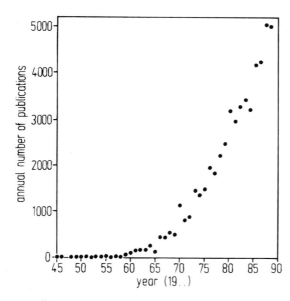

Fig. 6.1. Number of annual scientific publications in the field of Surface Science plotted vs time, for the period 1945–1990. The diagram is based on the number of entries in the journal Physics Abstracts under the topic: Surface Science. It does not include those listed under the topic: Surface Tension, which are concerned with liquid surfaces. After Jahrreiss [2]

from informing himself about the potential overlap of his own discipline with neighboring scientific areas.

Possibly, very important cross-relations among adjacent scientific fields will no longer be recognized, which in fact will mean a change for the worse, as far as mutual communication is concerned. Are there any means against this potential development? To be optimistic, there is the self-discipline of the individual scientist, who does not publish every minute detail of his work, but consolidates and carefully considers his results and the appropriate experiments before publishing, as well as keeping a careful eye on developments in neighboring sciences.

At the end of Chapter 4, we attempted to point to some developments that are likely to occur in surface analysis in the near future. It became clear that methods are seldom independent of the actual research problems – conversely, first there are problems and then scientists are motivated and attempt to solve them by suitable means.

Many more subjects could be listed in which surface science has significantly contributed to the progress made during the past decades, or is at least expected to have impact on further developments; these include information technology, materials science, corrosion, metallurgy, energy storage technology, heterogeneous catalysis, optics, high-temperature superconductivity, membrane biochemistry. Probably most evident is the increasingly complicated electronics of daily life, from video cameras, color copying machines, and telefax equipment, to extremely powerful microprocessor chips that have enabled worldwide availability of high-speed computers. This only became possible by large-scale, high-precision production of semiconducting devices. We recall and emphasize that the "single-crystal approach" introduced in Chapter 1 and pursued throughout this book has been most successfully employed in this area – real devices use well-oriented, ultra-clean single crystal samples of silicon, and the characterization of metal – semiconductor junctions requires and makes extensive use of surface analytical facilities, as do growth studies of multicomponent films, for example, by molecular beam epitaxy. Recently, the fabrication and characterization of semiconducting "null"-dimensional, so-called quantum-dot structures with exciting new physical properties have become possible [3], whereby the unique electron transport phenomena, conducting and luminescence behaviors of these dot structures have attracted considerable interest. It is, at present, impossible to estimate their impact on the design of novel devices.

Another fairly recent development in the field of semiconductors is the fabrication of gas sensors, and in order to optimize their functioning a thorough understanding of adsorption and surface-gas interaction phenomena on these materials must be attempted [4]. Novel electrical and optical devices are increasingly based on the so-called organic semiconductors which consist mostly of organic polymers such as polyacetylene doped with various compounds or elements that provide surprising conducting properties. Again, we refer to the literature and simply stress the impact of surface science in this exciting new area [5].

To reiterate, there is a mutual fertilization between instrumental developments and the fields of their application. Again, semiconductor technology is a striking example: the progress in fabrication procedures and the control of tolerance limits often hinges on the visibility of the product and the possibility to precisely measure its electrical and mechanical properties. Since the structural resolution of semiconductor devices is about to push forward to atomic dimensions, also the diagnostic tools must be able to routinely resolve atomic structures – and this despite the increasing susceptibility of these micro structures to damage caused by foreign atom impurities or mechanical strain and roughness. Scanning high-energy transmission electron microscopy and scanning tunneling microscopy have proven to be powerful diagnostic tools here, as has scanning Auger microscopy; it is believed that the consequent exploitation of these combined techniques will sooner or later enable tailoring of devices with the maximum possible fine structure and, hence, storage capacity. A helpful overview of problems solved, and future technical goals in semiconductor physics and technology can be obtained from several research reports and reviews collected recently (e.g., [6]) whereby theoretical and experimental facets are considered, including adsorption on semiconductor surfaces, Schottky barrier formation, molecular beam epitaxy of elemental and compound semiconductors, and surface chemistry of dry-etching processes.

Perhaps not so obvious is the role of surface science in energy technology. It is certain that the increasing pollution of the earth's atmosphere with carbon dioxide, caused by extensive burning of fossil fuels will unfavorably influence our climate via the greenhouse effect. A solution to this urgent problem would be the replacement of fossil fuels by hydrogen – a comparatively clean energy carrier; the storage of hydrogen, i.e., the development of "hydrogen batteries" may be a future goal. In this situation surface science can again play a beneficial role, since the hydrogen uptake of storage materials often depends on the rate of H_2 permeation through the surface of the respective solid (structure, chemical composition, cleanliness). In this context also the development of storage materials themselves can be supported by surface studies, that can help to elucidate, for example, hydrogen permeation, surface reconstruction, and heterogeneous hydride formation processes of the respective active materials, among which are intermetallic compounds containing lanthanum, magnesium, nickel, titanium or cerium. Another source of interest where surface science can have great impact is the development and improvement of fuel cells based on photocatalytic conversion of water into hydrogen and oxygen. Here we encounter photovoltaics and electrochemistry, battery technology, and related areas where, likewise, many of the essential elementary steps are not understood. Surface studies concentrating on adsorption and interaction of alkali metals (Li!) with metal chalcogenids (galena, for example) or related compounds can significantly help in developing new and powerful battery systems. The interested reader will find more information in a recent review article [7]. Also of interest will be solar cells, where photoelectric effects in the surface region of the solids are decisive for optimum operation. Again, very promising developments with regard to energy conversion efficiency of solar cells have been made.

A serious problem in materials science and metallurgy is hydrogen embrittlement and fracture, as well as corrosion processes in general, because they lead to tremendous replacement costs. To explore the responsible elementary (often electrochemical) processes is thus a prominent task of surface science, particularly in combination with electrochemistry. Only during the past 10 or 20 years has it become apparent (cf., Chapter 3) that chemisorption of active gases on solid (metallic) surfaces frequently causes relaxations and reconstructions of the entire surface region and, hence, opens reaction channels through which the attacking chemical agents can enter the bulk solid, thereby resulting in acceleration of the damaging process. This kind of corrosion can, for example, be stopped by appropriate surface coatings that resist the respective chemical agent.

We could easily continue listing problems whose solutions lie in surface science; instead we hold the well-justified hope that a concerted action of surface chemistry and physics along with the other (technical, biological) disciplines will steadily contribute to an improvement of man's quality of life. Undoubtedly, natural scientists will become increasingly interested in the problems pertinent to surfaces and interfaces, and we are confident that each scientist can help manage the growing body of research for our mutual benefit.

References

1. Somorjai GA (1978) Surface Science. Science 201:489–497
2. Jahrreiss H (1990) Zur Entwicklung der Oberflächenphysik. Vakuum in der Praxis 3:195–201
3. Sikorski Ch, Merkt U (1990) Magneto-optics with Few Electrons in Quantum Dots on InSb. Surface Sci 229:282–286
4. Göpel W (1985) Chemisorption and Charge Transfer at Ionic Semiconductor Surfaces: Implications in Designing Gas Sensors. Progr Surface Sci 20:9–103
5. Basescu N, Liu ZX, Moser D, Heeger AJ, Naarmann H, Theophilu N (1987) High Electrical Conductivity in Doped Polyacetylene. Nature 327:403–405
6. Jaegermann W, Tributsch H (1988) Interfacial Properties of Semiconducting Transition Metal Chalcogenides. Progr Surface Sci 29:1–167
7. King DA, Woodruff DP (eds)(1988) The Chemical Physics of Solid Surfaces and Heterogeneous Catalysis, vol 5: Surface Properties of Electronic Materials. Elsevier, Amsterdam

Subject Index

271